LIGHT AND COLOR IN NATURE AND ART

LIGHT AND COLOR IN NATURE AND ART

SAMUEL J. WILLIAMSON
New York University

HERMAN Z. CUMMINS
City College of the City University of New York

JOHN WILEY AND SONS
New York • Chichester • Brisbane • Toronto • Singapore

COVER ILLUSTRATION

HARBOR AT ST. TROPEZ by Paul Signac
From the Collection of Mr. Nathan Cummings, New York, NY.
(The lower section is a color reversed print)

COVER AND TEXT DESIGN BY LORETTA SARACINO

Library of Congress Cataloging in Publication Data:

Williamson, Samuel J.
   Light and color in nature and art.

   Includes bibliographies and index.
   1. Light.   2. Color.   I. Cummins, Herman Z., 1933–
II. Title.

QC355.2.W56   1983   535   82-11167
ISBN 0-471-08374-7

Printed in the United States of America

To Joan and Marsha

PREFACE

This book is an introduction to the science of light and color and its applications to photography, art, natural phenomena, and other related areas. It is intended primarily as a text for a one-semester or one-quarter college course for students with little or no background in science or mathematics.

The progress of science, especially since the sixteenth century, has provided a basic understanding of visual effects that has dramatically enriched our appreciation of what we see. From this knowledge we have increased our sensitivity to the interplay of light and objects around us. We have also gained a precision of language that permits subtle distinctions to be expressed. In this book we aim to explain the origin of phenomena that are commonly encountered in nature and art. Although we emphasize the physical aspects, all that we see cannot be explained solely by physical laws. Our investigation necessarily touches on aspects of physiology and psychology that directly influence how we perceive visual images.

The topics in this book are selected for their general interest, the role they played in the evolution of color science, or their relevance to professional activities. Most of the presentation is appropriate for the general reader. This includes the effect of mixing colors together (both light and paint), the notion of color spaces, how atoms and molecules affect light, how light can be "measured," the effect of using a lens, and many other topics. Some aspects of the subject are directly applicable to narrower interests, for instance stage lighting, art history, photography and television, perceptual psychology, and color printing.

The topics in this book could be presented in any of several different sequences. We have chosen to begin with phenomena that are familiar to everyone and that can be explained without an understanding of the detailed physical properties of light. These are subjects where perception plays the dominant role: the description and measurement of color and the laws that govern the effects of superimposed lights, of mixtures of pigments, and of reflecting surfaces. These topics introduce terminology that proves essential in understanding the rest of the book. Also, the relationship between color and the spectral composition of light is developed.

We are then in a position to examine the physical nature of light and the colorful effects that can be explained only by understanding what light is. These effects depend on the wave aspects of light or its quantum properties. The physical interaction between light and matter is discussed in order to explain the effect of pigments, dyes, and fluorescent materials. This first third of the book lays the foundation on which the rest of the text is built.

The topics that follow are relatively independent of each other but have been arranged so that earlier topics introduce concepts that are applied in later ones. Correspondingly, how light is measured precedes a discussion of how light is produced. The discussion involves practical considerations of various lamps and how much light they provide, including special features of the laser. In a parallel way, the concepts of optics and the way in which images are focused by a lens or mirror precedes a chapter on the human eye and visual system.

In the last third of the book we deal with topics of general interest in which the physical principles discussed earlier are apparent. These include the optical and colorful effects seen in the visual arts, the principles of photography and holography, a short section on television and, finally, a chapter on light and color in the atmosphere. In short, we have begun the book with a phenomenological description of light and color, proceeded to discuss the physical nature of light and its interaction with materials, and then applied these two levels of description to explain the visual effects that can be seen about us.

Readers will approach this book with specific interests in mind. For this reason we have organized our presentation to afford them considerable flexibility. For readers who prefer a qualitative overview we have set out a "fast track" sequence of topics, moving from one chapter to the next. A black square precedes and follows each fast track sequence. These topics are indicated in the table of contents by a black square preceding the number of each section. In these sections central topics are discussed with little emphasis on quantitative aspects. This fast track is appropriate as a text for a one-quarter course at the college level. It can be supplemented, as desired, by including additional sections not indicated by a black square.

These additional sections will interest the reader who wishes to go deeper into particular topics. Some of them emphasize a more quantitative approach, for instance, when we consider ways of measuring light and color. Here an appreciation for the magnitudes of physical quantities (e.g., the intensity of a light bulb) and perceived qualities (the chromaticity of the sun) is essential. Other topics, such as dealing with optical effects produced by differently shaped lenses, also rely heavily on a quantitative discussion. This enables us to take advantage of the predictive ability of mathematics, as when we wish to determine at what distance the image formed by a lens will be in focus. But even in these cases the mathematics goes no further than elementary algebra at the high school level. Generally we have minimized mathematics in order to concentrate on fundamental concepts. A combination of the fast track topics and most of the remaining sections would be suitable as a text for a one-semester course at the college level.

In addition to this main sequence of topics we have also included a selection of Advanced Topics. These deal with concepts that are central to an understanding of light and color, but they are appropriate only for a more technical course that extends beyond the conventional one-semester period. These Advanced Topics are designated by the letter "A" after the section number.

This book was developed for courses that the authors initiated at New York University and the City College of the City University of New York. We thank our many students whose perceptive questions have encouraged the preparation of this material. We have also benefited from numerous discussions with colleagues in many disciplines, and we regret that it is impractical to mention all of them individually. For their interest and help we extend our gratitude. In addition we have received valuable suggestions from readers of early drafts of chapters of the manuscript, and we sincerely thank Lloyd Kaufman, Charles Harbut, Lawrence Majewski, Allan Staley, and Jens Zorn for their careful attention and advice. Any remaining inadequacies are entirely our own responsibility. In addition we acknowledge the special help given by Joan Garey, Barbara Peterson, and Frances Tritt in typing the manuscript and organizing the illustrations. Last but of paramount importance we express our heartfelt gratitude to our wives, Joan and Marsha, whose encouragement made this project possible.

SAMUEL J. WILLIAMSON
HERMAN Z. CUMMINS

CONTENTS

LIGHT AND COLOR IN
NATURE AND ART

CHAPTER 1

INTRODUCTION

Light and color have an impact on virtually every phase of human experience, since we perceive the world largely through vision. Painting, film, books, theater, observation of natural phenomena such as rainbows and sunsets, and recognition of objects about us—all involve the subtle interplay of light and color. The quest for a scientific explanation of phenomena involving light and color has been a central topic during the past three centuries in the development of three independent disciplines. *Physics* analyzes the ways in which light is produced, how it can be measured, how it interacts with materials, and how it reaches the eye and forms an image on the retina. *Physiology* begins with the formation of the image and describes the action of the light on the eye's photoreceptors (the rods and cones in the retina), the subsequent processing of neural impulses produced by the receptors, and the transmission of the signals along the nerve pathways to the visual cortex of the brain. Finally, *perceptual psychology* deals with the processing of information and subsequent interpretation in the brain, which yields a perception.

In this book we shall develop a unified approach to an understanding of light and color. We shall concentrate on the physical aspects of light in building a framework for understanding the processes leading to the formation of images on the retina. But we shall also include enough visual physiology and perceptual psychology to establish important connections between the retinal image and the resulting visual perception.

■ 1-1 HISTORICAL PERSPECTIVE

Although not properly understood until the late Renaissance, optical effects are so commonplace it is not surprising that they were put to use early in the development of civilization. Archeological evidence from Mesopotamian cultures of 1500 B.C. shows that polished metal served as a mirror and a piece of glass became a "burning lens" to focus the sun's rays to kindle fire. In this way a working knowledge of elementary optics evolved several millenia ago; but the Western tradition that seeks an understanding of the physical principles that underlie such phenomena can be traced back only to the time of Classical Greek civilization.

During this period philosophers struggled to find a logical basis for optical phenomena and the process of vision itself. Their work was limited to speculation, because there was very little experimental evidence against which the predictions of theories could be tested.

With the growth of the Roman Empire the intellectual center in the Near East shifted from Greece to Alexandria. Increased emphasis was placed on experiments as a means of gaining knowledge. Hero and Ptolemy are particularly noted for this development. Ptolemy is credited with important discoveries about color, such as the rules of additive light mixtures, the effect of spinning a disk with sectors of various colors, and aspects of refraction whereby a light path is bent when entering a piece of glass at an angle. He was also the first to note that we have better perception of detail in the direct line of sight than to the side.

With the decline of the Roman Empire the center of intellectual activity gravitated to the Islamic world. The first great philosopher was al-Kindi who worked first in al-Basra in Mesopotamia and later in Baghdad (813–847). Respect for Greek thought stimulated him to translate, correct, and complete much that concerns optics. In the early eleventh century this work in the Near East was carried on by Avicenna (980–1037), perhaps the most influential natural philosopher of Islam.

The most significant figure in optics between antiquity and the seventeenth century was Alhazen (965–1039 A.D.), a contemporary of Avicenna. Alhazen demonstrated how anatomical, physical, and mathematical features could be integrated into a single theory of vision. His fundamentally new contribution was to adapt a suggestion by al-Kindi and show how the form of a subject is introduced into the eye. He imagined that the surface of a subject consists of closely spaced point sources, each of which sends forth its own rays in all directions. He believed these rays form a unique set that indicates the form of the subject, thus permitting the lens to obtain a clear impression.

In the thirteenth century the flood of Arabic and Greek literature in translation kindled in Europe a vigorous interest in optics. Roger Bacon, writing in Paris, drew on Alhazen and Aristotle for his material and elevated their ideas to prominence in the Western world. During the late Middle Ages and Early Renaissance, emphasis on realism by painters such as Giotto, Masaccio, and Brunelleschi stimulated an interest in perspective and optics. It was the Benedictine monk Francesco Maurolico (1495–1575) who recognized that the double convex lens of the eye causes rays to converge inside the eye. He believed that before the rays meet they enter the optic nerve and are conveyed to the brain. He could thus understand the cause of myopia (nearsightedness) as due to the excessive curvature and presbyopia (farsightedness) to insufficient curvature of the lens. Thus he was the first to provide a theory for the action of spectacles, then long in use.

Not until the seventeenth century was the problem solved of how the eye resolves a single point source since its light falls everywhere across the area of the lens. The explanation was offered by the mathematical physicist Johannes Kepler (1571–1630), working in Prague. Noted for formulating the laws describing the orbits of planets, Kepler apparently acquired an interest in optics during the solar eclipse of 1600. Using the law of refraction tabulated by the Dutch optician Willebrord Snell, Kepler worked out mathematically the directions taken by rays when they leave a point on a subject and are intercepted by a glass sphere. He found that the rays converge at another point behind the sphere at a location that depends on the position of their source. Arguing that the lens of the eye acts in the same way, Kepler solved the problem of multiple rays: rays from different parts of a subject will be focused by the lens at different positions on the retina. We now appreciate that the major focusing element of the eye is the cornea, and the lens serves for fine adjustments to the focus. The image on the retina is inverted, upside down and right to left. To account for how we perceive a proper image, Kepler was forced to conclude that optical concepts do not apply beyond the retina, and some other mechanism—yet to be explained—conveys an impression of the scene via the optic nerve to the brain. To Kepler goes the credit for explaining the essence of the optics of the eye and the retinal theory of vision. His achievement put the capstone on 2000 years of intellectual struggle.

The upswell of interest in optical effects during the Renaissance and the introduction of shaped pieces of glass for optical instruments, such as the microscope and telescope in the early part of the sixteenth century, awoke curiosity about the nature of light itself. It was for Isaac Newton in 1665, at the age of 23, to take the fundamentally important step with his series of experiments with glass prisms and reveal the spectral colors that constitute white light.

■ 1-2 ORIGINS OF THE SCIENTIFIC VIEW

Our modern ideas about light begin with the work of Isaac Newton. His ideas were modified in the nineteenth century by Young, Fresnel, and Maxwell, and were radically transformed in the twentieth century by Planck, Einstein, and the architects of the quantum theory. From these ideas there has emerged a comprehensive understanding of light and color powerful enough to explain even the subtlest physical aspects of optical phenomena.

Although the theory of light and color is rather complex in its entirety, virtually all of the remarkable phenomena of interest in nature and the arts can be understood by using concepts embodied in a simple version of the theory, the *classical wave theory of light.* This theory has been complete for more than 100 years; but historically the wave theory of light—both in its early Newtonian form and in its modern form—has had a long and difficult climb to acceptance by the general public and particularly by the art world. It is instructive to examine the source of this difficulty.

First, light and color have a long tradition of mystic and religious significance dating back to prehistoric times. The symbolic and decorative aspects of color are evident in the caves at Lascoux in France and Altamira in Spain, where pictoral representations of animals have been attributed to Cro-Magnon man of the period around 15,000 B.C. The symbolic value of color is found even today in the sand-painting of the Navajo in the Monument Valley area of the American Southwest, and in the association in Western popular culture of black with death, white with purity, and red with danger. Traditionally, then, light and color have been popularly viewed according to their esthetic and symbolic attributes.

Newton's scientific explanation for the nature of light was a dramatic break from this tradition. Newton asserted that the perception of color resulted entirely from the physical properties of the light and not from its aesthetic or symbolic attributes. His notion that passing white light through a glass prism will reveal differently colored components of the original light was not readily accepted for a second reason as well. The ideas of Aristotle were still widely viewed as authoritative, and Aristotle had proposed that all colors are derived from mixtures of black and white, in direct contradiction to Newton's ideas:

> Possible way of conceiving the existence of a plurality of colors besides black and white: juxtapose black particles and white ones. Those juxtapositions

involving simple ratios may be the most pleasing colors such as purple or crimson, like the concords in music. Irrational ratios may produce impure colors.

Aristotle, *Sense and the Sensible* (about 350 B.C.).

The third difficulty faced by the scientific explanation was its failure to distinguish between the *physical stimulus* and the *perception* that it produces. This has led to great confusion. In Newton's crucial experiment of 1665 in which sunlight passed through a prism and spread to form a spectrum, he concluded that the rays were being separated according to some physical property (which he called "refrangibility"):

The rays of smallest refrangibility are all disposed to exhibit a red color, while the most refrangible rays exhibit a deep violet color.

Newton, *Opticks.*

He was careful to distinguish between this *physical property* (refrangibility) and the *perception* that the light produces:

Indeed, rays, properly expressed, are not colored. There is nothing else in them but a certain power . . . to produce in us the sensation of this or that color.

His followers were often less careful to make the distinction and simply equated color and refrangibility.

Indeed the perception of color is an immensely complicated process, a fact well known to artists familiar with effects such as color contrast. Long before Newton, Leonardo da Vinci described many subtle effects whereby the color of a portion of a surface may be altered if a contrasting color surrounds it. Newton's theory that the composition of the light from each portion of the surface should dictate the color of that area simply cannot explain the contrast effect on physical grounds alone. Thus, the oversimplification of equating color with refrangibility led many people to violently reject Newton's ideas and the subsequent wave theory of light.

In 1810 Johann Wolfgang von Goethe, poet, playwright, and student of the arts, published his *Theory of Colours* in which he vigorously rejected Newton's work. Goethe reported many elegant experiments in which color effects illustrate the complexity of the relation between the physical stimulus and the resulting perception. Some of these relationships have only recently been explained by psychologists. But Goethe had attempted to relate them to the nature of the stimulus alone. Not really understanding Newton he reverted to an Aristotelian approach and dealt in vague poetic terms that do not stand up to careful and objective scrutiny. Yet his book is important, for its careful observations showing the highly complex relationship between physical stimulus and perception served as a forerunner of the modern field of perceptual psychology.

The Aristotelian view held that white light is the purest light, and when light appears colored it has been contaminated. Goethe in like manner supposed that white light has to be mixed with darkness to produce color. He argued for the unity and indivisibility of light and contemptuously referred to Newton's prism demonstrations as "the epoch of a decomposed ray of light."

This resistance to science by artists has gradually disappeared since Goethe's time. This is due partly to the development of the psychology of perception, which has succeeded in probing the relationships between the nature of physical stimuli and the resulting perception. Partly it is due to the scientific triumphs of the nineteenth and twentieth centuries, which have firmly established the validity of the wave theory of light and Newton's interpretations. And partly it is due to practical considerations arising from the growing impact of technology on art. Ironically, with the development in the 1800s of premixed paints and organic dyes and the introduction of mass-produced textiles, printing ink, and papers the need for an objectively correct color theory became urgent.

The first crucial step toward such a theory was taken in 1802 by the English physician Thomas Young. Through a series of experiments he demonstrated that a wide range of the colors we know can be reproduced by superimposing blue, green, and red lights in proper proportion. This fact suggests that the human visual system has analogous characteristics, and Young proposed that the eye has three different types of color receptors that are selectively sensitive to light in these three regions of the spectrum. The brain yields a perception of color on the basis of the relative strengths of the three signals sent by the receptors when an image is formed on the retina.

A way to measure color was introduced some 50 years later by the brilliant Scottish physicist James Clerk Maxwell. He showed how each color could be specified by the amounts of three standard red, green, and blue lights that must be added to reproduce the color of interest. He thereby initiated the quantitative science of colorimetry. Today the colors of paints, color films, theatrical lights, dyes, and the like are specified by methods derived from this approach. Artists and manufacturers have learned to follow these scientific systems to specify colors and understand optical effects. For the artist, however, this is only the beginning. How the materials will be used and what esthetic quality will result remain questions of art alone.

Young's suggestion of three color receptors in the eye was subsequently made more specific by the noted physicist and psychologist Hermann von Helmholtz in the late 1800s. From color matching experiments Helmholtz concluded that the three types of sensors have their peak sensitivities in different positions in the spectrum, but each sensor will also respond to light over a fairly broad range of the spectrum on either side of the peak. The Young–Helmholtz three-color theory of color vision was not confirmed until 1964 when two groups working independently (William Marks, Walter Dobelle, and Edward MacNichol at Johns Hopkins University and Paul Brown and George Wald at Harvard University) showed through light absorption measurements on individual cones of the human retina that indeed there are three types of cones that absorb light in different regions of the spectrum.

While Young's three-color receptor theory is correct, additional physiological work on animals such as the macaque monkey and the horseshoe crab indicates that the electrical pathways between the cones and the brain are more complicated than he envisioned, and considerable processing of sensory information occurs within the retina itself. This retinal processing leads to many subtle phenomena when viewing contrasting colors, some of which were described by Goethe. It is also consistent with the modern version of an "opponent" color theory that was introduced in the late nineteenth century by Ewald Hering to explain relationships between various colors as revealed by psychological studies. The way in which the "neural switchboard" handles color information in some ways bears a close resemblance to the method chosen by engineers to encode picture information for color television broadcasts.

Since the human plays the essential role in perception of light and color, we should add a word of caution that great differences are found between the range of colors perceived by various individuals. The overwhelming majority (99.5% of women and 92% of men) enjoy the maximum range of color perception ("normal vision"). But "color blind" individuals have less discrimination as a result of their more limited span of color perception. For example, objects that appear red for a normal person may instead appear gray for certain conditions of color blindness. The reason for color blindness in most cases is physiological and has been explained by the absence of either one or two types of color receptors in the retina. In this book we shall restrict our attention to phenomena as they would be seen by an observer with normal vision.

■ 1-3 OUTLINE OF THE PRESENTATION

Our book begins with several aspects of color familiar to people concerned with the effects of superimposing colored lights or mixing paints. We develop a precision in thought and language that shows these two processes to be distinctly different. The rules describing the outcome of color mixing differ for each process, and we shall establish the logical basis for this difference. As a result, the reader will gain appreciation of the relationship between how energy is distributed throughout the spectrum and the resulting perceived color. This establishes the connection between a precise characterization of the stimulus and our resulting perception.

The third chapter builds from these concepts to introduce the notion of a color space in which the position of a color gives an unambiguous specification of its qualities. As one useful example we explore aspects of the chromaticity diagram that provides a framework within which we can visualize the relationship between various colors. It also permits us to predict quite precisely the result of adding together two or more colors, something that cannot be done with more qualitative, psychological color spaces.

Some phenomena that we encounter every day have no explanation in color mixing theory. Examples include iridescence and fluorescence. Such remarkable effects have their basis in the wave or the quantum aspects of light itself, and we shall deal with these topics next. This discussion provides an understanding of the basic nature of radiant energy.

We make the important distinction between "light" and "radiant energy" in the chapter on photometry. This chapter deals with how light is measured and how light sources and reflecting surfaces are best described. The various means of producing light are then examined in a separate chapter. It also describes why the light that different sources emit has characteristic features that depend on the source.

The last half of this book deals with applications of the concepts that were laid down in the first half. We start with the principles of geometrical optics to explain how the propagation of light can be influenced by shaped pieces of glass. How images are formed and the notion of magnification are described when we consider common optical instruments such as the telescope, binocular, and microscope. Optical principles affecting the color of paints, dyes, stained glass, and the like are explored. In a digression from our discussion of the optics of the human eye, we extend our interest to encompass the physiology of the retinal receptors and the organization of the visual system. This includes the psychological effects that influence our perception of color in ways that depart from the simpler concepts established in the first two chapters of this book.

Some of the final chapters discuss in detail three important areas of color applications: photography, holography, and television. Our emphasis is on the technical side in order to explain the underlying principles. In photography this includes not only the optics of the camera but also photochemical aspects of the film. In holography, we explore the many ways in which holograms can be recorded and shown, including the whitelight hologram that has generated an explosion of interest in using this medium as an art form.

In the last chapter we turn to light and color in the atmosphere. Drawing on the principles of optics and colorimetry developed in earlier chapters, we explore the origins of mirages, rainbows, the blue sky, red sunsets, and other familiar effects. ■

FURTHER READING

General References on Light, Color, and Vision

C. Rainwater, *Light and Color* (Golden Press, New York, 1971) (paperback). A delightful short book covering in a brief, clear fashion all of the major topics discussed in depth here.

H. Varley, Ed., *Color,* An Architectural Digest Book (Viking Press, New York, 1980). A big book, profusely illustrated, giving a nontechnical discussion of color, its symbolism, its origin, and how it is used for decoration.

G. Agoston, *Color Theory and Its Application in Art and Design* (Springer Verlag, New York, 1979). A small book of remarkable clarity, accessible to everyone.

Historical References

D. C. Lindberg, *Theories of Vision from Al Kindi to Kepler* (University of Chicago Press, Chicago, 1976). A splendid account of the early development of concepts that have led to our modern notions about vision. Essentially no mathematics is used.

D. L. MacAdam, *Sources of Color Science* (M.I.T. Press, Cambridge, Mass., 1970). Extracts from original papers that are hallmarks in the development of our concepts of color. Highly recommended, especially for reading after completing Chapters 2 and 3.

I. Newton, *Opticks* (Dover Publications, New York, 1952).

J. W. von Goethe, *Theory of Colours.* English translation by C. L. Eastlake (M.I.T. Press, Cambridge, Mass., 1970).

Journals of Interest

Color Research and Application (Wiley-Interscience, New York). An international journal reporting on the science, technology, and application of color in business, art, design, education, and industry.

CHAPTER 2

ORIGINS OF COLOR

Most of us were first introduced to color theories through an experience with painting. We were told how to mix two colors from the palette to produce a third. And we were probably told that all known colors can be produced by appropriate mixtures of three "primary" colors—identified as blue, yellow, and red. An artist when conveniently describing one of the "primary" paints as blue realizes that the color is more accurately characterized as blue-green, otherwise known as cyan or perhaps turquoise. A similar shorthand is used for the "primary" red in place of the more accurate description of bluish-red, or magenta. This lack of precision in describing the primary colors suggests a color theory that is logically inconsistent and conflicts with the physical laws governing the reflection of light by pigments.

This chapter explores the world of color phenomena to examine the nature of color and to show how the rich variety of color mixing effects applied for example in painting, stage lighting, printing, and television can be understood from a few general principles. Many of the applications will be discussed in greater detail in subsequent chapters. These color mixing laws provide a satisfying base on which to deal with color in a way that is logically consistent. We can then appreciate that the color of an object depends not only on the type of molecules that form it but also on their physical arrangement—for example, the size of the particles in a pigment and the smoothness of a surface.

■ 2-1 NEWTON'S EXPERIMENTS

The foundation of color science as now accepted was laid 300 years ago by the great mathematician and physical scientist Isaac Newton. In 1666, shortly after he received the Bachelor of Arts degree at Cambridge University at the age of 23, the University was forced to close for 18 months to escape the rampage of the Great Plague. During this enforced vacation from teaching obligations, Newton's inventive mind led him to devise the calculus, theory of gravitation, and notion of the color composition of white light. Any one of these would have ensured his fame, but the investigation of light was in a sense unique because it relied on experiments that Newton himself conducted. His interest was probably stimulated by the curiosity of contemporary scientists about the peculiar effects observed when light passes through shaped pieces of glass. Only five years earlier in 1661 the Dutch biologist Anton van Leeuwenhoek had invented a powerful microscope that was a major advance over an earlier instrument devised in 1590 by Zacharias Jensen. And the importance of the telescope, invented in Holland in about 1608, was widely appreciated as a consequence of Galileo Galilei's publication of his discoveries in 1610 of the satellites of Jupiter, sunspots, and mountains on the moon. The colorful effects produced when sunlight passes through a glass prism were well-known to Newton's contemporaries, but it was his own genius that carried the experiments one step further and thus demonstrated a fundamental aspect of light.

One of Newton's most crucial experiments was to allow a narrow ray of light, formed by a hole in a windowshade, to pass through a triangular glass prism. On entering the prism, and again on leaving, the path of light is bent. This effect is now

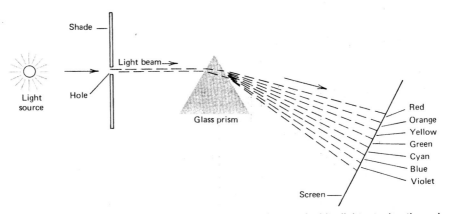

Fig. 2-1. Newton's experiment in which a narrow beam of white light passing through a prism is refracted and spread, with different colors of the spectrum being observed when the beam shines on a white screen (see Color Plate 1).

called *refraction*. A careful observer also notes that at each of these surfaces a small amount of light is reflected without penetrating the glass. Newton was primarily interested in the refracted light, and to study it he directed the beam refracted by the prism toward a distant white cloth, or screen. He then saw the full spectrum of colors spread out from violet through red (Fig. 2-1). A modern version of this experiment is shown in Color Plate 1, in which white light from a lamp is refracted by a prism to form the spectrum of colors. This effect was known before Newton began his studies, and one popular explanation held that as the prism refracted the light, it must have also affected the light's color by "staining" it in varying amounts.

But Newton overturned this notion with a further experiment shown in Fig. 2-2. He used a hole in a second shade placed beyond the prism to allow only the yellow portion of the spectrum to pass through to a second prism and observed the color when the emerging beam struck a white screen. It was still yellow! He reasoned therefore that a prism did not "stain" light or add color to light that was originally uncolored. Rather, he suggested, the components of light of various colors revealed by the first prism were originally constituents of the incident white light. The action of the first prism was to separate the white light into its components of different colors.

Spectral Colors

This was an observation with far-reaching consequences, because it provided a new basis on which to explain many color phenomena. The notion that light from a narrow portion of the spectrum could not be further subdivided into light of different colors led Newton to describe it as "homogeneal." Today we call it *monochromatic*. This experiment showed that monochromatic light in one region of the

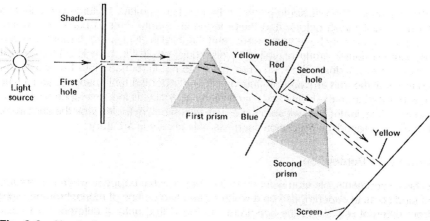

Fig. 2-2. Newton's experiment with two prisms showing that spectral colors (in this example, yellow) cannot be further subdivided into components of other colors.

spectrum differs from monochromatic light in another. Newton called the difference "refrangibility," meaning the ability to be refracted by a prism. Figure 2-1 shows that the blue monochromatic light is more refrangible than red light, because it is refracted more than the red. But Newton at this time had no clear notion about what light is, and his terminology has been replaced by the modern notion of *wavelength* to distinguish one monochromatic component of the spectrum from another. This comes from our present understanding that light travels through space with wavelike properties. Blue monochromatic light has a shorter wavelength than red, so we describe the prism experiment by saying that light of shorter wavelengths is refracted more by a glass prism than light of longer wavelengths. In Fig. 2-3 the colors of the spectrum are indicated together with the wavelength scale. The shortest wavelength of light—at the violet end—is about 400 nanometers, and the longest—at the red end—is about 700 nanometers. Since the *nanometer* is only one-billionth of a meter (about one-billionth of a yard), the wavelengths of light are incredibly short.

As soon as one looks at a spectrum like that in Color Plate 1, it becomes obvious that the number of colors that can be distinguished depends on one's degree of

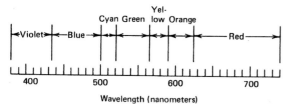

Fig. 2-3. Spectrum of light showing the colors corresponding to various wavelength regions.

sophistication. There is some evidence that certain primitive cultures use only two terms to distinguish colors, and these terms invariably refer to black and white. Other societies may have three terms, and the third is always red. If four terms are recognized, either green or yellow is distinguished. If six, blue is added; and if seven, brown. Indeed, the term "orange" was not introduced into European language until the fruit arrived from the Middle East after the tenth century, and it was not liberated from exclusive association with the fruit until the seventeenth century. Figure 2-3 indicates that we sometimes use seven colors to describe the spectrum, but as the preceding discussion suggests, this choice is arbitrary.

Color as a Perception

In his experiments, Newton went on to investigate what happens when mirrors are placed so as to superimpose on a white screen two beams of monochromatic light from different regions of the spectrum. He found that quite a different third color was seen! For instance, superimposed green and red beams produced a yellow color. This can be seen in Color Plate 2 where the area illuminated by the green and red beams produces a yellow closely resembling the region of yellow in the spectrum of Color Plate 1. This intermingled light was not monochromatic, because Newton found that when it passed through another prism two beams emerged with the two original monochromatic colors, green and red. Thus a yellow color may be produced either by monochromatic light from the spectrum or by a superposition of two different but appropriately selected monochromatic lights. The character of the light in the two cases differs, as revealed by the spectra produced when passed through a prism. Yet the perceived color of the light is the same.

From these observations Newton deduced a fundamental principle that we always must keep in mind: the notion of "color" applies only to the perception we have, not to the light itself. We should not confuse a perception with the stimulus that produces it. When we talk about the "color of light," we really mean the color it evokes in our mind. Light of a variety of different spectral compositions, appropriately chosen, can evoke the same color perception; therefore, describing light by the color it produces is ambiguous and often misleading. For convenience we may talk about a "yellow light," but we really should say a "light that we perceive as yellow."

All the colors in the universe which are made by light, and which do not depend on the power of imagination, are either the colors of homogeneous light [spectral colors] or are compounded of these. . . .

Newton's Proposition VII, Theorem V from *Opticks,* **Book 1, Part 1**

Newton's ideas were by no means immediately accepted by the scientific and intellectual community. As late as 1810 the famous German writer and natural philosopher Johann Wolfgang von Goethe published in the *Farbenlehre (Theory of Colour)* a systematic attempt to prove Newton wrong. As noted in Chapter 1 he argued for the unity and indivisibility of light and referred contemptuously to Newton's prism demonstrations as "the epoch of a decomposed ray of light." In effect he fell into the same trap as had thinkers in the pre-Newtonian era. He believed that visual phenomena, such as our perception when viewing two contrasting colors, could reveal the nature of the light producing each color. However, investigations during the last century have shown this to be incorrect. Our visual system has been found inadequate for the task of accurately describing the physical composition of light. Newton was in fact right when he concluded that a prism reveals uniquely the spectral composition of a light and that without the prism this composition is not accessible to human perception.

Spectral Power Distribution of Light (SPD)

The *spectroscope* is a modern instrument that works on the same principle as the prism to indicate the power (in watts) of the components at various wavelengths. Light entering the spectroscope is spread by a prism much as in Newton's experiment. As a detector moves across the screen, it measures for each wavelength of light the rate at which energy is received from the light source. The resulting curve giving the power at each wavelength is called the *spectral power distribution.* The spectral power distribution will play such an important role in our future discussions that we shall introduce an abbreviation for it: SPD. We gain a more discriminating appreciation of various colors if we keep in mind the shape of their SPDs. In Fig. 2-4 we illustrate the particular SPDs for the blue, green, and red portions of the spectrum. For blue the power is found concentrated at short wavelengths; for green, at intermediate wavelengths; and for red, at long wavelengths. We shall see later that although the SPD cannot be determined from the visual response it produces, our visual response to a particular SPD *can* be predicted on the basis of the properties of the human visual system, once the SPD is known. It is by this process that color can be "measured." Therefore the first step in determining the visual perception that a particular light source will produce is a measurement of the SPD. The *spectrophotometer* is a modern instrument that measures the SPD automatically.

■ 2-2 SUBJECTIVE SPECIFICATION OF COLOR

Because color is a perception we must rely on human observers to describe it. But even the most attentive and discriminating observer has no way of specifying color in an absolute or quantitative sense. That is, there is no way to "put a number" to color. Have you ever tried to describe the color of an object, such as a new brown coat to a friend over the telephone? It can be done only in a *comparative* sense, by comparing the attributes of one color with those of another. To communicate effectively with your friend, you both need previous exposure to the same color, or colors, that can be used for reference.

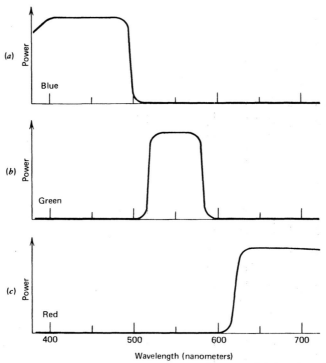

Fig. 2-4. SPDs for the (*a*) blue, (*b*) green, and (*c*) red portions of the spectrum of white light, as separated by a prism.

Hue, Saturation, and Brightness

Numerous psychological studies reveal that three attributes specify a color: *hue, saturation,* and *brightness.* We say these are "attributes" because they describe a perception, not physical properties of the light itself. *Hue* is the attribute that we denote by such adjectives as "red," "yellow," and "green." It is an obvious feature of monochromatic light that varies with wavelength and therefore is used to denote various regions of the spectrum. *Saturation* refers to the lack of "whiteness" in a color, or more precisely, how much a color differs from white (or gray). "Pink" is an example of unsaturated red. *Brightness* is the attribute that describes the perceived intensity of light. A common example will illustrate the use of these terms: the sun is seen at noon to have a yellow hue, which is strongly unsaturated and very bright. But at sunset the hue has shifted to red and the color is more saturated and dimmer.

The psychological description of color does not correspond simply to the spectral composition of the light evoking the perception. True, monochromatic light is seen as the most saturated example of a particular color. Yet an observer does not judge monochromatic colors from different regions of the spectrum to be equally saturated. Monochromatic violet or monochromatic red near the ends of the spectrum appear more saturated than monochromatic yellow near the center of the spectrum. (We shall see later that this curious effect can be explained in terms of the three color receptors of the eye, each of which responds to light over a broad region of the spectrum, with some overlap near the center.)

A particularly simple color to describe is gray. Any shade of gray represents a completely unsaturated color in which hue is totally lacking. Black and white are extreme examples of gray. The various grays can therefore be described by only one attribute—brightness. They range from black (the absence of any light) through the just-visible gray at very low light levels, to dazzling bright white. Because it lacks hue, a gray color is called *achromatic*.

Like whiteness, justice has no degrees. When they are qualified by "more or less," they owe such qualifications to the things with which they are mixed.

Dante
De Monarchia, c. 1300 A.D.

It should be emphasized that the description of a color in terms of hue, saturation, and brightness applies to our perception of light. The situation we envisage is a simple one in which we view a single light in the dark, or perhaps see light from a stagelight reflected from a large screen, or even view a white wall uniformly illuminated by a colored lamp. The color in these situations is determined exclusively by the nature of the light entering our eye. Psychologists know that describing the color of a surface in more common situations is more complicated. Our judgment of the color of a bookcover may be influenced not only by the type of light illuminating it but also by the color of other objects in view. We defer consideration of the interplay of surface properties and nature of the illumination until later in this chapter, and the subtle psychological features that go by the terms "lightness" and "color constancy" will be discussed in Chapter 10. For the moment let us restrict attention to investigating the color properties of light itself.

■ 2-3 ADDITIVE COLOR MIXING

In his 1704 edition of *Opticks* Newton described a number of experiments in which he arranged mirrors to direct two different monochromatic components of the spectrum to the same location on a white screen, and as we mentioned earlier he observed the color of the intermingled beams. Although he could produce many different colors, Newton noticed that he could not always produce white light by mixing two monochromatic colors together. His systematic studies revealed general rules of color addition, which we shall investigate in this section.

The rules discussed in this section are solely for *additive color mixing*, where one light is added to another. The spectral power distribution (SPD) of one light is added to the SPD of the other to produce the net SPD. This happens when two stage lights are simultaneously directed toward the same position. The more lights that are added, the brighter is the net effect. This differs from *subtractive color mixing,* which pertains to painting and color printing, where the more colors that are mixed together the dimmer becomes the surface. Subtractive color mixing will be discussed in the next section.

The rules of additive color mixing are valid whether or not the light that produces a given color is monochromatic. In this section we shall emphasize primarily the rules as they pertain to hue, and will not be especially concerned with the effect on saturation and brightness. For simplicity we shall assume that the brightness of each original color is about the same.

Experiments show that mixing green light and red light produces a yellow light. To summarize this mixture symbolically we write

$$G + R \rightarrow Y \qquad (2\text{-}1)$$

Of course, if we vary the relative brightness of G and R lights, the resulting hue will range from pure green through olive green, yellow, orange, and finally red. Many different colors can thus be formed by adding G and R. But, if G and R have nearly identical brightness, then Eq. 2-1 is correct.

Another way to think about this color rule is to refer to the SPDs for the green and red portions of the spectrum shown in Fig. 2-4. Then it is clear that by adding G and R together the SPD includes contributions from both intermediate and long wavelengths of the spectrum. Nevertheless experiment shows that this light is perceived as yellow.

A second example of addition is when blue and red lights are superimposed to produce magenta (a purple):

$$B + R \rightarrow M \qquad (2\text{-}2)$$

As the brightness of B varies relative to R the range of hues that may be produced spans the range from blue to reddish blue, magenta, bluish red and red. In terms of the resulting SPD we realize that important contributions from both the short wavelength and long wavelength regions of the spectrum produce the color magenta.

For a third example we note that when blue and green are added we obtain the color cyan (or turquoise):

$$B + G \rightarrow C \qquad (2\text{-}3)$$

Consequently when the SPD contains contributions from both short and intermediate wavelengths the color seen is cyan.

With the *three* colors B, G, and R mixed additively in various proportions a wider range of colors can be produced, including all of the preceding examples. If the three have similar brightness, when added they produce white:

$$B + G + R \rightarrow W \qquad (2\text{-}4)$$

Here the resulting SPD includes contributions from short, intermediate, and long wavelength regions of the spectrum.

Primary Colors

Experiments first reported by the English physician Thomas Young in the early 1800s show that virtually all colors can be produced from a set of three lights whose colors are found at widely separated regions of the spectrum. The three colors are called *primary colors* if they also can be added together to produce white. The three primary colors need not be monochromatic although they can be. There is no unique set of primaries, although some choices when mixed will provide a wider range of colors than others. For this reason a useful set of primaries has hues that are close to blue, green, and red. Therefore, blue, green, and red are usually considered to be the primary colors for additive color mixing.

Any two colors that produce white when added together are called *complementary*. The complement of a primary color is called a *secondary color*. Referring to Eq. 2-4 we see that the complement of R is B + G, which is the same as C. Therefore, cyan is a secondary color and is the complement of red. Similarly, Eq. 2-4 indicates that the complement of G is B + R, or M, so magenta is also a secondary color. Finally, the complement of B is G + R, or Y. Thus yellow is a secondary color and is the complement of blue. All of these results can be seen by carefully examining Color Plate 2.

The rules of additive color mixing are easily understood when we organize our thinking in terms of primary colors. When adding two colors we first determine which combination of primaries describes each of the two and then deduce the effect of adding the constituent primaries. We thereby predict the resulting hue and gain a notion about the degree of saturation.

Example 2-1   Adding Colored Lights Together

Suppose we wish to predict the color when a yellow spotlight and blue spotlight are directed onto a white floor. We can proceed by thinking about the primary colors that are represented. We recall that yellow can be produced by a mixture of G and R. Then it is clear that the addition of B will produce white, if all three primaries are present in nearly equal proportions. (Of course we might also have remembered that yellow is the complement of blue!)

Now for a more difficult case, suppose that a magenta and a yellow light are superimposed on the floor. We proceed by recalling that M is composed of B + R, and Y is composed of G + R. Adding the contributions gives:

$$M + C = B + G + R + R = W + R$$

We recognize that the combination of B + G + R is white, but we still have some R left over. The hue would be red, but the presence of the white makes it unsaturated, so it would appear pink.

Newton's Color Circle

An effective way to summarize approximately the additive relationships between colors was devised by Isaac Newton and is known as *Newton's Color Circle* (Fig. 2-5a). The three primary colors are arranged about the circumference of a circle with the corresponding complementary colors diagonally opposite. The colors from R ranging through Y, G, C to B are in the same relative positions as they fall in the spectrum. We note that violet is a color of the extreme end of the spectrum, but it is included with blue in this rendition of the color circle. It is a common belief that magenta (or the purples) are a part of the spectrum, but this is not true. Magenta is one perceived color that cannot be produced by monochromatic light. It can be seen only when the SPD contains contributions from both short and long

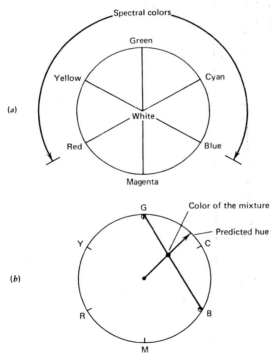

Fig. 2-5. (*a*) A modern version of Newton's Color Circle showing the range of spectral colors. (*b*) The result of adding two colors (here, blue, and green) is represented by a point on the line joining the two colors; its hue is indicated by where a line from the center through the point meets the circle.

wavelengths. Newton closed the color circle with magenta because it logically fits there, diagonally opposite its complement green, not because this position represents a physical aspect of light.

Newton recognized that this arrangement has the virtue that it allows us to roughly predict the result when two colors are added together. We need only draw a line that connects the positions of the two colors (Fig. 2-5b). The resulting color lies somewhere on that line, depending upon the relative brightness of the two colors. The point on the line in Fig. 2-5b represents one example when adding B and G of nearly equal brightness, for the resulting color is nearly halfway between the two. The hue of this color is represented by a position on the circumference that is deduced by drawing a second line from the center of the circle outward through the point. Where this line intersects the circumference designates the hue of the resulting color, in this example C. Evidently when complementary colors are added in proper portions the result is represented by a point lying at the center of the circle. Thus with white represented at the center and spectral hues (plus the purples) on the circumference, Newton deduced that a measure of the saturation of a color is how far it lies from the center. Newton's color circle is a kind of map, showing saturated colors on the circumference and progressively less saturated colors closer to the center. It is a useful way to visualize color relationships, because it includes information about both hue and saturation. However, it cannot accurately predict the results of additively mixing two colors, since it does not assign numerical values to colors. Examples of more accurate color maps will be described in Chapter 3.

Additive Color Effects

Many examples of additive color mixing are found in our contemporary world. They can be classified by the method used in adding colors. The first is *superposition* in which two or more colored light beams simultaneously illuminate a surface. This is common in stage productions and outdoor advertising and is a direct example of the preceding discussion. Especially interesting effects come from two different colored lights placed on opposite sides of the stage; for then shadows of one light take on the color of the other, and the interplay of original and additive colors lends vitality and depth to the presentation.

A second method of additive color mixing depends on spatial fusion and is called *partitive* mixing. When the sources of differently colored lights are viewed directly and are sufficiently small and close together that they cannot be seen individually, the perceived color is the additive mixture of the individual colors. One example of this is color television. The face of the TV screen is covered with an array of tiny spots 0.2 mm in diameter of three different kinds of phosphors (Color Plate 4). They emit B, G, and R light when three separate electron beams directed toward them from the back of the tube strike the phosphors. As there are more than 200,000 triads of dots, it is impossible to see them individually at normal viewing distances and only the net effect of their light is perceived. The relative brightness of the three colors at each location on the screen is determined by the relative intensity of the three electron beams when they scan through that location; therefore the full range of colors can be produced by the electrical signals that govern the beams. The principles of television will be discussed in detail in Chapter 13.

A second example of partitive mixing is found in the paintings of the Pointillists, who flourished at the end of the nineteenth century. Artists such as Georges Seurat were fascinated by the optical effect of putting small dots of different colors close together on the canvas. Seen from a sufficient distance the light from the dots mixes additively on the retina; on approaching the painting one receives a kinetic impression as the partitive aspect is lost and the individual dots come into view. In most Pointillist paintings the dots are not sufficiently small nor close together to yield partitive mixing at normal viewing distances. Paul Signac's painting *"Harbor at St. Tropez,"* which appears on the cover of this book, illustrates the vibrant effect achieved by the Pointillists.

Shortly after the Pointillists, the Lumière brothers developed autochrome, a process of color photography in which small colored grains of starch do achieve partitive mixing. Textiles give a more common example of partitive mixing whereby cloth finely woven from differently colored threads takes on a new color from that of the individual threads. Mosaics, which were popular in Byzantine churches, when viewed from afar also may display partitive mixing. Interspersed red and green glazed chips are seen as dark yellow or ochre; whereas blue and green are seen as dark cyan. Blue and yellow, complementary colors, are seen as gray.

Partitive mixing of light *reflected* from a surface such as a mosaic differs from the effect of light that is *produced* by the surface itself, as in color television where a full range of colors can be achieved including bright white. In the first case the inclusion of more colored dots does not necessarily increase the overall brightness. Some dots may represent a dark color and consequently reflect very little light. Including more of these dots would diminish brightness. Thus, blue and yellow tiles are perceived as gray, not white, in partitive mixing of reflected light. It is impossible to produce a dazzling white by a partitive mixture of reflected blue, green, and red lights, because too much light is absorbed at the surface by chips that reflect just blue or green or red.

The third method of adding color is through *visual persistence*. When two colors are flashed quickly in alternation on a screen, the human visual system cannot follow the individual flashes, and the perception is similar to that of two colors mixed by addition. Analogous effects are obtained when a disc with sections painted in various colors is spun rapidly.

■ 2-4 SUBTRACTIVE COLOR MIXING

The notion of primary colors, which proves useful for discussing additive mixing of light, can be applied to a different set of situations where the SPD (spectral power distribution) is altered as light passes through a medium that absorbs differing amounts of power across the spectrum. This occurs when a beam of light passes through one or more colored filters (such as a piece of colored glass or a plastic containing a dye). The more colored filters we insert into the path of the beam, the more light is absorbed and the dimmer is the emerging light. A similar effect is produced when paints of various colors are mixed together and light penetrates the mixture before being reflected. These are examples of *subtractive color mixing.*

The rules of subtractive mixing are more complicated than those of additive mixing. If several filters are placed in a beam of white light only the first filter receives all of the white light. The others receive variously colored light depending on the filters that precede them. To figure out what happens we must follow the path of light step by step.

Selective Absorption

By convention a filter is named by the color of the light that emerges from it when it is placed in a white beam. A red filter allows long wavelengths to pass and absorbs the rest. A common yellow filter allows medium through long wavelengths to pass and absorbs short wavelengths. The process whereby a medium absorbs light in only certain regions of the spectrum is called *selective absorption*. It is the primary cause of most colors that we see. In terms of our additive primaries, a red filter allows only the R component of the incident light to pass. A common yellow filter transmits G and R. Similarly a magenta filter transmits B and R; and a cyan filter transmits B and G. Figures 2-6*a*, *b* and *c* and Color Plate 3 show the effect of these individual filters on white light.

With these transmission properties in mind we can consider what happens when two filters are placed one behind the other in a white beam. If we unhappily chose a red filter and blue filter, we must conclude that *no* light emerges, and the resulting color is black. This happens because in the spectral region where one filter transmits light the other absorbs it. A more interesting example occurs when a magenta filter and a yellow filter are inserted into a white beam (Fig. 2-6*d*). The magenta filter allows B and R to pass. The yellow filter would transmit any incident G and R, but since G is absorbed by the magenta filter, only R emerges. The combination of the two filters therefore has the same effect as a red filter. It makes no difference which of the two filters is inserted into the beam first.

Subtractive Primary Colors

Following a similar train of logic, a combination of yellow and cyan filters transmits green light (Fig. 2-6*e*); and cyan and magenta give us blue light (Fig. 2-6*f*). Various combinations of two or three filters (magenta, yellow, and cyan) produce essentially all colors including red, green, or blue from originally white light. Therefore magenta, yellow, and cyan are called the *primary colors for subtractive color mixing.* Color Plate 3 summarizes the results that we have just deduced when two or three filters of these subtractive primaries are superimposed. The reader will perceive that the subtractive primary sometimes loosely called "red" is actually magenta, since it contains some blue; and the primary often referred to as "blue"

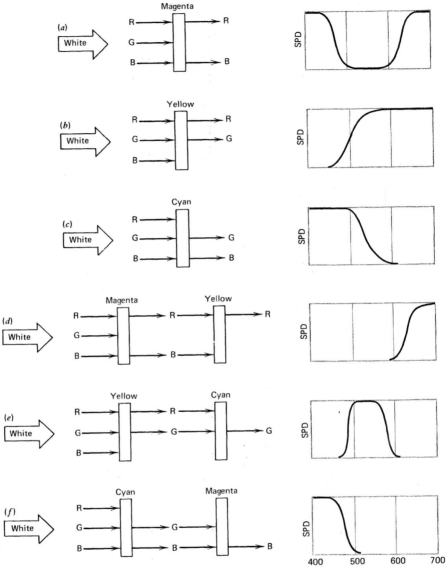

Fig. 2-6. Effect of magenta (*a*), yellow (*b*), and cyan (*c*) filters on white light. Also shown (*d–f*) is how the three additive primary colors can be produced from white light by using two filters in succession. On the right are shown the SPDs of the emerging light in each case.

is actually cyan. These subtractive primary colors are easy to remember because they are also the complements of the primary colors for additive color mixing. The basic explanation given above for the rules of subtractive mixing was first put forth by Herman von Helmholtz in 1850.

Rather than learn a new set of rules for subtractive color mixing, it is simpler and perhaps more satisfying to apply our knowledge of the additive primaries and how filters affect light. This is shown in the following example.

Example 2-2   Subtractive Mixing

What light emerges when yellow light is incident on a cyan filter? The incoming light has red and green primary components. Because the filter passes only green and blue, the red component is absorbed and only the green emerges.

What is the color of light coming from a magenta filter when green light is incident? Since only red and blue are passed by the filter, no light emerges. The color is therefore black.

Spectral Transmittance Curves

Now that the fundamental principles of subtractive color mixing have been established let us return to consider in greater depth the effect of a filter. The description in Fig. 2-6 is oversimplified, because specifying the emerging light as containing red and blue, or red and green, for example, does not provide details as to exactly what percentage of the light at each wavelength is transmitted. In other words, we do not know the details of how the SPD is affected. Given knowledge of the SPD of the incident light, we would like to predict the SPD of the transmitted light. A more accurate description of a filter's effect is given by its *spectral transmittance curve* (Fig. 2-7). The curve specifies the fraction of incident light that is transmitted for each wavelength across the spectrum. This percentage at a given wavelength is called the *spectral transmittance* for that wavelength. The rest is absorbed by the filter, or perhaps is reflected back from one or both surfaces. More precisely, the value of the spectral transmittance at a given wavelength is defined as:

$$\text{spectral transmittance} = \frac{\text{transmitted power}}{\text{incident power}} \tag{2-5}$$

The spectral transmittance therefore can range from 0 to 1 or, when expressed as a percentage, from 0% to 100%. Figure 2-7a illustrates a red filter for which the curve is high only in the red portion of the spectrum. The curve in Fig. 2-7b is for a blue filter, which has appreciable transmittance only in the blue portion of the spectrum. What color results if this red filter and blue filter are both inserted in a beam of white light? Since the portion of the spectrum transmitted from either filter is not transmitted by the other, no light emerges, and the color is black.

We can now combine two important concepts to understand how a particular light source and a particular filter will influence the color of the emerging light. The

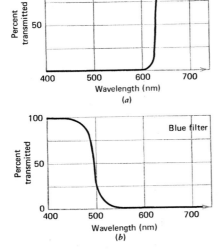

Fig. 2-7. Spectral transmittance curves for a red filter and a blue filter, showing the percentage of light at each wavelength that is allowed to pass through the filter. These are idealized curves, because a typical filter will pass somewhat less than 100% of the light in the spectral regions where it transmits.

light source is characterized by its SPD, and the filter by its spectral transmittance curve. When the filter is placed in front of the light source, the amount of power produced by the light source at each wavelength will be reduced by an amount determined by the value of the transmittance curve of the filter at that wavelength. To obtain the SPD of the light transmitted by the filter, we simply multiply the SPD of the source by the value of the filter's transmittance curve at each wavelength.

An example is shown in Fig. 2-8. The shape of the SPD for the emerging light exhibits features that are characteristic of both the SPD of the source and the transmittance curve of the filter.

Transmittance

Manufacturers of filters provide a spectral transmittance curve for each filter and may also give the overall percentage of light that is transmitted. This is called simply the *transmittance*. A filter designated by a transmittance of 32% will transmit 32% of the light intensity if white light is incident. The transmittance takes into account the varying sensitivity of the eye to different regions of the spectrum, as will be described in Chapter 6. In cases where a filter is inserted into a beam that already has a distinct hue, the value of the transmittance given by the manufacturer no longer applies.

Spectral and Monochromatic Filters

The common practice of labeling a filter by the hue it produces from a white beam, although useful, is very imprecise and sometimes misleading. By contrast, the transmittance curve of the filter completely describes its effect. As an example of the advantage of considering the transmittance curve, Fig. 2-9 gives three curves for different filters that could all be called "yellow." The curve describing a high spectral transmittance that extends over a broad range of wavelengths—from green through red—is typical of common filters used in stage lighting and displays. But yellow could also be achieved by a filter that transmits a narrow range of wavelengths centered at the yellow region of the spectrum. This is said to describe a *spectral filter*. If light emerging from this filter were analyzed by a spectroscope, it clearly would not reveal any green or red components, only yellow. A spectral filter that allows one of the additive primary colors to pass is sometimes called a *primary filter*.

For some purposes an even more narrow range of wavelengths is desired. Filters with this characteristic are said to be *monochromatic*. The distinction between spectral and monochromatic filters is arbitrary because they merely differ in the width of the transmitted spectrum. Generally, we can justify calling a filter monochromatic if its width is less than, say, 10 nm.

The colors from a common, spectral, and monochromatic filter usually differ in saturation. The spectral filter produces much greater saturation than a common one, because much less of the spectrum is passed. A monochromatic filter pro-

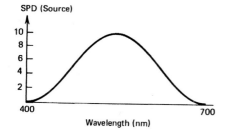

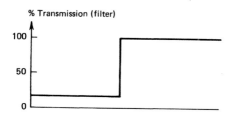

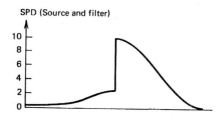

Fig. 2-8. The SPD of a light source, transmittance curve of a filter, and the SPD of the light emerging from the filter.

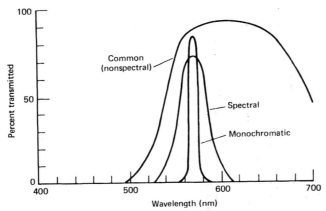

Fig. 2-9. Transmittance curves for three different types of yellow filters.

duces even greater saturation. However, spectral and monochromatic filters are not commonly used on the stage or in displays. They pass such a narrow portion of the spectrum that even if the transmittance curve is high in that region, the over-all transmittance is very low. Therefore, unless a particularly high-powered light is available, the colored beam is too dim. The usual desire for a bright beam compels the use of nonspectral filters with unsaturated colors.

The subtractive color mixing laws summarized in Fig. 2-6 are applicable only for the common, nonspectral filters. For instance a spectral yellow filter does not transmit red and, therefore, can only yield black in combination with a magenta filter. Similar reasoning leads us to conclude that some of the illustrations in Fig. 2-6 are not correct if the cyan filter is spectral cyan. On the other hand, magenta is not a color of the spectrum and so a spectral magenta filter cannot be made; a magenta filter always passes light in the blue and red portions of the spectrum. Color Plate 5 gives examples of the transmittance curves and colors of some filters commonly used in stage lighting. Owing to the limitations of the color printing process, these colors are slightly different from those that would be produced on stage.

A filter that is particularly simple to characterize has the same transmittance for all wavelengths. It is called a *neutral density filter.* Of course it does not affect the shape of the SPD of light passing through it but merely reduces the brightness by decreasing the SPD by the same percentage at each wavelength. Neutral density filters are rated according to their transmittance. A 39% filter transmits 39% of the intensity of the original beam at each wavelength, and 39% of the overall light intensity.

Paints and Inks

The coloring effects of paints and inks applied to a surface are also described by the laws of subtractive color mixing. The color comes from fine particles called a *pigment* suspended in a clear fluid. With inks the pigment is left attached to the paper as the fluid evaporates. For oil painting the fluid medium is traditionally linseed oil, which is sufficiently viscous to keep the pigment in suspension and dries to provide a protective cover. As shown in Fig. 2-10, the layer of paint, which is seen in cross section, contains many pigment particles at varying depths below the surface. Light shining on the surface passes through the dried linseed oil and strikes the particles of pigment. A portion of the light is reflected while the rest is transmitted to the next particle lying below. The reflected portion has its color little affected, because it has not penetrated far into the particle, while the transmitted portion undergoes considerably more selective absorption. The cumulative effect of encountering many particles leaves the light that finally emerges appreciably affected by the pigment. The effect when light is transmitted through one pigment and then another is identical to the effect achieved by placing a succession of colored filters in the beam of light. Pigments of the artist's palette behave like nonspectral filters with few exceptions. Thus subtractive color mixing laws that successfully describe how light is altered by nonspectral filters also describes how light is altered by pigments.

Enhancement of color by successive reflections is a very real effect that is apparent in many ways. Interior decorators know that in a narrow hallway, light reflects many times from one wall to another, and this accentuates the saturation of the prevailing color beyond what happens on a single reflection. To avoid unpleasant coloration and darkening, decorators recommend that paints with unsaturated colors be applied in confined public areas.

The Impressionists of the nineteenth century understood very well the distinction

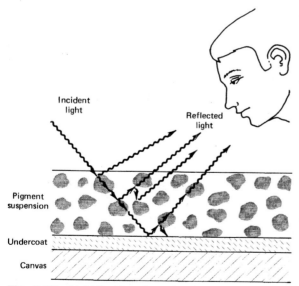

Fig. 2-10. Light that is reflected from paint has passed through many of the pigment particles and has also been reflected by many other particles before it leaves the surface of the paint.

between subtractive color effects achieved by mixing paints and additive effects when colored dots are placed side by side on a white reflecting surface. With the latter, where the eye produces the color mixing, each area of the painting reflects more light than if the paints were mixed, because only one pigment is present. This gives greater reflectance and brightness to better convey the effect of intense sunlight. Even when the individual dots of variously colored paints are not sufficiently fine to produce true partitive color mixing, the intensity and juxtaposition of colors produce a dramatic effect. Conventional painting, achieved by mixing paints or applying them in successive layers, causes a diminution of brightness as absorption takes place over a broader span of the spectrum within each small area of the painting.

Substitute optical mixture for the mixture of pigments. In other words: break down tones into their constituent elements because optical mixture creates much more intense light effects than the mixture of pigments.

Camille Pissarro, in a letter to Durand-Ruel, 1886

■ 2-5 PRINTING

Color printing in books and magazines provides an example where both subtractive and additive color mixing are important. Before taking up this topic, let us consider the simpler process of reproducing a black-and-white picture by applying black ink to white paper. A black-and-white photograph is not really black and white; it is black and gray and white. It offers a continuous range of brightness, and in the printing trade it is called a *continuous tone.* By contrast, the *halftone* process commonly used for illustrations in magazines consists entirely of black dots on a white background (Fig. 2-11). To print a region of medium gray the paper is printed with black dots of such a size and spacing that the area is half dots and half white space. If the dots are sufficiently small and numerous, the optical illusion produced by visual fusion makes them undetectable at a normal viewing distance, and the area is seen as a uniform gray.

In a halftone reproduction of a photograph the dots are equally spaced but they vary in size. The pattern of dots is made by photographing a negative of the original picture through a halftone screen. The negative has light areas where the original picture was dark, and dark areas where the original was light. A halftone screen is made from two plates of glass, each scribed with closely spaced parallel lines, and one plate is rotated so that its lines are at right angles to the other's. The scribed lines are about as wide as the spaces between them. Typically 50 lines fall within a distance of one centimeter. When the photograph is made, light passes through the negative and then through the small square spaces of the screen whereupon it spreads to produce dots on the film proportional in size to the amount

Fig. 2-11. (*a*) A halftone reproduction of a photograph of foxtail pines in the High Sierra Wilderness area of California. (*b*) An enlargement of the rectangular area of the reproduction showing the individual dots that comprise it. (Forest Service, U.S.D.A.)

of light coming from the negative. Light gray areas of the negative make larger dots, and dark areas make small dots. Therefore the large dots correspond to regions that are dark on the original positive picture and small dots to regions that are light. This developed film containing the pattern of dots is used to prepare the printing plate on which ink will be spread.

For moderate press runs of a few hundred thousand copies, quite decent reproduction is possible with *lithography*. This process was invented by Aloys Senefelder of Munich in about 1798. He found that if he drew a design with an oily substance on the smooth surface of a stone, when he wet the stone's surface with a solution of gum arabic and water the oil repelled it but elsewhere the stone absorbed it. Then, when Senefelder rolled an oil-based ink (made of carbon particles, oil, soap, and wax) on the stone, the ink adhered to the oily design but was repelled from the water. Pressing the stone to paper transfers the ink and imprints the design. In modern lithography the plate is a thin sheet of aluminum coated with a light-sensitive emulsion. When the plate is exposed to ultraviolet light passing through the film containing the pattern of dots, the molecules of the emulsion are altered so that ink cannot adhere to them. This leaves the dots free to retain ink, which is pressed onto paper when the presses run. The surface of a lithographic plate is quite smooth; it is the chemical difference between regions of the emulsion that determines where the ink is retained. In offset lithography, introduced just after 1900, the pattern of ink is "offset" from the plate onto a rubber roller before being transferred to the paper (Fig. 2-12). This saves wear on the plate, and the soft rubber makes a more even impression on the paper.

A *letterpress plate* is also used to reproduce halftones but has a raised printing surface, the remainder having been etched away. It is a simple process and is appropriate for making a comparatively small number of copies. By contrast, *gravure* is used for magazines of wide circulation that must maintain high quality for

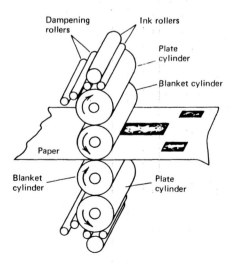

Fig. 2-12. Arrangement of a press for offset lithography. The rotating plate comes in contact first with water then with oily ink. The inked image is transferred to a rubber blanket that deposits it on the paper. The press shown here simultaneously prints on both sides of the paper.

a large number of copies. The ink is carried in etched recesses in the plate, which is then pressed onto paper. In fact it was by a process similar to gravure that the German James Le Blon discovered the basis for modern color printing in the mid 1700s: a wide range of colors could be produced by overprinting combinations of magenta, yellow, and cyan inks. Le Blon engraved his design in the surface of blocks, one for each of the three colors. The depth of the incision determined the amount of ink and thus the density of the pigment applied by that block at each location. The blocks were inked and pressed in turn onto the paper, with care taken to insure proper alignment; in that way the first color printing was achieved.

Four-Color Printing

In modern printing the same principle is applied as in Le Blon's process, but four colors are used and each image is composed of dots. Although a number of procedures are in use, essentially they all end up by producing four screened positive films of the picture: one for the magenta parts, one for the yellow, and one for the cyan. Originally these individual *color separations* were made by photographing the original picture through magenta, yellow, and cyan filters, but, now they are produced electronically with a laser beam that scans the picture. Printing inks do not perfectly represent subtractive primaries (each pigment transmits somewhat in spectral regions where it should be completely absorbing) so when the three colors are superimposed they do not produce black but rather a dark brown. To overcome this defect a black separation is also made and printed over the others (Color Plate 6).

When making the screened separations, the lines of dots are rotated by about 30° between one color and the next, so that dots overlap in some areas and are juxtaposed in others. This blends the colors more uniformly and makes the need for alignment of the individual printing plates less exacting. Because of the partial overlap of the dots both subtractive and additive color mixing play a role. Where dots overlap the color is predicted by subtractive rules; where they are adjacent the partitive mixing of spatial fusion brings additive mixing into play. For instance, where magenta and yellow dots overlap a saturated red is produced; but where they are adjacent, an unsaturated red is seen (recall Example 2-1 of Section 2-3). The color is made even less saturated in the last case because of additional white from the paper that shows between the dots. The process of color printing is a demanding art in which the proper size and alignment of the dots and the correct balance and stickiness of the inks to match the paper must be realized. During the press run these and other adjustments must be made to ensure uniformity and accuracy in rendering the colors.

■ 2-6 SPECTRAL REFLECTANCE CURVES

The way in which a surface affects the SPD when reflecting light can be precisely characterized by its *spectral reflectance curve*. This curve is analogous to the spectral transmittance curve for a filter but, instead, gives at each wavelength the percentage of incident light that is reflected. An example for three common farm products is shown in Fig. 2-13. It may seem surprising that butter reflects a higher percentage in the red portion of the spectrum than a ripe tomato; but of course butter has a yellow hue when illuminated by white light because it also reflects in the green portion of the spectrum, where a tomato does not. Because butter reflects an appreciable percentage of light over the entire spectrum, its color includes a high proportion of white; therefore the yellow is highly unsaturated. By contrast a red tomato reflects over a much more restricted portion of the spectrum and consequently the color is more saturated. These reflectance curves suggest that butter has a brighter color than tomatoes, because it reflects more light. This is actually a correct conclusion in this case, but it is not safe to make statements about brightness on the basis of a reflectance curve and the spectrum of the incident light alone. The relative sensitivity of the eye to light of different wavelengths must also be taken into account (as will be shown in Chapter 6). This factor will emphasize the contribution from intermediate wavelengths more than from short and long wavelengths.

Reflectance

If we wish to know just the *total* percentage of reflected light from a given surface, then the value of the *reflectance* is of prime interest. This number, analogous to the *transmittance* for a filter, is given for a specific SPD from the illuminating light source. Usually the SPD corresponding to white light is of greatest interest. The value of the reflectance quoted for a surface takes into account the relative sensitivity of the eye to different wavelengths. Consequently under the same light, surfaces with higher reflectance appear brighter than those with lower reflectance.

The general problem of predicting how a particular light source and a particular colored surface will interact is very similar to the problem of colored filters discussed in Section 2-4. The SPD of the light source is multiplied, at each wavelength, by the value of the spectral reflectance of the surface. The resulting curve, which gives the SPD of the light leaving the surface, can then be used to predict the visual perception that will be produced, as will be discussed in Chapter 3.

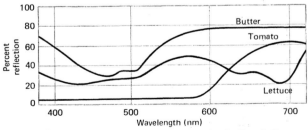

Fig. 2-13. Spectral reflectance curves for butter, lettuce, and tomato, giving the percentage of the incident light reflected at each wavelength. (From *Light and Color,* General Electric Company Publication TP-119.)

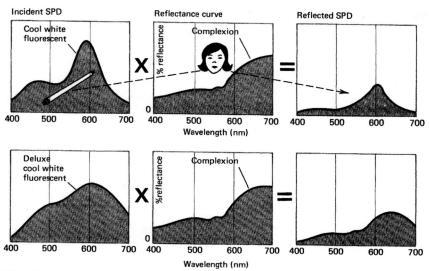

Fig. 2-14. How different fluorescent lamps affect skin color. Under the deluxe lamp much more red is incident on the skin and reflected to the eye; therefore the appearance is warmer and closer to the color seen under sunlight or incandescent lights. (From *Light and Color,* General Electric Company Publication TP119.)

Figure 2-14 illustrates how the reflectance curve of a surface (human skin), multiplied by the SPD of each of two different light sources, gives rise to reflected light with distinctly different SPDs that can be expected to appear quite different. In fashion magazines much attention is given to the effects of illumination on the color of lipstick, eyeshadow, and dress color, for instance. Color Plate 7 illustrates how strongly colored illumination produces particularly dramatic effects.

...The colors of all natural bodies have no other origin than this, that they are variously constituted so as to reflect one kind of light in greater plenty than another. And this rule I have tested in a dark room by illuminating those bodies with uncompounded light of different colors. For by that means any body may be made to appear of any color. When so illuminated they do not exhibit their daylight colors but always appear of the color of the light cast upon them, but yet with this difference, that they are most bright and vivid in the light of their own daylight color.

Isaac Newton
"New Theory About Light and Colors," *Philosophical Transactions of the Royal Society of London, 80,* 3075 (1671/72)

Example 2-3   Color of a Surface

What color is reflected from a surface appearing yellow under white light when cyan light is used to illuminate it? We think in terms of the primary components that constitute a light, since in common lights the colors are not spectral. As the incident light contains both blue and green, and the surface reflects in the green and red portions of the spectrum, only the green component is reflected. That is the color of the surface.

What color is seen with magenta illumination? The incident light has blue and red components, whereas the surface reflects in the green and red; therefore only the red component is reflected.

What color appears with blue illumination? As this color is not reflected, the surface is black.

■ 2-7 TRANSMISSION THROUGH FILTERS

Filters alter the spectral composition and intensity of light by a combination of two effects: reflection and absorption. Both processes remove light from an incident beam. Some filters may rely almost exclusively on one kind of light alteration. A very effective neutral density filter can be produced by evaporating onto glass a thin layer of a suitable metal, such as Inconel, an alloy of iron, nickel, and chromium. The thickness of the layer determines the transmittance, and it affects both the absorption of light and the amount reflected at the surface. But other types of filters, particularly colored ones, depend mainly on absorption.

The light absorbing material in a filter can be either fine particles of a *pigment* or dispersed as individual molecules, called a *dye.* The supporting medium is transparent and is commonly a glass plate or sheet of plastic or celluloid. The series of Wratten filters produced by Eastman Kodak consists of a mixture of organic dyes suspended in clear gelatin, which is sandwiched between glass plates for protection. Filters fall into two classes (Fig. 2-15). Those with a texture that deflects light

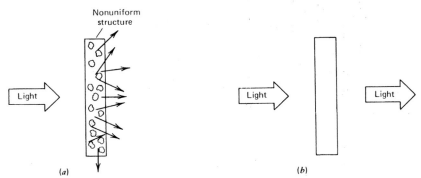

Fig. 2-15. (*a*) Diffuse filter from which light emerges in all directions and (*b*) transparent filter from which light emerges in the same direction that it originally had.

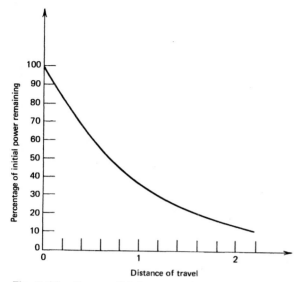

Fig. 2-16. Exponential decrease of the light intensity at one wavelength with distance of travel inside an absorbing filter.

into random directions as it passes through are said to be *translucent* or *diffuse.* Those that allow light to pass straight through and just absorb a portion of it are called *transparent.* The texture of a diffuse filter may be internal, as in filters containing pigments or in translucent objects such as a thin porcelain dish, jade, or alabaster. Or the texture may be on the surface, as for "frosted" or "ground" glass.

As light passes through an absorbing, transparent filter its intensity continuously decreases as it penetrates deeper. This is shown in Fig. 2-16 for the intensity at one wavelength. This effect was known as early as 1728 from the work of Pierre Bouguer, a French scientist who carried out pioneering investigations of the passage of light through the atmosphere. The curve in Fig. 2-16 is said to have an *exponential* shape, because it is described by the decaying exponential function of mathematics. One feature of the curve is that it decreases by equal percentages for equal successive distances of penetration. Note that this same form of decay (the exponential) occurs whenever the decrease of a quantity in a given distance, or interval of time, is proportional to its value at the beginning of that interval.

Bouguer's Law

Transparent filters placed in a light beam have predictable effects, because the light emerging from one filter passes on to the next one. This is not true for translucent filters because much of the diffusely emerging light does not strike the next filter. Nor is it true for filters that depend upon reflection for their major effect. Let us therefore focus our attention on the consequence of placing a number of identical transparent filters in the path of a light beam. If the transmittance of each filter at that wavelength, for example, is 50%, the emerging beam on passage through one filter is one half as intense at that wavelength as the original beam. If two filters are used (Fig. 2-17) the beam coming from the first is reduced by another factor of $\frac{1}{2}$ on passage through the second; therefore the net transmittance is: $\frac{1}{2} \times \frac{1}{2} = (\frac{1}{2})^2$ or $\frac{1}{4}$. For three filters it is $\frac{1}{2} \times \frac{1}{2} \times \frac{1}{2} = (\frac{1}{2})^3$ or $\frac{1}{8}$. The progression of these numbers indicates that if the symbol N represents the number of filters in the beam and T represents the transmittance (expressed as a decimal or fraction) at a wavelength of interest to us, the net transmittance is T^N, or T multiplied by itself N times. The same is true if instead of N filters we had one filter that is N times thicker than the one with a transmittance of T. The T^N prediction is known as *Bouguer's law*.

It follows from Bouguer's law or the exponential law of Fig. 2-16 that the shape of the spectral transmittance curve for a filter can be severely altered if the filter is made thicker. For example, if there is a difference in transmittance between two separate regions of the spectrum, the difference is accentuated if a thicker filter (or additional identical filters) is used. Thus the color of the emerging light may change. This effect is known as *dichroism*, which means "two colors."

Dichroism of a Dye

A similar effect is commonly observed with dyes. If the concentration of the colorant molecules in a solution is increased, there are more molecules to provide selective absorption—with the same result as if the light had instead passed through a correspondingly greater thickness of the original solution. If T is the spectral transmittance at a given wavelength for the original concentration, increasing the concentration N-fold yields a spectral transmission of T^N. This is known as *Beer's law* and is the equivalent for concentration to Bouguer's law for thickness.

Example 2-4 Shift of Hue

Figure 2-18 shows the accentuation of differences between various regions of a transmittance curve when the concentration of a dye in a solution is increased. At the concentration for curve *(a)* the peak in the blue is about three times

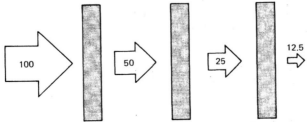

Fig. 2-17. Light with an original intensity of 100 units passing through identical filters in succession. The transmittance of each filter in this example is 50%.

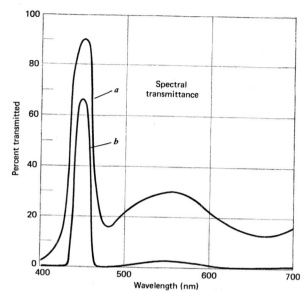

Fig. 2-18. (*a*) A dye at low concentration may appear green because the broadness of the peak in the center of the spectrum more than compensates for the higher transmittance in the blue. (*b*) The same dye at a fourfold increase in concentration will have a greater suppression of transmittance in the green than blue because it was originally lower in the green. The dye will now appear blue.

greater than the peak in the green. But the peak in the green is sufficiently broad to make this dye appear green. If the dye concentration is increased by four, the peak in the blue of the curve *(b)* becomes $3^4 = 3 \times 3 \times 3 \times 3 = 81$ times greater than the peak in the green. This is sufficient to outweigh the broader peak in the green, and the dye now appears blue. The shift from green through cyan to blue with increasing dye concentration is an example of dichroism.

A filter may be characterized by its *density* instead of transmittance. This is particularly common in characterizing a photographic negative. The definition of a filter's density is given in Advanced Topic 2A.

One-Way Mirrors

A "one-way mirror" enables a person to see into another room without being seen. Its effect is based on reflection and a marked difference in the level of illumination between the two rooms. There is no special feature that affects light moving in one direction differently from light moving in the other direction. The surfaces of the glass are thinly coated with a highly reflecting metal so that much but not all of the incident light is reflected. If the room of interest is brightly lit compared to the dark

Fig. 2-19. One-way mirror allows the observer on the left to see into a room without being seen.

space in which the observer resides, he cannot be seen from the brightly lit room because his image is dim in comparison to the reflection of other objects in the room he is observing.

This is shown in Fig. 2-19. The observer is illuminated by only 10 units of light and thus reflects at most these 10 units toward the mirror. If the transmittance of the mirror is only 10%, then only 1 unit emerges into the room being observed. Within this room of interest, the illumination is arranged to be at a higher level, say with 100 units. The subject of interest reflects at most these 100 units toward the mirror, only 10 units of which are transmitted with the remaining 90 units being reflected. If the subject should glance at the mirror, he would see 90 units of his own image and only 1 unit of the observer's; therefore the latter would not ordinarily be seen. The one-way mirror is all the more effective if the observer's room is made even dimmer. However, if the observer should light a match, it could indeed be seen from the bright room!

■ **2-8 COLORS OF SMOOTH AND ROUGH SURFACES**

The color of a surface seen by reflected light is affected in part by the color of the illumination and in part by the nature of the surface. In this section we investigate the interplay of these two aspects that provide a rich visual experience when we view the surfaces of objects around us.

Specular Reflection

Whenever light passes from one medium to another, a fraction of it is reflected at the interface of these media. The remainder is refracted and enters the second medium where it will be absorbed in accordance with the optical properties of the medium. If the surface is smooth, light is reflected from it in a specific direction, determined by the angle at which the incident light strikes the surface (Fig. 2-20). The angle of the reflected light is identical to the angle of the incident light. The light is said to reflect *specularly*. Although the color of specularly reflected light depends on the nature of the second medium, in general it is not very different from that of the incident light. It has not penetrated deeply into the medium nor has it been reflected many times by pigment particles.

Diffuse Reflection

If the surface on the other hand has a pronounced texture—that is, if it is rough— light is reflected diffusely (Fig. 2-21). In addition, if the light penetrates the medium and is affected by irregularities in the structure, as when it is reflected from pigment particles, it emerges in all directions and contributes to the diffuse reflection. In such a case diffusely reflected light is more strongly influenced by selective absorption at the surface than is specularly reflected light.

Matte and Glossy Surfaces

A surface that provides predominantly diffuse reflection is called *matte*. Felt, an asphalt road, and newly fallen snow are matte. All have extremely rough surfaces. A surface that exhibits predominantly specular reflection is said to be *glossy*.

The reflection of light from any surface is partly diffuse and partly specular (Fig.

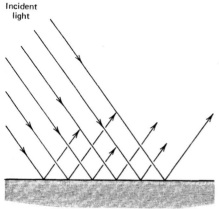

Incident
light

Fig. 2-20. A smooth surface specularly reflects light; that is, it comes off the surface in a specific direction that depends on the angle of the incident light in relation to the surface.

2-22). The interplay between the two can be fascinating, because the proportions change with the direction of illumination and viewing angle. Artists appreciate this when painting and include areas suggesting specular reflection to emphasize form, enhance depth, and suggest the color of the light illuminating the subject. A striking example is found in the paintings of El Greco, who applied generous areas of white to highlight robes and faces in portraits. Specular reflection also is exploited by photographers who backlight subjects to accent hair. This works because specular reflection is particularly effective when light is only deflected by a small angle during reflection.

Wearing apparel made from pile fabrics, such as velvet where fibers stand perpendicular to the surface, achieve their striking appearance from the contrast between diffuse and specular reflection. When looking into the pile the surface is matte, but when viewing light reflected from the side of the fibers the surface is glossy.

The nature of specular reflection determines the difference between "flat" black paint and "glossy" black paint. The latter has a suspension medium that forms a hard, smooth surface, and this medium is very effective in producing specular reflection in one direction, depending on where the source of illumination is. You can often feel the difference between flat and glossy paint, the latter being distinctively smoother, which also makes for easier cleaning. Glossy paints in general are more saturated than matte, except when reflecting light directly toward the viewer. Matte paint achieves its dullness through an irregular surface texture. Nevertheless there are always small regions that specularly reflect light toward the viewer, no matter what his or her vantage point, and this desaturates the color produced by the diffusely reflected component. The most effective matte surfaces may reflect as much as 50% of the incident light. Artists realize that the texture of a surface can be suggested by the degree of saturation of its color.

Desaturation of colors in ceramic mosaics, pictographs, colorful rocks, and dry

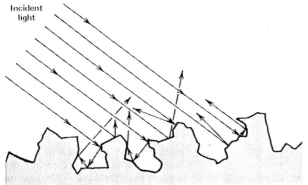

Fig. 2-21. In diffuse reflection from an irregular surface reflected light comes off the surface in all directions.

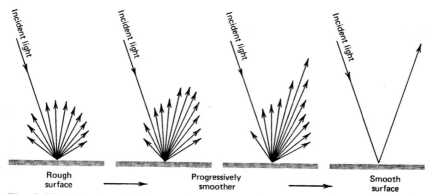

Fig. 2-22. Interplay of diffuse and specular reflection from representative surfaces. The length of each arrow indicates the relative intensity of the light reflected in that direction.

wood surfaces comes from the component of specularly reflected light provided by small facets of surface having just the right orientation. Photographers know that moistening a surface with water immediately darkens the area and provides more saturated colors. A similar effect is seen when wood is varnished, where the rich colors of the grain texture are considerably enhanced. In these and similar examples one factor is most important: the reflectance at the surface of a solid covered by a thin layer of liquid (or wax or varnish) is considerably less than if the liquid were absent. Thus more light penetrates into the solid before being reflected out, and it is more affected by selective absorption. As a consequence, color differences are enhanced. How reflectance is influenced by the media on each side of a boundary will be taken up in more detail in Section 11-9. A second effect of a smooth liquid surface on top of a colored undersurface comes from multiple reflections, where a portion of the light reflected by the colored surface is re-reflected back when it comes to the liquid surface. Each time the light is reflected back to the undersurface there is a corresponding increase in selective absorption.

Although the importance of surface conditions was emphasized in the preceding discussion, the nature of the illuminating light source also influences the observed proportion of specularly and diffusely reflected light. A single, small lamp establishes highlights on a subject from surfaces that specularly reflect light toward us. Surfaces characterized by diffuse reflection provide more saturated colors. But, if an array of lamps illuminates a subject uniformly from both sides, much more of the surface can specularly reflect light toward us. The nearly uniform specular reflection removes highlights and desaturates the color, leaving the subject with a dull, lifeless appearance. Artists recognize that these two extremes are also experienced on a sunny, clear day and on an overcast day. Direct sunlight provides strongly directed light from a single source, whereas an overcast sky gives a more diffuse illumination coming from various regions of the sky.

Dichroic Mirrors

The notion of dichroism mentioned in Section 2-7 refers to the change in color of certain dyes as the concentration of the colorant is varied. A distinctly different "two-color" effect is associated with a few pigments, printing inks, and nonabsorbing materials used as the reflecting surface of special mirrors. The reflectance curve of the surface of these *dichroic materials,* not the light's absorption within the bulk, is primarily responsible for their color effects. A dichroic pigment that transmits blue (and consequently is labeled with that adjective) reflects elsewhere in the spectrum. In reflected light, the pigment may appear yellow or bronze.

A set of dichroic mirrors is an essential part of a modern color television camera. Red light alone is reflected from one mirror to a light-sensitive detector where the optics form a red image of the scene, the transmitted light passing on to a second dichroic mirror that reflects blue light to a second detector. The remaining center portion of the spectrum continues on to a third detector for recording the green image. In this way the three detectors can produce separate signals representing the red, blue, and green portions of the scene.

SUMMARY

Newton's discovery that white light is separated by a prism to form a spectrum led him to associate color with a physical property of light, now recognized as its wavelength. The amount of power that the light contains at each wavelength determines the spectral power distribution (SPD) of the light, from which we can predict the perception of color it produces in our visual system. The subjective specification of color includes three attributes, hue, saturation, and brightness. Hue for spectral colors is the perceptual quantity most closely connected with wavelength.

Newton's experiments led him to formulate laws of additive color mixing based on three additive primaries—blue, green, and red—and to construct the color circle that permits a qualitative estimate of the hue and saturation produced by additive mixing. Additive mixing can occur in two different ways: as superposition (in stage lighting, for example) or as partitive mixing (in the array of tiny dots in the picture tube of color television). In additive color mixing, the final SPD is just the sum of the SPDs of the individual lights.

Subtractive color mixing occurs when light passes through a series of pieces of colored glass (filters) or when different colored paints or dyes are mixed together. Each filter, pigment, or dye absorbs some portion of the incident light, and the SPD of the remaining light that is not absorbed determines the color of the mixture under the particular illumination involved. Colored filters are characterized by a spectral transmittance curve that specifies the percentage of incident light that will be transmitted at each wavelength. When similar filters are stacked together or when the concentration of a dye is increased, the hue of the emerging light may change, an effect called dichroism. Dichroism results from the dependence of absorption on increasing filter thickness or dye concentration (governed by Bouguer's law and Beer's law, respectively), which accentuates differences between absorption at different wavelengths.

The laws of subtractive color mixing are based on the three subtractive primaries—magenta, yellow, and cyan—which are the complements of the three additive primaries. In some situations both additive and subtractive color mixing occur. In four-color printing, separate magenta, yellow, cyan, and black images are printed as patterns of dots that partially overlap. Where they overlap, subtractive mixing occurs; where they do not overlap, partitive additive mixing occurs.

A surface can be characterized by its spectral reflectance curve, which shows the fraction of incident light that will be reflected from the surface at each wavelength. The spectral reflectance curve and the SPD of the light incident on the surface together determine its apparent color. An object with high reflectance in the red and zero reflectance in other spectral regions will appear red under white illumination but black under blue illumination. ■

QUESTIONS

2-1. Can a cyan light always be decomposed by a spectroscope into blue and green components?

2-2. Describe as accurately as possible the color of a yellow school bus on a sunny day.

2-3. What is the color of monochromatic light with a wavelength of 600 nm?

2-4. Which of the following appears most saturated, and which least saturated: sunlight, red stop light, mint ice cream?

2-5. Beams of red and green lights are directed onto a white floor. What color is seen where they overlap?

2-6. What color light should be superimposed on a green light to produce white?

2-7. What color should be mixed additively with cyan to produce white?

2-8. Can you produce black by directing a proper combination of colored lights onto a white wall?

2-9. Draw an example of an SPD representing magenta light. Also draw an example for yellow light produced by superimposing red and green lights. Finally draw the SPD when the magenta and yellow lights are superimposed. What is the color?

2-10. Suppose that a yellow lamp is in a stage light, but you want a green color. If only magenta and cyan filters are available, which should you use? Explain.

2-11. When a yellow piece of chalk is illuminated by magenta light, what color does the chalk appear to be?

2-12. Does an oil painting become brighter or darker when more pigments are mixed together? Explain why.

2-13. Why is four-color printing more popular than the three-color process that relies only on magenta, yellow, and cyan inks?

2-14. Explain how four-color lithography depends on both additive and subtractive laws.

2-15. What qualities of a surface enhance specular reflection?

2-16. Can a more saturated color be produced with an enamel or a flat paint? Why?

2-17. Why is frosted glass an example of a translucent filter? What accounts for its translucent quality?

2-18. If the purple filter and the yellow filter whose transmittance curves are shown in Fig. 2-11 are placed one in front of the other in a beam of white light, draw the SPD of the emerging light.

2-19. If you wished to exhibit a tomato in a shop, would it look better under cyan or under yellow illumination? Why?

2-20. In a psychology experiment the intensity of a lamp must be adjusted in a precise manner; for this reason two neutral density filters, each of 50% transmittance, are placed one in front of the other in its beam. What percentage of the light emerges from the two?

2-21. If a certain amount of dye is dissolved in a large beaker of water and the spectral transmittance at a particular wavelength is found to be 30%, what is the spectral transmittance if an equal amount of dye is added to the original solution? What is the spectral transmittance if three times the original amount is added to the original solution?

FURTHER READING

C. Rainwater, *Light and Color* (Golden Press, New York, 1971) (paperback). A short, comprehensive description of phenomena associated with light and color. A valuable introduction, containing many effective illustrations.

W. D. Wright, *The Measurement of Color* (Macmillan, New York, 1958). Well-written introduction to important notions concerning color. Includes discussions of Beer's law and optical density.

R. W. G. Hunt, *The Reproduction of Colour,* Third Edition (Fountain Press, London, 1975). Authoritative explanation of color reproduction techniques in printing, photography, and television.

E. A. Dennis, *Lithographic Technology* (Bobbs-Merrill, Indianapolis, Indiana, 1980). Comprehensive presentation, from the preparation of material to the finished product.

R. S. Hunter, *The Measurement of Appearance* (Wiley, New York, 1975). An examination of the interplay of specular and diffuse reflection at a surface, perception as affected by the mode of viewing, measurement techniques to characterize light reflected from various surfaces, and methods for specifying color. Moderately advanced.

F. Garritsen, *Theory and Practice of Color: A Color Theory Based On Laws of Perception* (Van Nostrand–Reinhold, New York, 1975). Lavishly illustrated account of color mixing and how the mode of viewing affects color. This visual delight is spoiled by loose wording and poorly developed presentation. Well worth reading once you already understand the material.

J. Beck, *Surface Color Perception* (Cornell University Press, Ithaca, New York, 1972). Psychological aspects of perceiving surface color and form—the notion of lightness, effects of adaptation, importance of contrast. For the more advanced reader.

R. W. Burnham, R. M. Hanes, and C. J. Bartleson, *Color: A Guide to Basic Facts and Concepts* (Wiley, New York, 1963). A compendium of facts pertaining to aspects of color; in outline form with some illustrations. Particularly useful as a reference.

CHAPTER 3

MEASURING COLOR

How can we measure color? The science of colorimetry was created to find an answer. Because color is a human perception, colorimetry is based on the visual response of observers, and its laws are subject to the inherent characteristics of the visual process. Our exploration of color science will reveal surprising aspects of color that are not generally appreciated.

One's first thought on how to measure color accurately might be to specify the SPD (spectral power distribution) of the light. This physical description is useful for gaining a general notion of the color, but it is not a convenient method for indicating precise aspects of color, except for spectral hues. We would find it difficult to form an idea of a color if the spectrum had a complicated shape, as can easily be confirmed by examining Color Plate 5.

For a colorimetric system to be useful it should maintain contact with our everyday notions of color attributes—a spectrum alone will not do. Furthermore the system should have predictive capability, the feature whereby we can predict the result of adding two colors together. Several systems have been devised, but only one has been generally accepted. In this chapter we shall first become acquainted with the notion of a color space and with two systems that provide a means of specifying a color by reference to the best-matching member in an array of standard samples. With this background we can see the advantages of an alternative method—the *chromaticity diagram*—as a convenient and precise way to specify color. For this reason the chromaticity diagram has been adopted by an international technical body called the Commission Internationale d'Éclairage (International Commission on Illumination, or C.I.E. for short). The predictive features of the C.I.E. chromaticity diagram for color mixing and its relation to other color spaces will be discussed in some detail. Advanced Topic 3A at the end of the book explains the mathematical basis for the C.I.E. chromaticity diagram.

■ 3-1 COLOR SPACES

We describe our perception of a color in terms of three attributes: hue, saturation, and brightness. As we mentioned in Chapter 2, Newton introduced a convenient way to visualize relationships between hue and saturation with his color circle, which provides a kind of two-dimensional color map (Fig. 2-5). The next logical step is to add the third dimension to include brightness (Fig. 3-1). This forms a three-dimensional *color space.* A color is specified by the position at which its three attributes would place it in this space. These are the three color coordinates. The achromatic colors ranging from black to white lie on the vertical line through the center. Colors with increasingly greater saturation are found increasingly farther from this line, and hue is indicated by the direction about the line. We can visualize relationships between colors by their respective positions in this color space. These relationships, in turn, will depend on how the coordinates of the color space have been selected. In two commonly used systems described below the color space is determined by a set of standard samples, and a color is specified by comparison with these standards. These systems provide a workable though

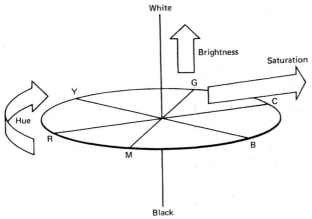

Fig. 3-1. A three-dimensional color space indicating the attributes of color perception. Spectral colors and purples are located on the circumference, less saturated colors closer to the center, and brighter ones at higher positions.

subjective means of specifying a color that is of considerable importance to workers in painting, printing, textiles, and decorating. We refer to the Munsell and Ostwald systems for specifying colors, although there are also other systems. For any color space the principal question is what units should be used to indicate hue, saturation, and brightness. This is the nub of the problem in colorimetry.

■ 3-2 MUNSELL SYSTEM

The rapid development of industry at the turn of the present century created a need for accurate and reproducible color standards. In response to this need Albert Munsell, a painter and art teacher, devised a color atlas in 1905 that has become known as the *Munsell Book of Color.*[†] It has subsequently been revised many times and is now perhaps the simplest subjectively based color system and the most widely used. It contains over 1500 color samples prepared from stable pigments, with either matte or glossy finish. They should be viewed with white illumination representing north skylight. The samples are arranged in a three-dimensional color space, as shown in Color Plate 8. It is impractical to include all colors, because about 10 million surface colors can be distinguished by a normal observer under optimal conditions. Therefore, representative samples were selected so that adjacent samples differ by approximately equal perceptual intervals.

The *Hue scale* is divided into 100 equally spaced Hues. They are grouped under

---

†Published by Munsell Color Company, Inc., 2441 North Calvert Street, Baltimore, Maryland, 21218.

families denoted by five "principal" Hues P, B, G, Y, and R and five "intermediate" Hues PB, BG, GY, YR, and RP, arrayed as shown in Fig. 3-2. Each of the 10 Hue families contains 10 subdivisions, labeled numerically from 1 to 10, with 5 being the midpoint. For example, the notation "7YR" refers to a hue 2 steps from the midpoint of yellow-red toward yellow. Since there is no absolute scale for the psychologists' notion of saturation, Munsell introduced one that he denoted by the term *Chroma*. The scale for Chroma extends from 0 representing a neutral gray out to 10, 12, 14 or perhaps as far as 18 for complete saturation, depending on the Hue. Each step in saturation appears to cause equivalent change. The scale Munsell chose for brightness is called *Value* and has 9 steps between black and white. Black has a Value of 0, and white has 10. A shorthand notation for specifying a color gives, in order: Hue, Value, and Chroma in the form H V/C. For example a nearly saturated yellow of moderate brightness could be designated 4Y 6/10. All grays have no Hue and zero Chroma. These achromatic or neutral colors are written N V/. Thus a gray visually halfway between white and black would be N 5/.

A cutaway view of the Munsell color space illustrating the different V/C for Hue 5Y is shown in Fig. 3-3. Munsell did remarkably well in choosing a wide variety of color samples and spacing them in nearly equal increments. In 1943 recently devel-

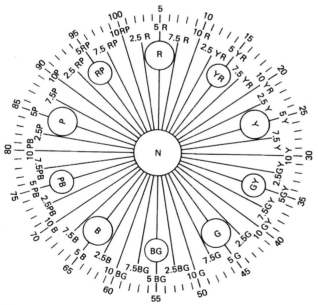

Fig. 3-2. Munsell's division of the Hue scale into 100 parts, illustrating his notation in terms of five principal and five intermediate hues. The numbers around the outside ranging from 1 to 100 provide an alternative numerical way to specify Hue.

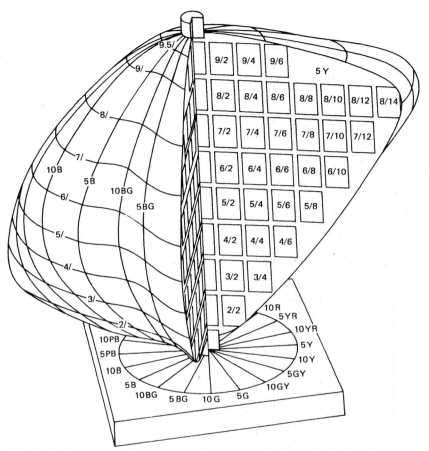

Fig. 3-3. A cutaway of Munsell's color space showing a leaf of the Hue 5Y, with Chroma increasing outward from the central vertical axis and Value increasing with height. The directions of leaves with other Hues are indicated on the base. (From A. Stimson, *Photometry and Radiometry for Engineers,* Wiley, New York, 1974.)

oped spectrophotometric techniques enabled minor improvements to be made in regularizing the spacings, and these revised samples are denoted by the present Munsell notation.

With its Hue scale based on a division into equal differences, the Munsell System fails to place complementary colors on opposite sides of the achromatic axis. In other words, geometrical relationships are not significant in this color space. The distance separating two points in color space has no simple relationship to differences in their perceived Hue, Chroma, and Value.

One fascinating aspect of the Munsell color space is its irregular shape. It bulges outward toward the blues and purples at low Values and toward the yellows at high Values. Of course for the lowest Value only black is represented. At somewhat higher Values purples can be distinguished with a larger range in Chroma than for other hues, so the color space bulges outward with its greater number of standards in the region of PB and P (see Fig. 3-3). Since at intermediate Values nearly all hues can be represented by the same range in Chroma, one side is not favored more than another (although the boundary of the color space is not strictly circular). For high Value we cannot perceive strongly saturated colors except for yellows and, consequently, the color space extends outward in the yellow region to accommodate the greater range of Chroma. Finally the highest Value representing white stands alone at the top. The form of the Munsell color space enables us to visualize the variety of colors that are possible at various levels of Value (or brightness).

■ 3-3 OSTWALD COLOR SYSTEM

Wilhelm Ostwald, a German chemist who won the 1909 Nobel prize for his basic research in physical chemistry, introduced a new system of color specification in 1917. Ostwald's color system is based on his theories about the nature of color, which are not widely accepted, but the system itself has proven to be very useful and is favored by many artists and others working with practical color problems, such as interior decoration and color printing.

Unlike the Munsell System, whose adjacent samples are chosen to represent equal perceived differences, the Ostwald System was developed from an analysis of the reflectance curve for each color. The details of how this is done will be explained in subsequent sections of this chapter that describe the C.I.E. color system. The results of the analysis provide a characterization of each color in terms of its *dominant wavelength, purity,* and *luminance.* In a very approximate sense these measured attributes of the light reflected from a surface correspond to the perceived attributes of hue, saturation, and brightness.

The Ostwald color space fills the interior of a double cone shown in Fig. 3-4a. The outer circle represents the colors of greatest purity attainable at the same luminance (called "full color"), while the vertical axis down the center is neutral (achromatic) with white at the top and black at the bottom. Complementary colors are on opposite sides of the neutral axis.

All colors having a given dominant wavelength lie on a triangular leaf bounded by the neutral axis and a full color point as shown in Color Plate 9 and diagrammed in Fig. 3-4b. The location of a particular color on this triangle is determined by its purity and luminance: colors of increasing purity lie farther away from the neutral axis, while colors of increasing luminance (under illumination by standard white light) lie further up on the triangle (see Color Plate 9). Ostwald color sets are commercially available in boxes of colored paper samples.

The Ostwald notation for each sample was based on studies with a disc colorimeter. This is a simple device that consists of a flat disc mounted on the end of

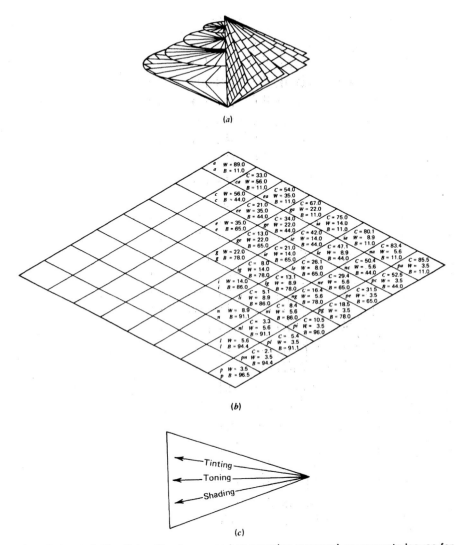

(a)

(b)

(c)

Fig. 3-4. (*a*) The Ostwald color space has samples arranged on separate leaves for each dominant wavelength; (*b*) two different notations for samples on one leaf are shown: one gives the percentages of full color (*C*), white (*W*), and black (*B*) for each sample, and the other indicates the sample by two letters, whereby colors of the same tint have the same first letter and colors of equal shade the same second letter; (*c*) directions of change in the color of a sample when a full color is tinted, toned, or shaded.

an axle. The disk has angular segments of a full color, white, and black; when the axle is spun rapidly the individual colors appear to blend. When the areas of the segments are adjusted so that the disk's color matches a sample's, the proportions of the total area devoted to full color, white, and black specify the color of the sample. Thus each Ostwald sample is labeled by three numbers representing the proportion of full color (C), white (W), and black (B), and the three numbers add to 100%.

A color specified by 34, 22, 44 has additive mixtures of 34% full color, 22% white, and 44% black. The values of luminance and purity represented by samples in Ostwald's color space are chosen so that on proceeding away from full color, samples lying on the same upward-sloping diagonal have identical black content and those on downward-sloping diagonals have identical white content.

Although a color is technically described in the Ostwald system by three num-bers, another qualitative designation is often employed for describing a mixture of paints or inks when white and black are mixed with a full color. A full color mixed with white is called a *tint* of the full color; a full color mixed with black is called a *shade*; and a full color mixed with gray (both black and white) is called a *tone* (Fig. 3-4*c*).

It is often found that pleasing color combinations can best be achieved by restricting the range of dominant wavelengths (which produces a dominant color scheme) and using colors with similar locations within their respective Ostwald tri-angles. These are referred to as *isotones*, meaning colors with the same tone (that is, with equal black contents and equal white contents). Colors that are isotones will have identical Ostwald numerical designations but different dominant wave-lengths.

The Ostwald system enjoys popular use in printing where, for example, in the halftone process a color is produced by the fusion of separate black and colored dots. It also appeals to artists when mixing black, white, and colored pigments and to interior decorators seeking harmonious color combinations. Variations of this system are widely used in industry, such as the Color Harmony Manual produced by Container Corporation of America. Another related systematic arrangement of color samples is found in the Maerz and Paul Color Dictionary, which includes sam-ples produced by color printing.

■ 3-4 THE C.I.E. DIAGRAM

Although the Munsell and Ostwald color systems provide very useful ways of spec-ifying colors, they suffer from two serious problems when applied to the measure-ment of color. Comparison of an unknown sample (or light source) against the reference samples in a color atlas or Ostwald set is a highly subjective process. Furthermore, if the reference samples change with time because of fading, the measurement process will become less and less reliable.

During the early part of the twentieth century, the need for an objective system of color specification, independent of comparison standards and subjective judgment, became increasingly evident. Therefore the C.I.E. (Commission Internationale d'Eclairage) was charged with selecting a new objective system of colorimetry. The problem the commission faced can be roughly summarized as follows: consider a colored light source (or colored sample illuminated by a standard source of white light) whose SPD has been measured. From the SPD derive one quantity called the *luminance,* which determines its brightness, and two quantities, which together are called the *chromaticity,* that determine its hue and saturation.

The procedure adapted by the C.I.E. was so successful that it has become the universally recognized system for specification and measurement of color from the lighting industry to pigment manufacturing to psychological research in color perception. The C.I.E. procedure, which will be discussed in detail in the remainder of this chapter, converts the SPD into three quantities Y, x, and y by the use of standardized tables based on the judgment of the "average" human observer. The value of Y gives luminance, which is a quantitative measure of the intensity of light leaving a surface. The exact definition of luminance will be discussed in Chapter 6. The perceptual attribute of "brightness," which is not quantitative, is closely related to luminance and is frequently used in its place, though strictly speaking incorrectly. The two other coordinates x and y are the *chromaticity coordinates,* which locate the color (with respect to hue and saturation) on a two-dimensional color map called the *C.I.E. chromaticity diagram.* In this section we shall focus our attention on this diagram, since hue and saturation of a color are our prime interest in this chapter.

Color Plate 10 shows a C.I.E. diagram with a rough approximation to the actual colors that each location represents. At first glance the C.I.E. diagram closely resembles Newton's color circle (Fig. 2-5). The pure spectral colors are arranged along a horseshoe-shaped perimeter, indicated by the appropriate wavelengths (from 380 nm at the extreme violet end to 780 nm at the extreme red end). The perimeter is closed across the bottom by a straight line representing the nonspectral purples. Every possible chromaticity is specified by a pair of numbers x and y giving its coordinates. The x value of a specific point on the diagram is found by moving straight down vertically from the point to the x-axis, and the y value by moving straight across horizontally to the y-axis.

The special pair $x = \frac{1}{3}$, $y = \frac{1}{3}$ (whose chromaticity is denoted "equal energy" in the diagram) locates the *achromatic point.* This represents the grays and white. It is also called the *equal power point* because achromatic colors can be produced by an SPD that has the same amount of power at every wavelength. Colors whose chromaticity lies near the achromatic point are unsaturated, while those lying near the perimeter of the diagram are saturated. Complementary colors lie on opposite sides of the achromatic point.

Dominant Wavelength of a Color

Although the chromaticity coordinates x, y precisely specify the chromaticity of a color, their values do not immediately convey to our mind an image of that color. The notion of *dominant wavelength* was introduced to help our intuition. Dominant wavelength is determined by where a straight line drawn from the achromatic point through the point representing the chromaticity of interest intersects the perimeter of the diagram. For example, the chromaticity labeled *D* in Fig. 3-5 has a dominant wavelength of 548 nm. As suggested by the hue of monochromatic light at this wavelength, the color *D* is yellowish-green.

The dominant wavelength for a chromaticity lying below the achromatic point does not exist if the line from the achromatic point intersects the line of purples. In this case, the notion of *complementary dominant wavelength* of the color is introduced by drawing a line in the opposite direction, from the chromaticity of interest *through* the achromatic point to the perimeter on the far side. The complementary dominant wavelength for the point *P* is thus 548 nm.

Purity of a Color

The notion of the *purity* of a color is also defined by the position of its chromaticity. Purity is a measure of how close the chromaticity lies to the perimeter of the diagram. With reference to Fig. 3-5, if *a* is the distance from the achromatic point to the chromaticity of interest, say *D*, and *b* is the distance to the perimeter, the purity

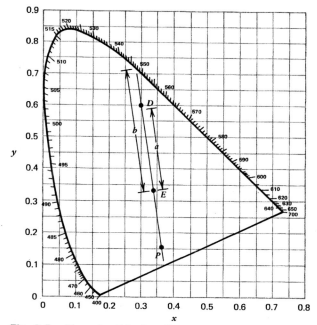

Fig. 3-5. The color *D* is described by a dominant wavelength of 548 nm. The color *P* has no dominant wavelength, but a color that is complementary to *P* has a dominant wavelength of 548 nm in this example. The purity of *D* is defined as the ratio determined by dividing the distance *a* by the distance *b*.

of the color is defined as the ratio of these distances:

$$\text{purity} = \frac{a}{b} \tag{3-1}$$

Thus the purity of colors ranges from 0 for the achromatic point to 1 for saturated colors at the perimeter.

A color can therefore be described by its two chromaticity coordinates x and y, or by its dominant wavelength and purity. Either set of numbers precisely locates a point on the C.I.E. diagram. The measures of dominant wavelength and purity let us better visualize the color in question, but the chromaticity coordinates are more useful when dealing with additive mixtures of colors.

Color Addition

Although Newton's color circle suggests qualitatively how to predict the result of additively mixing two colors, the C.I.E. diagram provides a way of precisely and quantitatively predicting the result. This is a source of its great importance. Consider, for example, two colors G and R whose chromaticities are shown in Fig. 3-6. An additive combination of G and R will produce a color with a chromaticity that lies on the straight line joining G and R. If the brightness (Y value) of each light source is the same, the chromaticity of the mixture will lie about halfway between G and R. Otherwise, it will lie closer to whichever point has the greater brightness (Y value).

. . . Accordingly, taking the luminance of the light into account also, we find that the quality of every color impression depends on three variable factors, namely, luminosity [brightness], hue, and saturation. There are no other differences of quality in the impression made by light. This result may be expressed as follows:

The color impression produced by a certain quantity of mixed light x can always be reproduced by mixing a certain amount of white light a with a certain other amount of some saturated color b (spectrum color or purple) of definite shade.

Hermann Ludwig Ferdinand von Helmholtz
Physiological Optics, (1924 English translation: Dover, New York, 1962)

If a third light source such as B is mixed with G and R, the chromaticity of the mixture can lie anywhere within the triangle bounded by the 3 points B, G, and R. The range of chromaticities available through additive combinations of these three colors is called the *gamut* of the three colors. If the gamut includes the achromatic

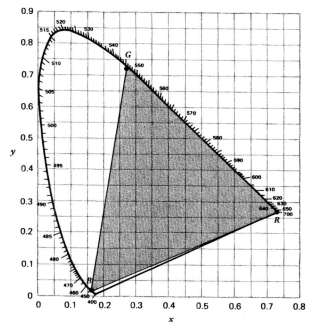

Fig. 3-6. C.I.E. chromaticity diagram indicating the chromaticity of three light sources, B, G, and R. The chromaticities obtainable by mixing any two lie on the straight line joining them. The full gamut of chromaticities available by mixing the three sources in varying amounts lies within the triangle joining the three points.

point, as in this example, the three colors can be considered a set of additive primaries. It is important to realize, however, that because the perimeter of the C.I.E. diagram is curved, it is impossible for *any* set of three primaries to have a gamut that includes all colors.

The C.I.E. diagram is a useful format for comparing the range of colors that can be reproduced by various media. Figure 3-7 shows this by the triangular gamut for the additive primaries of color television and the irregular boundary enclosing the chromaticities achieved by commercial four-color printing. Color printing relies on both additive and subtractive mixing effects, and that is the reason why its boundary is not triangular. Figure 3-7 shows that color television achieves a wider range of chromaticities than printing, and has colors of higher purity. But neither is capable of producing high-purity colors with dominant wavelengths between about 490 and 510 nm (the cyans).

The conversion of an SPD into chromaticity coordinates *(x, y)* is a somewhat technical procedure that we will discuss fully in later sections. Nowadays, however, the conversion is usually carried out automatically by a device called a *colorimeter* (a computer built into a spectrophotometer). An operator can insert a sample (a painted surface, for example) into the instrument, which measures the

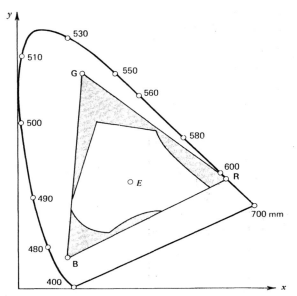

Fig. 3-7. C.I.E. diagram showing the chromaticities for the blue (B), green (G), and red (R) phosphors of a color television set, together with the gamut of colors they can produce in additive combinations. The irregular boundary encloses the range of chromaticities reproduced by commercial four-color printing.

SPD and computes the chromaticity coordinates and the Y value. Since the Munsell and Ostwald reference samples have also been measured this way, the results for the unknown sample can then be readily translated into the equivalent designations for either of these systems. The C.I.E. system thus provides a method for measuring and specifying colors objectively, without comparison to reference samples, and without subjective judgment. ▪

3-5 TRISTIMULUS VALUES

The historical developments leading to the C.I.E. diagram began with Newton's color circle. It was the first successful method that summarized consistently several important relationships between colors (e.g., complementarity, progression of spectral hues, variation in saturation, and additivity of red and blue to yield purple). Yet it failed to provide a *numerical* measure for color and thus lacked accuracy and reproducibility. After Newton, there was considerable interest in the possibility of establishing a color theory based on the additive properties of three primary colors. A great conceptual advance was made in 1802 by an English physician, Thomas Young, who suggested a threefold character of color *perception*. His notion was that three different sets of receptors in the eye respond to light in three different regions of the spectrum. The perceived color could then be described by the intensity of sensation in each of the three receptors. This is why, he argued, only three properly selected lights comprise a set of primaries. It turns out that Young's theory is close to the truth, but it was not until 1965 that detailed study of cones in the retina of the eye showed three types with differing spectral sensitivity.

The modern science of colorimetry was created by the Scottish mathematical physicist James Clerk Maxwell about 60 years after Young. His experiments with spinning discs on which are mounted adjustable lengths of colored paper enabled him to explore quantitatively the color mixing laws. He found that there is no unique set of primary colors whether or not they are monochromatic. Many combinations of three colors whose hues are chosen from a sufficiently broad span over the spectrum can match a wide range of colors including white. However, none of these sets can be used to match *all* known colors! If the primaries of a set represent colors that are close to spectral hues widely distributed across the spectrum, the gamut is large; if not, it is small. Color matching studies are the basis of the science of colorimetry.

Color Matching

The modern version of Maxwell's color matching measurements is performed under controlled conditions in which an observer views a split field (technically, a "bipartite screen") (Fig. 3-8), half of which is illuminated with the sample color and the other half with the chosen mixture of the three primary colors. The illuminated region subtends a visual angle of 2°, which is sufficiently small so that only receptors in the fovea at the center of the retina are stimulated when the observer looks directly at the center of the screen. The notion is that only cones are then responding, and the results obtained should be characteristic of daytime vision. This avoids complications that might be introduced by peripheral vision where rods play a more important role. Similar tests with a 10° field of view yield slightly different results, but they are not sufficiently altered to warrant our considering the differences in detail.

Notice that these tests rest on what is termed the *local theory* of color vision. This theory is based on the assumption that the color we perceive is determined by the light

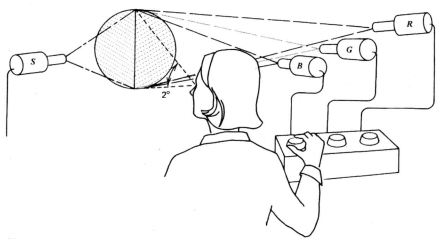

Fig. 3-8.   Color matching measurements with a small-field bipartite screen.

that falls on a localized region of the retina. For broad areas of uniform color this property describes what we see, but there are exceptions. Contrast phenomena clearly show that the apparent color of a subject is affected by the color of the surrounding field. See Chapter 10 for examples. As of now no way has been found to incorporate such effects into a general theory of color measurement. In the presently accepted method for color matching, a neutral color that introduces the least influence on the results surrounds the split field. This simplification should be kept in mind when the notions of colorimetry are applied in situations where contrast effects may be important.

The science of *psychophysics* is concerned with the measurement of human perception. Color matching described above is an example of a psychophysical measurement. Such measurements do not rely on a qualitative description by the observer, as would be required for specifying the psychological variables of hue, saturation, and brightness of a color. The observer need only indicate when two colors appear identical. The results are therefore found to be reasonably accurate and reproducible. Chapter 6 describes methods for carrying out psychophysical measurements for matches of brightness without regard to hue and saturation.

Matching Relations

A color match implies that hue, saturation, and brightness of the particular combination of primaries are identical to those of the sample. If we designate unit amounts of our primary lights by the symbols **B, G,** and **R,** and if we let the symbols B, G, and R denote the number of units of the primaries needed to match the color of the sample **S,** the condition for the match is economically summarized by the equation:

$$\text{S} \blacktriangleleft B\text{B} + G\text{G} + R\text{R} \tag{3-2}$$

The special symbol of the eye $\blacktriangleleft$ is used instead of an equals sign to remind us that this expresses a perceptual match, in which the two sides of the split screen appear to be indistinguishable. The numbers B, G, and R are the "color coordinates" of the sample in the color space determined by **B, G,** and **R.** They are called the *tristimulus values* of the color **S** with respect to the particular set of primaries **B, G, R.** By using a different set of primaries we would obtain a different set of tristimulus values. But if we employ the same set of lamps to match against a wide variety of sample colors, the three values of B, G, R for each color give a reproducible and accurate specification. Tristimulus values are a psychophysical measure of color.

For convenience the relative intensities of unit amounts of the primary lights **B, G,** and **R** are chosen so that they match white when added together:

$$\text{W} \blacktriangleleft 1\text{B} + 1\text{G} + 1\text{R} \tag{3-3}$$

Thus the tristimulus values for white are all equal. For matches to other samples we gain a rough intuitive appreciation for the hue of the color by noting which tristimulus value is largest. And we can gauge the approximate desaturation by how nearly equal to each other they are.

One possible set of primaries that has proven useful in the past is composed of monochromatic blue at 436 nm, green at 546 nm, and red at 700 nm. Experiments show that for this set to match white the power reflected into the eye from these three must be in the proportion 0.014:0.019:1. Consequently unit amounts of **B, G, R** appearing in Eq.

3-3 could be chosen as 0.014 microwatt at 436 nm, 0.019 microwatt at 546 nm, and 1 microwatt at 700 nm.

It has been generally found that increasing or decreasing the power of each primary by the same factor will change the overall brightness but not affect a match for hue and saturation. This means that if B, G, and R are each multiplied by (or divided by) the same number, then an observer will find that the two halves of the screen still match in hue and saturation, although they no longer match in brightness. We therefore can describe a color by its *chromaticity*, which depends on the relative magnitudes of B, G, and R, but not on their absolute numerical values. Since chromaticity is determined by hue and saturation alone, it can be specified by two numbers rather than three. The specification of chromaticity will be discussed in detail later in this chapter.

The theory which I adopt assumes the existence of three elementary sensations by combination of which all the actual sensations of color are produced. I will show that it is not necessary to specify any given colors as typical of these primary sensations. Young has called them red, green, and violet; but any of the three colors might have been chosen, provided that white resulted from their combination in proper proportions.

James Clerk Maxwell
"Theory of the Perception of Colors," *Transactions of the Royal Scottish Society of Arts, 4,* 394-400 (1856).

Maxwell discovered that *any* color can be matched with a primary set **R, G, B,** *provided* that subtraction as well as addition is allowed. That is, we know it is impossible to match most saturated samples because the addition of two or more primary colors is known in most cases to produce a desaturated color. But if one of the primaries is redirected so that it mixes additively with the sample instead of with the other primaries, a match can be made. For example, suppose that **R** must be added to our sample **S** to achieve a color match:

$$\textbf{S} + R\textbf{R} \blacktriangleleft B\textbf{B} + G\textbf{G} \tag{3-4}$$

This expression can be formally rearranged to give us the standard form displayed in Eq. 3-2 if we subtract $R\textbf{R}$ from both the left and right sides:

$$\textbf{S} \blacktriangleleft B\textbf{B} + G\textbf{G} - R\textbf{R} \tag{3-5}$$

This is what Maxwell meant by "subtraction." The negative tristimulus value $(-R)$ does not mean that this light was physically subtracted from the light of the other primaries (for that of course is impossible). It simply means that the primary had to be added to the sample.

Color Matching Functions

Color matching experiments of the type described above provide the basis for an objective system of colorimetry. If we could measure the SPD of an unknown light source (or of a colored sample illuminated by white light) and convert the SPD into its equivalent tristimulus values *B*, *G*, and *R*, then its color would be specified. Any two colors with the same *B*, *G*, and *R*, no matter how different their SPDs, must look alike.

The key to this conversion lies in the *color matching functions.* During the 1920s perceptual psychologists (notably John Guild and W. D. Wright in England) performed color matching experiments with a large group of observers in which the tristimulus values corresponding to matches of the primaries **B, G,** and **R** against *monochromatic* test samples were determined. The amount of power (in watts) falling on the test half of the screen was kept constant, and the resulting tristimulus values at each wavelength were recorded. The results, suitably averaged over minor differences among observers, were tabulated as $\bar{b}_\lambda$, $\bar{g}_\lambda$, $\bar{r}_\lambda$. The bars over the letters indicate that they characterize test samples of equal power; the subscript λ indicates that the specified amounts of the primaries match monochromatic light at wavelength λ. The values of these color matching functions at 20 nm wavelength intervals are listed in Table 3-1. The very small values that occur near both ends of the spectrum result from the poor sensitivity of the eye to light of the corresponding wavelengths. Thus, much less *B*, *G*, *R* are required to match one watt of deep red or blue light than one watt of green or yellow light. How the values vary with wavelength is graphically shown in Fig. 3-9.

The tristimulus value *B* for a given SPD is obtained by summing the numbers $\bar{b}_\lambda$ for all wavelengths across the spectrum, each weighted by the actual amount of power P_λ in the SPD at that wavelength. That is, for a given wavelength λ we multiply $\bar{b}_\lambda$ by P_λ, and do this in succession one wavelength at a time; then we add all of the individual products to arrive at the value for *B*. Similarly *G* or *R* can be separately calculated by using the appropriate color matching function. If the values of the color matching functions in Table 3-1

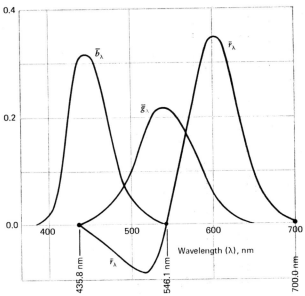

Fig. 3-9. Color matching functions for the set of monochromatic primaries **B** (at 435.8 nm), **G** (at 546.1 nm), and **R** (at 700.0 nm). (D. B. Judd and G. Wyczecki, *Color in Business, Science, and Industry,* Third Edition, Wiley, New York, 1975.)

TABLE 3-1 Average Color Matching Functions $\bar{b}_\lambda$, $\bar{g}_\lambda$, and $\bar{r}_\lambda$.[a]

Constant Radiance Test Stimulus at Wavelength λ(nm)	Color Matching Functions		
	$\bar{r}_\lambda$	$\bar{g}_\lambda$	$\bar{b}_\lambda$
380	0.00003	−0.00001	0.00117
390	0.00010	−0.00004	0.00359
400	0.00030	−0.00014	0.01214
410	0.00084	−0.00041	0.03707
420	0.00211	−0.00110	0.11541
430	0.00218	−0.00119	0.24769
440	−0.00261	0.00149	0.31228
450	−0.01213	0.00678	0.31670
460	−0.02608	0.01485	0.29821
470	−0.03933	0.02538	0.22991
480	−0.04939	0.03914	0.14494
490	−0.05814	0.05689	0.08257
500	−0.07173	0.08536	0.04776
510	−0.08901	0.12860	0.02698
520	−0.09264	0.17468	0.01221
530	−0.07101	0.20317	0.00549
540	−0.03152	0.21466	0.00146
550	0.02279	0.21178	−0.00058
560	0.09060	0.19702	−0.00130
570	0.16768	0.17087	−0.00135
580	0.24526	0.13610	−0.00108
590	0.30928	0.09754	0.00079
600	0.34429	0.06246	−0.00049
610	0.33971	0.03557	−0.00030
620	0.29708	0.01828	−0.00015
630	0.22677	0.00833	−0.00008
640	0.15968	0.00334	−0.00003
650	0.10167	0.00116	−0.00001
660	0.05932	0.00037	0.00000
670	0.03149	0.00011	0.00000
680	0.01687	0.00003	0.00000
690	0.00819	0.00000	0.00000
700	0.00410	0.00000	0.00000
710	0.00210	0.00000	0.00000
720	0.00105	0.00000	0.00000
730	0.00052	0.00000	0.00000
740	0.00025	0.00000	0.00000
750	0.00012	0.00000	0.00000
760	0.00006	0.00000	0.00000

[a]The functions are for observers with normal color vision viewing a two degree visual field. The primaries are monochromatic **B** (at 435.8 nm), **G** (at 546.1 nm), and **R** (at 700.0 nm) with relative intensities so that when added they match white.

Source. From C.I.E. Publication No. 15, *Colorimetry*, 1971.

are chosen for this calculation, a rather coarse-grain sample is taken, since we consider only the values of the functions and consequently P_λ at 20 nm intervals. The relevant power P_λ that we use to multiply $\bar{b}_\lambda$, $\bar{g}_\lambda$, or $\bar{r}_\lambda$ for a particular λ is then the amount found in the SPD within a 20 nm range centered at that λ. If we are interested in the chromaticity and not in the brightness, then the units in which the SPD is measured are unimportant since they affect the magnitude but not the relative values of the computed B, G, and R. Example 3-1 shows an even coarser calculation of tristimulus values in which for brevity we sample the SPD at 50 nm intervals. In this example we concentrate on chromaticity.

Example 3-1 Calculating Tristimulus Values

Figure 3-10 shows the reflectance curve for a sample of fresh butter. It has the same shape as the reflected spectral power distribution when illuminated by a white lamp providing equal power at all wavelengths. Let us calculate the corresponding tristimulus values. To proceed in orderly fashion we make a table as shown below: first we list the wavelengths chosen evenly across the spectrum, and then enter the corresponding values of the measured SPD. The more wavelengths, the more accurate the determination of tristimulus values. The exact physical units in which P_λ is measured are irrelevant when characterizing chromaticity. Then we enter in Columns 3, 5, and 7 the corresponding values for the color matching functions as read from Table 3-1. The number of wavelengths included in this table is not sufficient for an accurate calculation of B, G, R, because this coarse sampling does not properly represent details of the power distribution. Usually entries of P_λ at 5 nm intervals or smaller are desired. As the column headings suggest, the next step in the calculation is to multiply the value of P_λ in Column 2 by the value of $\bar{b}_\lambda$ on the same line in Column 3 and enter the result in Column 4. Similarly, values in Columns 2 and 5 are multiplied to obtain entries in Column 6; and Columns 2 and 7, for entries in Column 8. The value of R is just the sum of all entries in Column 4; the value of G, the sum of entries in Column 6; and the value of B, the sum of entries in Column 8. Thus the color of fresh butter is given by the tristimulus values $+12$, $+23$, $+33$, for the set of primary colors **B, G, R.** Only two significant figures are retained for each value, if that is the accuracy of the SPD data.

Example Calculation of Tristimulus Values

1	2	3	4	5	6	7	8
λ (nm)	P_λ	$\bar{r}_\lambda$	$P_\lambda \bar{r}_\lambda$	$\bar{g}_\lambda$	$P_\lambda \bar{g}_\lambda$	$\bar{b}_\lambda$	$P_\lambda \bar{b}_\lambda$
400	56	0.0003	0.02	−0.0001	−0.01	0.0121	0.68
450	31	−0.0121	−0.38	0.0068	0.21	0.3167	9.82
500	35	−0.0717	−2.51	0.0854	2.99	0.0478	1.67
550	70	0.0228	1.60	0.2118	14.83	−0.0006	−0.04
600	78	0.3443	26.86	0.0625	4.87	−0.0005	−0.04
650	78	0.1017	7.93	0.0012	0.13	0.0000	0.00
700	77	0.0041	0.32	0.0000	0.00	0.0000	0.00
750	76	0.0001	0.01	0.0000	0.00	0.0000	0.00
		R = sum =	33.84	G = sum =	23.01	B = sum =	12.09

At the risk of overstressing a point and to avoid a popular misconception, we emphasize that the curves in Fig. 3-9 are *not* the spectral power distributions of the three pri-

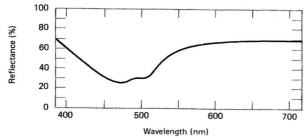

Fig. 3-10. Reflectance curve of butter in the visible portion of the spectrum.

maries **B, G, R.** The primaries that we have chosen are monochromatic colors. What the curves indicate is how many units of each of the three primaries are needed to match one unit of monochromatic power at any wavelength. Where one of the color matching functions is negative, that primary must be added to the monochromatic sample and not to the other two primaries to achieve a match.

3-6 THE MAXWELL TRIANGLE

Maxwell realized that chromaticity of a colored surface is relatively insensitive to its brightness, or luminance, and this is why chromaticity matches (for hue and saturation only) are useful. Furthermore, since only two variables describe chromaticity, using all three tristimulus values is superfluous. We need only *two* numbers to specify chromaticity. Many choices can be devised, but some offer more advantages than others. Maxwell's approach was to notice first that luminance differences are eliminated from consideration if each tristimulus value is divided by the sum of the three tristimulus values. Accordingly, he defined new variables b, g, and r in this way:

$$b = \frac{B}{B + G + R}, \qquad g = \frac{G}{B + G + R}, \qquad r = \frac{R}{B + G + R} \qquad (3\text{-}6)$$

The numbers b, g, r have an important feature: they are unchanged if samples differ only in luminance but not chromaticity. For example, if one sample has twice the luminance but the same chromaticity as another, B, G, and R are each twice as large. However, since this increase in the numerators in Eq. 3-6 is compensated by an equal increase in the denominator, b, g, and r do not differ. To specify chromaticity we need quote only two of these numbers, say r and g. The third adds no new information because it can be deduced from these two numbers. This fact is a result of the mathematical equality: $b = 1 - r - g$. The reader can verify that this is an equality by substituting for b, r, and g their expressions from Eq. 3-6.

The clever definition in Eq. 3-6 provided Maxwell with a way to introduce a color map on which the chromaticity of colors could be represented. Knowing geometry he realized that all possible combinations of three positive numbers that add to unity can be represented by different points within a triangle whose sides have equal lengths (Fig. 3-11). The variables b, g, r are called *chromaticity coordinates* of the sample's color.

The Maxwell triangle was the first two-dimensional color space based on psychophysical measurements. It can be generalized to include the third dimension for luminance.

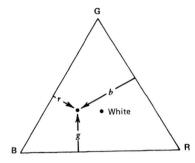

Fig. 3-11. Maxwell triangle showing that the chromaticity of a color can be indicated by the coordinates *b, g, r*, representing distances from each side.

And it allows us to predict the result of adding two colors, say for simplicity of equal luminance (and therefore equal brightness). Because of the additive nature of color matching as expressed by Eq. 3-2, when two colors are added, their respective tristimulus values add. A little thought about the meaning of Eq. 3-6 should reveal the implication of this statement: when two colors are added, the coordinates of the result are appropriate *averages* of the coordinates of the original colors.

Figure 3-12 gives an example of a nearly saturated red that is added to an unsaturated cyan. The resulting color will have coordinates that place it between the original colors. If the red and cyan have equal luminance, the result lies midway between them. Should the luminance of the red be increased, the resulting coordinates are shifted downward along the line to lie closer to the red. Similarly, if the cyan is made more intense, the coordinates shift upward along the line to lie nearer to the cyan. Prediction of the result of adding any other two colors can be carried out in the same way. This color map is analogous to Newton's color circle in the way that the effects of color mixing are predicted. The important difference is that geometrical relationships and distances between colors in Maxwell's triangle have a precise significance based on psychophysical measurements, whereas only general notions are conveyed by relationships in Newton's color circle.

Need for Curved Boundaries

Using Maxwell's triangle we run into trouble when we look for the location of a saturated yellow or saturated cyan. In the latter case for instance, adding **B** and **G** yields a color that falls on the straight edge of the triangle. But experiments show that such a color produced by addition is not as saturated as a spectral hue. Thus a range of spectral hues

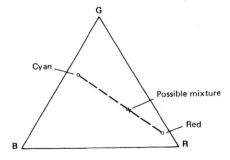

Fig. 3-12. The color resulting from an additive mixture of the red and cyan shown on Maxwell's triangle falls somewhere on the line joining them, depending upon the relative brightness of the two colors.

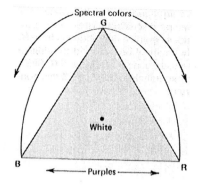

Fig. 3-13. Generalization of the Maxwell triangle to include all spectral colors in the diagram. Colors between the triangle and curved boundary cannot be matched by positive additive mixtures of the primaries.

near cyan must fall *outside* the triangle. This is true also for yellow: the point must lie outside the line joining **R** and **G**. Hermanm von Helmholtz explained that this inadequacy of a triangular map comes from Maxwell's inclusion of only positive tristimulus values in his color space. A proper map containing all known colors must have boundaries that include some of the region outside the triangle. A schematic illustration is given in Fig. 3-13. Spectral colors lie on the curved boundary, and all colors must be embraced between this curve and the straight line representing the purples. On this map the gamut of a set of primaries represented by the colors **B, G, R** has a more limited area: the gamut includes all colors lying within the triangle formed by the three points designated **B, G, R**. It is clear from the curved shape of this color space that any set of three primary colors cannot contain all known colors in its gamut. (This result is a direct consequence of the occurrence of negative numbers in the color matching functions in Table 3-1.)

3-7 C.I.E. COLOR SPACE

Before describing the color space adopted by the Commission Internationale d'Eclairage, it is worthwhile reviewing the motivation behind the use of tristimulus values. The notion that a color can be specified by the three numbers B, G, R is satisfying because this suggests a parallel with the fact that our perception of color is based on signals from the three sets of photoreceptors in the retina. But the analogy is not exact even if the spectral power distributions of **B, G, R** are chosen to match the spectral sensitivities of the three photoreceptors. As explained previously, not all colors can be described by positive amounts of **B, G,** and **R** regardless of what primary colors are used. Thus the visual system must operate in a more sophisticated manner than by simply determining color on the basis of three positive outputs from the photoreceptors. If this is so, there is no reason to retain the **B, G, R** primaries if another set offers more advantages.

C.I.E. Primaries

In 1931 the Commission Internationale d'Eclairage did indeed adopt a new set of primaries. They are called **X, Y, Z** and are defined mathematically by positive and negative combinations of **B, G, R.** In choosing the appropriate combinations, the C.I.E. ensured that the corresponding tristimulus values X, Y, Z are always positive. This was principally for computational convenience. The combinations were selected so that a match to white (or other achromatic color) is represented by equal values for X, Y, Z, as was the case with the **B, G, R** system. The combinations were also arranged so that the value of *Y* determines the luminance of the sample color. (This will be discussed further in Chapter 6.) Because the definition of the **X, Y, Z** primaries includes negative contributions from **B, G,** or **R**, the C.I.E. primaries are not colors; that is, they cannot be matched to additive mixtures of **B, G, R** no matter what type of lights or filters are used for **B, G,** and **R**. Since in practice we never need to carry out color matching experiments with **X, Y, Z** this non-physical nature of the C.I.E. primaries should not worry us.

TABLE 3-2 Color Matching Functions for the **X, Y, Z** Set of Primary Colors Adopted by the C.I.E. in 1931

Wavelength λ (mm)	$\bar{x}_\lambda$	$\bar{y}_\lambda$	$\bar{z}_\lambda$
380	0.0014	0.0000	0.0065
390	0.0042	0.0001	0.0201
400	0.0143	0.0004	0.0679
410	0.0435	0.0012	0.2074
420	0.1344	0.0040	0.6456
430	0.2839	0.0116	1.3856
440	0.3483	0.0230	1.7471
450	0.3362	0.0380	1.7721
460	0.2908	0.0600	1.6692
470	0.1954	0.0910	1.2876
480	0.0956	0.1390	0.8130
490	0.0320	0.2080	0.4652
500	0.0049	0.3230	0.2720
510	0.0093	0.5030	0.1582
520	0.0633	0.7100	0.0782
530	0.1655	0.8620	0.0422
540	0.2904	0.9540	0.0203
550	0.4334	0.9950	0.0087
560	0.5945	0.9950	0.0039
570	0.7621	0.9520	0.0021
580	0.9163	0.8700	0.0017
590	1.0263	0.7570	0.0011
600	1.0622	0.6310	0.0008
610	1.0026	0.5030	0.0003
620	0.8544	0.3810	0.0002
630	0.6424	0.2650	0.0000
640	0.4479	0.1750	0.0000
650	0.2835	0.1070	0.0000
660	0.1649	0.0610	0.0000
670	0.0874	0.0320	0.0000
680	0.0458	0.0170	0.0000
690	0.0227	0.0082	0.0000
700	0.0114	0.0041	0.0000
710	0.0058	0.0021	0.0000
720	0.0029	0.0010	0.0000
730	0.0014	0.0005	0.0000
740	0.0007	0.0003	0.0000
750	0.0003	0.0001	0.0000
760	0.0002	0.0001	0.0000
770	0.0001	0.0000	0.0000
780	0.0000	0.0000	0.0000

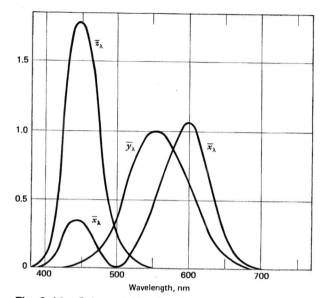

Fig. 3-14. Color matching functions for the **X, Y, Z** set of primary colors that define the color matching properties of the C.I.E. 1931 Standard Observer. (From D. B. Judd and G. Wyczecki, *Color in Business, Science, and Industry,* Third Edition, Wiley, New York, 1975.)

The tristimulus values X, Y, Z can be calculated by using the color matching functions $\bar{x}_\lambda$, $\bar{y}_\lambda$, $\bar{z}_\lambda$ listed in Table 3-2 and shown in Fig. 3-14. Notice that all of the values are positive, as must be the case if all possible tristimulus values are positive. Each tristimulus value for a sample of interest is calculated by proceeding from one wavelength to the next and multiplying the appropriate color matching function by the power P_λ at that wavelength as specified by the sample's SPD. Then all these numbers are added together. This procedure is similar to the one described in the previous section and shown in Example 3-1 for calculating B, G, R. The color matching functions $\bar{x}_\lambda$, $\bar{y}_\lambda$, $\bar{z}_\lambda$ shown in Table 3-2 were originally calculated from the psychophysically determined $\bar{b}_\lambda$, $\bar{g}_\lambda$, $\bar{r}_\lambda$ by using the same formulas that were chosen to relate **X, Y, Z** to **B, G, R**. A modern instrument called a colorimeter measures the SPD of a color and then by computer calculates X, Y, Z using tabulated values of $\bar{x}_\lambda$, $\bar{y}_\lambda$, $\bar{z}_\lambda$ stored in its memory.

C.I.E. Chromaticity Diagram

If we focus attention on the chromaticity of a color and disregard luminance, only two numbers are required for a complete specification. It makes sense to follow Maxwell's lead and introduce new numbers that have the overall strength of the primaries divided out:

$$x = \frac{X}{X + Y + Z}, \qquad y = \frac{Y}{X + Y + Z}, \qquad z = \frac{Z}{X + Y + Z} \tag{3-7}$$

The three new descriptive numbers *x, y, z* are called the *C.I.E. chromaticity coordinates.* Their values are independent of the luminance according to this definition. Together they comprise what is termed the C.I.E. *chromaticity* of a color.

Relationships between the chromaticities of colors can be displayed on a map. The C.I.E. avoided the clumsy triangular coordinates used by Maxwell and instead introduced coordinates arranged at right angles (Cartesian coordinates). This system of rectangular coordinates is usually the way relationships are shown in all branches of science. It is also found on road maps. Because a map requires only two coordinates to denote position, only *x* and *y* defined by Eq. 3-7 were chosen to be shown. The value of *z* provides no new information, since $z = 1 - x - y$.

All possible colors on the C.I.E. chromaticity diagram are represented by chromaticities that fall within an area of the map bounded by a horseshoe shaped curve called the *spectrum locus,* where monochromatic hues are found, and a straight line that joins its ends called the *line of purples* (Color Plate 10). Values are given for wavelengths at various positions on the spectrum locus. Achromatic colors, which differ only in luminance, have equal values for *X, Y, Z* and therefore have the same chromaticity with coordinates $x = \frac{1}{3}$ and $y = \frac{1}{3}$.

Figure 3-6 shown earlier illustrated the gamut of three additive primaries (equivalent to Maxwell's triangle) embedded within the chromaticity diagram. The primaries **B, G, R** for this triangle were chosen to be monochromatic colors at wavelengths of 440, 546, and 700 nm. The overall similarity between the chromaticity diagram and our generalization of Maxwell's triangle (Fig. 3-13) is apparent. In both cases the area within the triangle represents the gamut of the primaries. In neither case are all possible colors included within the gamut. We shall later return to discuss the reasons for this in more detail.

Standard Sources

Recall that the color of a surface is affected by the SPD of the source illuminating it. For this reason the C.I.E. has designated several SPDs approximating common sources as *standard illuminants* and has tabulated the values of each. Once the spectral reflectance curve of a surface is measured, the SPD of the reflected light from any one of the standard illuminants can be calculated. Then the chromaticity of the reflected light can also be calculated.

The SPDs of three standard illuminants A, B, and C are given in Table 3-3, and the curves will be shown in Fig. 7-4 of Chapter 7. The positions of their colors on the C.I.E. diagram are shown in Fig. 3-15 and Color Plate 10. The chromaticity of the C.I.E. standard illuminant A closely matches the chromaticity of a 100 watt incandescent lamp commonly used in lighting fixtures. Other incandescent sources operating at various temperatures are indicated in Fig. 3-15 by their "color temperatures." Roughly speaking, light of a given color temperature has the same chromaticity as light emitted by an object heated to that temperature. The value of the temperature is conventionally expressed in degrees on the Kelvin scale, and this is denoted by the symbol K following the value for the temperature. A precise definition of color temperature and an explanation of the Kelvin scale is given in Section 7-3.

A low color temperature indicates a reddish appearance; a medium color temperature, white; and a high color temperature, bluish. The 100 watt lamp has a color temperature of about 2800 K. Direct sunlight is described by a color temperature of 4870 K and is closely matched by C.I.E. standard illuminant B. Light from an overcast sky having a color

TABLE 3-3 Relative Spectral Power Distributions of the C.I.E. Standard Illuminants A, B, and C[a]

Wave-length (nm)	A	B	C	Wave-length (nm)	A	B	C
380	9.79	22.40	33.00	580	114.44	101.00	97.80
385	10.90	26.85	39.92	585	118.08	100.07	95.43
390	12.09	31.30	47.40	590	121.73	99.20	93.20
395	13.36	36.18	55.17	595	125.39	98.44	91.22
400	14.71	41.30	63.30	600	129.04	98.00	89.70
405	16.15	46.62	71.81	605	132.70	98.08	88.83
410	17.68	52.10	80.60	610	136.34	98.50	88.40
415	19.29	57.70	89.53	615	139.99	99.06	88.19
420	21.00	63.20	98.10	620	143.62	99.70	88.10
425	22.79	68.37	105.80	625	147.23	100.36	88.06
430	24.67	73.10	112.40	630	150.83	101.00	88.00
435	26.64	77.31	117.75	635	154.42	101.56	87.86
440	28.70	80.80	121.50	640	157.98	102.20	87.80
445	30.85	83.44	123.45	645	161.51	103.05	87.99
450	33.09	85.40	124.00	650	165.03	103.90	88.20
455	35.41	86.88	123.60	655	168.51	104.59	88.20
460	37.82	88.30	123.10	660	171.96	105.00	87.90
465	40.30	90.08	123.30	665	175.38	105.08	87.22
470	42.87	92.00	123.80	670	178.77	104.90	86.30
475	45.52	93.75	124.09	675	182.12	104.55	85.30
480	48.25	95.20	123.90	680	185.43	103.90	84.00
485	51.04	96.23	122.92	685	188.70	102.84	82.21
490	53.91	96.50	120.70	690	191.93	101.60	80.20
495	56.85	95.71	116.90	695	195.12	100.38	78.24
500	59.86	94.20	112.10	700	198.26	99.10	76.30
505	62.93	92.37	106.98	705	201.36	97.70	74.36
510	66.06	90.70	102.30	710	204.41	96.20	72.40
515	69.25	89.65	98.81	715	207.41	94.60	70.40
520	72.50	89.50	96.90	720	210.36	92.90	68.30
525	75.79	90.43	96.78	725	213.26	91.10	66.30
530	79.13	92.20	98.00	730	216.12	89.40	64.40
535	82.52	94.46	99.94	735	218.92	88.00	62.80
540	85.95	96.90	102.10	740	221.66	86.90	61.50
545	89.41	99.16	103.95	745	224.36	85.90	60.20
550	92.91	101.00	105.20	750	227.00	85.20	59.20
555	96.44	102.20	105.67	755	229.58	84.80	58.50
560	100.00	102.80	105.30	760	232.11	84.70	58.10
565	103.58	102.92	104.11	765	234.59	84.90	58.00
570	107.18	102.60	102.30	770	237.01	85.40	58.20
575	110.80	101.90	100.15	775	239.37	86.10	58.50
580	114.44	101.00	97.80	780	241.67	87.00	59.10

[a]Figures 7-2 and 7-4 show the SPDs for these three sources.

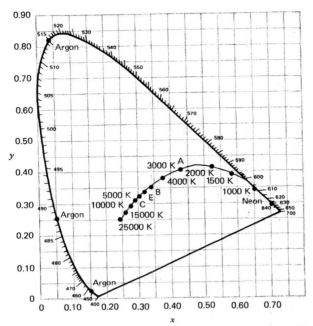

Fig. 3-15. The C.I.E. chromaticity diagram for the **X, Y, Z** set of primaries. The inside curve shows the chromaticities of incandescent sources at the indicated color temperatures. The C.I.E. Standard Illuminants are denoted by A, B, and C. The chromaticities for laser light from a neon laser (633 nm) and the green, blue, and indigo lines from the argon laser (515 nm, 488 nm, and 458 nm) are also shown.

temperature of 6770 K is closely matched by standard illuminant C. Also in the C.I.E. diagram are shown the chromaticities of light from the "equal power source" (E) and several common lasers. Manufacturers of dyes, textiles, and cabinets, for example, can provide precise descriptions of their colors by specifying the chromaticity coordinates under illumination by one of the standard sources.

Additive Color Mixing

Because of the additive nature of the color matching condition (Eq. 3-2), when two colors are mixed additively their respective tristimulus values add. This is true whether we consider the set **B, G, R** or the set **X, Y, Z** of primaries. As proved in Advanced Topic 3A, it follows that the "straight-line" rule also applies in the C.I.E. diagram: the resulting chromaticity lies somewhere on the line connecting the two original chromaticities. Thus a color with the chromaticity *T* in Fig. 3-16 can be produced by adding white light to monochromatic light of the same dominant wavelength of 500 nm, if proper proportions are chosen. Increasing the brightness of the monochromatic light shifts the resulting chromaticity closer to the spectrum locus. This enhances its purity without affecting its dominant wavelength. On the other hand, increasing the brightness of the white light shifts the resulting chromaticity toward the achromatic point, thereby decreasing its purity without change in dominant wavelength.

To determine exactly where the resulting chromaticity lies we first need to characterize the amount of light coming toward our eye from each of the original sources. This is related to the "luminance" of the source, which will be defined in Chapter 6. In the C.I.E. system, the value of the luminance L_v is simply $L_v = 683 Y$.

This is the general rule for color addition: If x_1, y_1 and x_2, y_2 are the C.I.E. chromaticity coordinates of the original colors with Y tristimulus values Y_1 and Y_2, respectively, the chromaticity coordinates of the resulting color x_3, y_3 are given by

$$x_3 = \left(\frac{Y_1}{Y_1 + Y_2} \right) x_1 + \left(\frac{Y_2}{Y_1 + Y_2} \right) x_2$$

$$y_3 = \left(\frac{Y_1}{Y_1 + Y_2} \right) y_1 + \left(\frac{Y_2}{Y_1 + Y_2} \right) y_2$$

(3-8)

These relations are valid only if y_1 and y_2 are reasonably close to each other in value. The full version of these equations will be discussed in Advanced Topic 3A. The expressions in Eq. 3-8 are intuitively reasonable. They say that each of the resulting chromaticity coordinates (say, x_3) is the average of the respective coordinates of the components (x_1

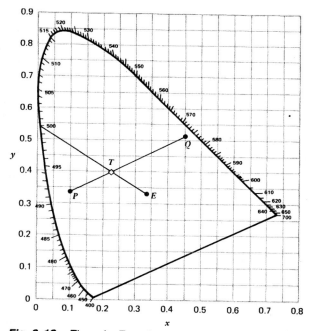

Fig. 3-16. The color T can be produced by adding monochromatic light having a wavelength of 500 nm to achromatic light of the appropriate luminance. As explained in Example 3-2, the color T can also be produced by adding a color P and a color Q whose luminance is half that of P.

and x_2) *weighted according to their relative contributions to the total luminance.* This is sometimes called the "center of gravity" rule.

Example 3-2   Superimposed Stage Lights

For a theater stage production a blue-green light and yellow light are mixed in proportions so that the blue-green light has twice as many units of luminance as the yellow. This can be achieved by using a blue-green filter in front of one lamp having twice the transmittance of a yellow filter in front of a second identical lamp. The spectral transmittance curves for the two filters have been used to calculate the tristimulus values and from those values the chromaticity coordinates. The blue-green (P) has $x_1 = 0.12$, $y_1 = 0.33$ whereas the yellow (Q) has $x_2 = 0.45$, $y_2 = 0.51$. See Fig. 3-16. Because the blue-green has twice as many units of luminance as the yellow, we let $Y_1 = 2$ and $Y_2 = 1$ in Eq. 3-8. Then the chromaticity coordinates for the resulting color are given by:

$$x_3 = \frac{2}{2+1}\,0.12 + \frac{1}{2+1}\,0.45 = 0.08 + 0.15 = 0.23$$

$$y_3 = \frac{2}{2+1}\,0.33 + \frac{1}{2+1}\,0.51 = 0.22 + 0.17 = 0.39$$

Thus the resulting chromaticity (labeled T) is closer to the blue-green than yellow because the blue-green has a greater luminance.

Equation 3-8 can thus be used to predict the chromaticity that results by adding two colors if the chromaticities and relative luminances of the two are known.

In the special case of equal luminances ($Y_1 = Y_2$) Eq. 3-8 predicts $x_3 = \frac{1}{2}x_1 + \frac{1}{2}x_2$. Then the resulting value for x_3 is the *average* of the original values. The same is true for y_3. The resulting chromaticity lies midway between the original colors.

Complementary Colors

The positions of complementary colors on the chromaticity diagram are now apparent: they lie on opposite sides of the achromatic point, so that the chromaticity of the sum according to Eq. 3-8 falls exactly on that point. Therefore it is possible for a color with low luminance to be the complement of another having much greater luminance provided the latter lies much closer to the achromatic point.

Gamut of a Set of Primaries

From the preceding law for additive color mixing we can see why the gamut of a set of primaries is the triangular area determined by their three chromaticities. Figure 3-6 showed an example for three monochromatic primaries, and Fig. 3-7 illustrated the gamut that would be spanned by a typical color television set. The boundaries are determined by the "straight-line" rule when only pairs of primaries are added in various proportions. The interior is spanned when the third primary is added in various proportions to any mixture of the other two.

We can see why it is impossible for the gamut of three primaries **B, G, R** to include all colors. Referring to Fig. 3-17, suppose that a sample **S** is monochromatic light with $\lambda = 500$ nm. Since any combination of **B, G, R** must lie *within* the triangle having apexes at **B, G, R,** the sample color can never be matched by any such combination. On the other

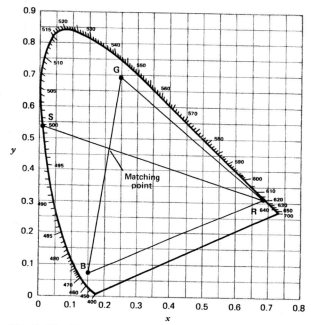

Fig. 3-17. The gamut of the primaries **B, G, R** lies within the triangle formed by their chromaticities. The sample **S** does not lie within this gamut. However, if **R** is added to **S** in just the right proportion, the resulting chromaticity lies on the line joining **B** and **G** and thus can be matched by proper proportions of **B** and **G**.

hand, in a color matching measurement with a bipartite screen, if the **R** light is redirected so that it adds to the sample, the resulting color will fall somewhere on the straight line joining **R** with **S**. By comparison the primary side of the bipartite screen will produce a color somewhere on the line joining **B** and **G**. The point where these two lines cross determines the matching condition.

Comparison with Psychological Attributes

A color space is most useful if it has a direct correspondence with how we perceive colors. Do equal distances between various points correspond to equal differences in their attributes? One way to examine this question is to measure how far we must go from a given chromaticity in various directions to just perceive a difference in color. Figure 3-18 shows the results for various positions in color space. The axes of the ellipses show how far from the center the chromaticity must deviate before a difference is perceived. For clarity, the size of each ellipse has been increased by a factor of ten.[†] Evidently, distance

---

[†]David MacAdam of Eastman Kodak Company established in 1942 how much scatter in the chromaticity occurs when observers attempt to match a given sample with the sum of three primary lights. The spread in results (standard deviation) determined the size of the ellipse that is centered on the chromaticity of the sample. These "MacAdam ellipses" provided a scientific basis for characterizing how close one color is to another.

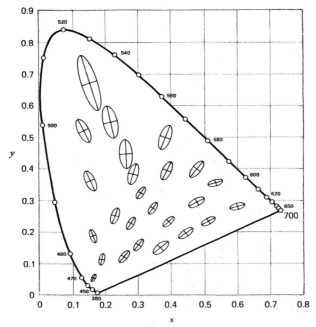

Fig. 3-18. Inaccuracy for color matches at various chromaticities. The lengths of the axes in each ellipse indicate the distance on the chromaticity diagram for which the color is just noticeably different from the color at the center of the ellipse. For clarity, the size of the ellipses has been expanded by a factor of ten; therefore the actual distances on the diagram would be ten times smaller than shown. (From MacAdam, *Journal of the Optical Society of America, 32,* 1942, p. 247.)

on the chromaticity diagram is not a fair measure of the noticeability of color differences. Small deviations in position are more easily perceived for chromaticities toward the bottom of the diagram; equal distances on the diagram do not mean equal visual differences.

A second comparison can be made with respect to chromaticity attributes as expressed, for example, by variables in the Munsell color system. Figure 3-19 shows curves of chromaticities of constant Munsell Hue radiating from the achromatic point. If Hue were identical to the psychophysical variable of dominant wavelength, these curves would all be straight lines. The diagram shows they are nearly straight, but not exactly. The region of yellows shows closest agreement. In fact the Munsell system was revised to obtain reasonably good agreement in this fashion after it was discovered that several of the original samples deviated considerably from these smooth curves. The presently accepted Munsell samples are described by Hue and Chroma in this revised notation.

If Munsell Chroma were identical with the psychophysical purity, a curve of constant Chroma would encircle the achromatic point and would have approximately the same shape as the perimeter (the spectrum locus plus the line of purples). This is clearly not the case in Fig. 3-19, because the contours do not extend sufficiently far toward the upper left and go too far toward the right. There is rather poor correspondence between Chroma and purity.

The lack of precise correspondence between visual attributes and colorimetric speci-

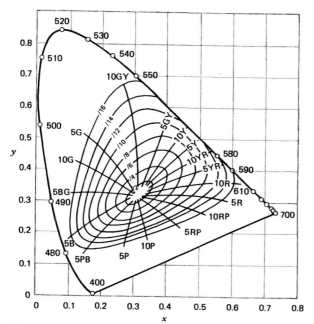

Fig. 3-19. Radiating from the achromatic point are curves denoting colors judged to have constant Munsell Hue at Value 5. Circling the achromatic point are curves of constant Munsell Chroma. Munsell notation is used to indicate Hue and Chroma. (After S. M. Newhall, D. Nickerson, and D. B. Judd, *Journal of the Optical Society of America 33*, 385, 1943.)

fications encourages continued research in color science to find a better arrangement of color space. In the meantime the 1931 C.I.E. chromaticity diagram is generally used for the precision it offers in color measurement. It also makes clear from the position of chromaticities and the "straight-line" rule for color mixing why psychologists have developed the following conclusions pertaining to additive color mixing:

1. A spectral hue has the maximum possible saturation for that hue.
2. The mixture of two spectral colors that are closer together in the spectrum than complementary colors yields a hue of an intermediate spectral color, but less saturated.
3. When two spectral colors are farther apart in the spectrum than complements, the hue may not be spectral at all but a purple.
4. Purples are the only saturated colors that can be made by mixing spectral colors.
5. A range of spectral colors from blue-green (493 nm) to yellow-green (567 nm) have no spectral color as a complement, but the complementary color can be produced by a mixture of red and blue.
6. All possible colors can be formed by additive mixing of various pairs of spectral colors in proper proportion.

The shape of the spectrum locus also explains why saturated red, orange, yellow, and yellow-green surfaces can be produced with a high reflectance, whereas saturated green and blue are much darker. Because of the straightness of the red-to-green locus, light from a large range of wavelengths can be added to produce a color in the yellow that is still almost completely saturated.

We now see why there is a bulge in the Munsell color tree in Fig. 3–3 having both high Value and high Chroma in the yellow part of the spectrum. This is not possible for the curved portions of the locus, where the result of adding chromaticities on the boundary produces a chromaticity inside the boundary, which therefore is unsaturated. To remain saturated the reflected light can contain power only within a narrow range of wavelengths; and consequently because of the very selective reflectance such a surface appears dark.

3-8 METAMERISM

Some readers may recall the experience of seeing the colors of two items of clothing match under the fluorescent lights of a store but not match in daylight. This is an example of *metamerism* (pronounced me·tam′er·ism) and the two colors seen as identical under fluorescent light are said to be *metamers* (met′a·meres). Evidently the spectral reflectance curves for the two items differ and, thus, for a given type of illumination the spectral power distributions of light coming from them differ. Two samples may happen to be metamers under one condition of illumination but generally are not under another. Because such colors appear identical they must have identical tristimulus values. A metameric match is defined as an exact psychophysical match of two stimuli with different SPDs.

The phenomenon of metamerism points out that color perception is determined by the *total effect* of contributions from all parts of the visible spectrum. Differences in details of the spectral distributions for two samples may have no effect on the perceived color. We cannot say just by noticing a difference in spectral power distributions whether the colors differ. An illustration of metamerism is shown in Color Plate 11. The orange skin and the orange filter have quite different reflectance curves, but they look alike in the illustration. Their common color is indicated by identical sets of tristimulus values. Consequently for each set of tristimulus values there is a family of SPDs that look alike and the colors are called "metamers." A point on the C.I.E. chromaticity diagram may represent a large number of metamers.

Metamerism illustrates the essential insight of colorimetry that we have explored in this chapter. The physical stimulus reaching the eye consists of the SPD that contains an extensive amount of information. However, the response of the eye and its associated neurological network produces a perception that can be characterized by three quantities such as hue, saturation and value, or chromaticity (x, y) and luminance (Y).

Since the perceptual response is far simpler than the physical stimulus producing it, many different physical stimuli (each with a different SPD) can produce identical perceptions. The methods of colorimetry embodied in the C.I.E. procedures thus allow us to bypass the unnecessarily large amount of information required to describe a physical stimulus (the entire SPD) and specify it by three quantities that predict its perceptual qualities. The key to this simplification lies in the ingenious psychophysical experiments which led to the color matching functions that summarize the perceptual response of the "average human observer."

SUMMARY

Three qualities can be perceived in a color: hue, saturation, and brightness. Two colors are identical when each of these three qualities match. However, there is no way to specify the hue or saturation or brightness of a given color except by comparison with a set of sample colors. A variety of color systems have been developed for this purpose. Among them is the Munsell system, in which the color samples were selected to appear evenly spaced for differing hue or saturation or brightness, and the Ostwald system, in which color samples are specified by the proportions of white, black and "full" color segments on a spinning disc that match each sample. Specifying color by comparing with samples in one of these color spaces is adequate for many purposes but suffers from problems of subjective judgment and fading of samples. The C.I.E. chromaticity diagram is an alternative method in which samples are not needed. It offers the advantages of precision, an easily visualized relationship between the colors, and the capability of predicting quantitatively the result of adding together two or more colors. The chromaticity of a color is specified by its coordinates x, y in the diagram and these coordinates can be calculated once the SPD of the color is known. Colors lying near the boundaries of the chromaticity diagram are saturated, while those near the center are unsaturated. If two colors are added together the resulting color lies on the straight line joining the original colors. This straight-line relationship makes the C.I.E. diagram particularly useful for showing the relationship between colors produced by additive mixing. Consequently the gamut of colors corresponding to a set of three primary colors lies within the triangle formed by the positions of the three primaries. For this reason the C.I.E. diagram is often employed to show the range of colors that can be achieved by various techniques for color reproduction. This system, based on the results of psychophysical experiments, allows the perceptual response to be predicted quantitatively from the SPD of the stimulus.

QUESTIONS

3-1. What are the three attributes that specify our perception of a color?

3-2. What are the three attributes of a color that specify its position in the Munsell color space? To which psychological attribute does each correspond?

3-3. What is an achromatic color? Give some examples.

3-4. What aspect of a color in Ostwald's color system corresponds to Hue in Munsell's?

3-5. How are the brightness and saturation of a color affected by tinting, by shading, and by toning?

3-6. Give an example on the C.I.E. diagram for the position of the chromaticity of a saturated color. What hue does the color of your choice have? Show where a less saturated color of the same hue is found.

3-7. What do the symbols x, y, and Y represent when describing a color in the C.I.E. color space?

3-8. Suppose that a color sample is measured on a colorimeter and the color is specified by $x = 0.1$, $y = 0.6$, and $Y = 0.9$. Where is the chromaticity found on the C.I.E. diagram? What is the hue of this color? Is the color completely saturated?

3-9. If a buyer for a chain of stores wishes to specify the color of a fabric that will be used in a series of dresses, the range of acceptable colors can be specified on the C.I.E. diagram. Show this range if the buyer says that the chromaticity must lie within a distance of 0.1 from the chromaticity at $x = 0.3$ and $y = 0.55$ when under northern skylight.

3-10. When a color specified by $x = 0.4$, $y = 0.3$ is added to another specified by $x = 0.2$, $y = 0.5$, what is the range of the resulting color when the two are added in various proportions? What is the chromaticity of the color if the two colors are equally bright?

3-11. If a color television system is designed with primaries having the chromaticities at $x = 0.65$, $y = 0.35$ and $x = 0.15$, $y = 0.8$ and $x = 0.1$, $y = 0.1$, show the gamut of the colors that they can produce.

3-12. What is the dominant wavelength and purity of a color whose position on the C.I.E. diagram is specified by $x = 0.4$, $y = 0.55$?

3-13. What are the possible complements of the color specified in Question 3-12? There is only one complement that has the same luminance (and same brightness). What is it?

3-14. Suppose that the color specified by $x = 0.1$, $y = 0.6$ is to be matched by adding a monochromatic color to white. What should be the wavelength of the monochromatic color? To achieve an exact match would the monochromatic color be brighter (have greater luminance) or dimmer (less luminance) than the white?

3-15. What is the wavelength of a monochromatic light that would be the complement of a monochromatic light with a wavelength of 490 nm?

3-16. Explain in terms of the C.I.E. diagram why a monochromatic color with a wavelength between about 494 nm and 570 nm has no spectral colors as its complement.

3-17. What are tristimulus values?

3-18. What is the significance of a negative number appearing as one of the tristimulus values for a color?

3-19. What are color matching functions? Were the color matching functions for the **X, Y, Z** primaries determined directly from color matching measurements? Explain.

3-20. Deduce the tristimulus values for the color of the tomato whose reflectance curve is given in Fig. 2-13 for the **R, G, B** primaries. Also deduce the values for the **X, Y, Z** primaries.

3-21. How is it possible for two colors having different SPDs to have identical chromaticities (that is, to be metamers)?

FURTHER READING

F. W. Billmeyer, Jr. and M. Saltzman, *Principles of Color Technology*, Second Edition (Wiley-Interscience, New York, 1981). A lucid explanation of the principles of measuring color and how color is controlled in various applications. Outstanding annotated guide to the literature.

D. L. MacAdam, *Color Measurement: Theme and Variations* (Springer Verlag, New York, 1981). An excellent comparison of various color measurement systems.

D. B. Judd and G. Wyszecki, *Color in Business, Science and Industry*, Third Edition

(Wiley, New York, 1975). Concepts of human color vision; specification, measurement, and control of color; definition of colorimetric terms now in use. A practical book for the more advanced reader.

R. S. Hunter, *The Measurement of Appearance* (Wiley, New York, 1975). Practical aspects of how to measure the SPD of a sample and detailed explanations of the advantages of various color spaces in relating color measurements to perception.

R. M. Boynton, *Human Color Vision* (Holt, Rinehart and Winston, New York, 1979). Excellent explanations of color vision, color spaces, and color matching procedures, as well as the physiology of the human visual system.

W. D. Wright, *The Measurement of Color,* Fourth Edition (Macmillan, New York, 1958). Very well written and clear discussion of color measurement techniques, tristimulus values, and the chromaticity diagram. Many useful illustrations, including color samples and their reflectance spectra.

R. W. Burnham, R. M. Hanes, and C. J. Bartleson, *Color: A Guide to Basic Facts and Concepts* (Wiley, New York, 1963). A compendium of facts pertaining to color, in outline form with some illustrations.

F. Grum and C. James Bartleson, Eds., *Optical Radiation Measurements,* Vol. 2: Color Measurement (Academic Press, New York, 1980). Colorimetry as treated by the C.I.E., color spaces, color measuring instruments, the practical aspects of color matching with a mixture of paints, and color aspects of light sources.

P. J. Bouma, *Physical Aspects of Colour,* Second Edition (St. Martins Press, New York, 1971).

G. Wyszecki, "Resource Letter OC-1: Colorimetry," *American Journal of Physics,* Vol. 37 (1969), p. 1201. An annotated list of books, review articles, and pedagogical articles on colorimetry, including relevant background topics.

CHAPTER 4

WAVE ASPECTS OF LIGHT

Light is familiar to us. Perhaps it is too familiar: we may no longer be intrigued by the question "What is this thing that we call 'light'?" Isaac Newton was not content merely to describe our perception of light by a set of color mixing rules derived from experiments. He sought to understand the underlying physical reality. Light has a wavelike property that gives rise to the colors of a soap bubble and the iridescence of a butterfly wing; it can be scattered by small particles, which accounts for the color of blue eyes; and it can be "polarized," as we can demonstrate with polarizing sunglasses.

Many of these effects were known to Newton, and to gain insight into why light behaves as it does he proposed a physical model. Some aspects of his model were vaguely described, because he had insufficient evidence to make them more precise. Newton supposed that light is a kind of disturbance conveyed by particles in motion. The straight-line motion, or raylike property, of these particles explains shadows from opaque objects and reflections from a mirror surface, which causes these particles to rebound. Newton's particles were endowed with a property called "refrangibility," which determined the angle by which they are refracted by a prism, but how this refraction takes place was not made clear. This was not the first particle model for light. Some 30 years earlier the French scientist René Descartes, in developing a theory for the rainbow, had advanced a similar notion, except that he added the conjecture that each particle also rotated about its axis, and the color of light is due to the rotational speed. Descartes thought that materials which change the color of light change this speed of rotation. This notion, however, was contradicted by Newton's experiments that revealed the laws of color addition.

Controversy ensued from the moment Newton proposed his ideas at a meeting of the Royal Society in London (published in his book *Opticks,* which appeared some 40 years later). An opposite point of view announced in 1665 by Robert Hooke, a fellow member of the Society, held that light was a wave. Hooke thought it much like a sound wave but, instead of a disturbance in air, it derived its unique characteristics as a wave in an invisible medium called the "ether," a medium permeating all space. Later the Dutch astronomer Christiaan Huygens developed Hooke's wave theory and in 1690 showed mathematically that if the effect of a prism or lens is to slow the speed of a wave, the fundamental laws of optical systems could be explained, at least as well as by Newton's competing theory of ray optics. The speed with which light waves moved through the ether must be considerably higher than for a sound wave in air, because it eluded all attempts in the seventeenth and early eighteenth centuries to measure it.

Huygens also proposed a mental picture of wave propagation that forms the basis for much of the subsequent work on diffraction phenomena that we consider in this chapter. Huygens' principle asserts that each point on a light wave acts as a source of secondary waves, and that the overlap of the spreading secondary waves determines how the light wave travels. Huygens' ideas were scorned by Newton, however, whose tremendous reputation guaranteed that the wave theory would be ignored for another century.

The decisive experiment that forced the scientific world to recognize the validity

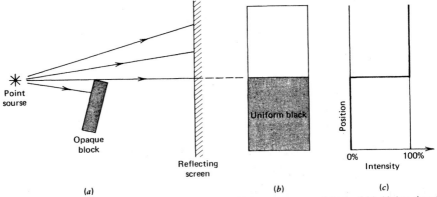

Fig. 4-1. (a) Shadow on a white screen cast by an opaque block. (b) Naive (and incorrect) view of how the light intensity varies across the screen. (c) Graph of intensity versus position suggested by (b).

of the wave theory was announced in 1802 by the English physician and scientist Thomas Young. Young's interest in the physiology of vision led him to investigate the nature of light as well, and to the discovery of the crucial effect of *interference* between two beams of light formed when light is made to pass through two narrow holes placed close together. The pattern of light and dark regions behind the holes could not possibly be explained by ray optics; instead the pattern was the natural result of waves that spread out from two closely spaced sources representing the holes. These observations were supported a few years later by a number of results announced by the French mathematician Augustin Fresnel and also by the German optical manufacturer Joseph von Fraunhofer. Their separate investigations of the unusual patterns created when light passes through a single small hole of varying shape, or close to the edge of an object, demonstrated the spreading of waves or *diffraction* effects. Many phenomena that we see everyday result from wave characteristics of light, but we (and Isaac Newton as well!) do not recognize them until they are pointed out. The rainbow colors of an oil slick on water and the iridescence of mother-of-pearl and peacock feathers are illustrations of interference; whereas the colored spectra seen when we view an electric lamp through a closely woven cloth or the light streaks seen when we squint at a lamp are examples of diffraction.

Many experimenters have attempted to reveal the presence of the invisible ether that Hooke believed to exist; the best known experiment was performed by the American scientists Albert Michelson and Edward Morley in 1887. But all attempts yielded negative results. Today we know that an ether is not needed for the propagation of light waves. Light has an existence of its own as a combined electric and magnetic phenomenon. This origin in electromagnetism was predicted by the Scottish physicist James Clerk Maxwell in the late nineteenth century. At the end of this chapter we shall show that it is the electrical property of light that permits it to be polarized.

■ 4-1 THE EDGE OF A SHADOW

Look closely at the edge of a shadow and you will see that it is not truly sharp. Of course there may be blurring when several lamps provide the illumination. That blurring comes simply when light is incident on the edge of an object from several directions. But of more fundamental interest is the situation where the light is provided by a source so small that it can be considered as a point (the sun will do). If we accept Newton's ray model for light, Fig. 4-1a shows what we expect if an opaque block is interposed between the source and a white screen. The portion of the screen above the block should be illuminated whereas below the block's edge it should be absolutely dark (Fig. 4-1b). The expected light intensity falling on different portions of the screen is portrayed by the curve in Fig. 4-1c. The curve shows 100% of the original intensity above the edge and an abrupt drop to 0% below.

However, when the edge of a shadow is closely examined we see that this simple idea does not represent the physical reality! The actual situation is much more

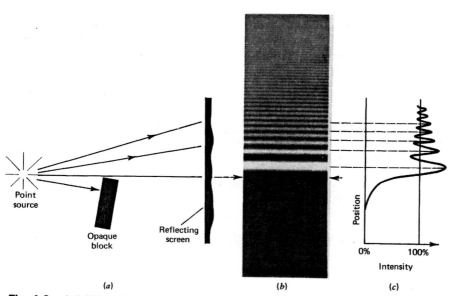

(a)   (b)   (c)

Fig. 4-2. (a) Diffraction at the edge of an opaque block (b) Photograph of an actual diffraction pattern near a straight edge whose position is indicated by the two arrows. (c) Curve showing the observed intensity at various positions near the edge of the shadow. (From M. Cagnet, *Atlas of Optical Phenomena*, Springer-Verlag, 1962.)

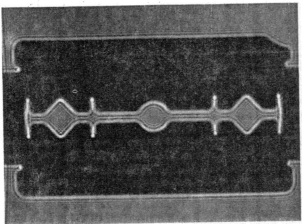

Fig. 4-3. Diffraction around the edges of a razor blade produces the light and dark fringes on the outside. Interference between light diffracted at the nearby inner edges produces the patterns of crosses and circles in the center. (F. W. Sears, M. W. Zemansky, and H.D. Young, *University Physics,* Addison-Welsey, Reading, Massachusetts, 1982.)

interesting. Figure 4-2 shows that for monochromatic light the edge of the shadow is blurred and moreover displays a series of dark and bright bands; whereas from the ray model we had expected to see a uniform intensity.

Patterns of this sort had been studied by Francesco Grimaldi in Newton's time, but their origin was not understood. They cannot be explained by particles or rays of light traveling in straight lines from the source. Instead, they indicate that the edge of an obstruction causes the direction of light to deviate—an effect known as *diffraction.* The ripplelike pattern of intensity above the edge correctly suggests that light behaves as a wave. A striking illustration of diffraction at the edge of a razor blade is shown in Fig. 4-3. Diffraction is a common feature of wave motion, as we realize when watching water waves bend around the end of a breakwater or pilings of a pier. The visual consequences of light's diffraction will be discussed in more detail later in this chapter. First, however, some important properties of a wave must be noted.

■ 4-2 PROPERTIES OF WAVES

Light waves and water waves share some common features, although in other ways they differ. Let us consider their common properties. Figure 4-4a is a "snapshot" of a water wave, showing that the wave crests are regularly spaced. The distance between crests is called the *wavelength.* It is conventional to represent the value of the wavelength by the Greek letter lambda (λ). Of course the wavelength is also the distance between troughs or, for that matter, the distance between any two equivalent points on adjacent waves.

The "strength" of the wave is its *amplitude* (*A*), or the height of the peak above the undisturbed waterline. Since a wave of this type has the crests lying above the undisturbed waterline by the same amount that the troughs lie below, the total vertical distance between trough and crest is twice the amplitude. The *intensity* of a wave is related to the square of its amplitude, or A^2. Thus doubling the amplitude quadruples the intensity. More will be said later about intensity.

Waves travel at a *speed* (*v*). The speed depends on the type of wave and the nature of the medium through which it travels. Waves on the surface of deep water with wavelength of $\lambda = 1$ meter travel at a speed close to $v = 1$ meter per second (abbreviated '1 m/s'). Light, on the other hand, travels at a much faster speed of about 300 million meters per second, or 3×10^8 m/s. (Readers unfamiliar with this scientific notation for large numbers should refer to Example 4-1.) Light travels so fast that its speed is hard to measure. It travels 33 centimeters (cm) in only one billionth of a second (10^{-9} s). In 1 second it can go three-quarters of the way to the moon. When light travels through a vacuum, as in outer space, it always goes at the same speed; so a special symbol (*c*) is conventionally used to denote this speed. The speed of light in a vacuum, to an accuracy of better than 1%, is

$$c = 3 \times 10^8 \text{ m/s} \tag{4-1}$$

When light encounters atoms during its travel, as when passing through a transparent material, it goes more slowly. In glass it travels only two-thirds as fast as in

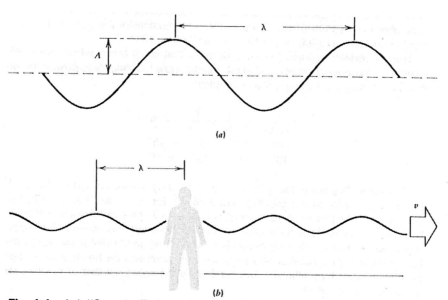

Fig. 4-4. (*a*) "Snapshot" of a water wave showing the wavelength λ and amplitude *A*. (*b*) Since waves travel at a speed *v*, it will take λ/v seconds for one complete wavelength to pass by.

a vacuum. Letting the symbol v denote its speed in a particular medium, we can relate this speed to its speed in a vacuum by the expression:

$$v = \frac{c}{n} \qquad (4\text{-}2)$$

The number n, which varies with the nature of the medium, is called the *refractive index* of the medium. Because n appears in the denominator of this equation, the larger the refractive index the slower is the speed of light. In addition to prescribing the speed of light in a given material, n determines the refractive effect when light enters that material, as will be explained in Chapter 5. For most solids and liquids the refractive index has a value lying between $n = 1.2$ and $n = 2$. For air the refractive index $n = 1.0003$ is so close to $n = 1$ that light travels through air essentially as fast as it travels through a vacuum.

Example 4-1   Powers of Ten Notation

The phenomena of light are described by quantities that range from the very small to the incredibly large. Powers of ten notation avoids the awkwardness of writing many zeros when dealing with such quantities. The basis for this notation is factors of 10. For example 1000 is written as 10^3 (read "ten to the third power") because it represents 10 multiplied by itself three times. The "3" is called the exponent. Fractions such as $\frac{1}{1000}$ or the equivalent decimal 0.001 are represented by negative exponents, 10^{-3}. This follows from the rule that when a number is moved from the denominator to the numerator the sign of the exponent changes: $1/1000 = 1/10^{+3} = 10^{-3}/1 = 10^{-3}$.

The convenient feature of powers of ten is that when two numbers are multiplied their exponents simply add algebraically. How to manipulate powers of ten can be seen by examining a few examples:

$$\frac{1}{100,000} = \frac{1}{10^5} = \frac{10^{-5}}{1} = 10^{-5}$$
$$10^3 \times 10^4 = 10^{3+4} = 10^7$$
$$10^7 \times 10^{-5} = 10^{7-5} = 10^2$$

Most quantities are not simply powers of 10; they are expressed by the product of a prefactor and a suitable power of 10, for example, 8.37×10^7. For convenience the power of 10 is usually chosen so that the prefactor is a number between 1 and 10. How a decimal number is converted into powers of 10 notation is shown in the following examples. When a decimal point is moved to the left (or right) a given number of places, we must compensate the change in value of the prefactor by multiplying the number by the same number of positive (or negative) powers of 10:

$$13,700 = 1.37 \times 10^4$$
$$0.0025 = 2.5 \times 10^{-3}$$

When two numbers in powers of ten notation are multiplied, prefactors are multiplied separately from the power of 10 that follows. Several examples of how numbers in powers of 10 notation are multiplied or divided are given below:

$$(3.2 \times 10^4) \times (5.6 \times 10^{-2}) = (3.2 \times 5.6) \times (10^{4-2})$$
$$= 17.92 \times 10^2$$
$$= 1.792 \times 10^3$$

$$\frac{7.2 \times 10^9}{2.4 \times 10^5} = \left(\frac{7.2}{2.4}\right) \times \left(\frac{10^9}{10^5}\right) = 3 \times 10^{9-5} = 3 \times 10^4$$

Now let us return to consider another aspect of waves. In Fig. 4-4b a sunbather on shore regards the waves moving past at a speed v. As one crest after another passes by directly in front of him, our observer sees the water level rise and fall in cyclic fashion. How long does it take one cycle (called the *period*) to pass by him? Since a full wave must travel a distance λ to pass by, the period is this distance divided by the speed: λ / v. Now we proceed one step further to determine the number of cycles that are seen each second, called the *frequency* of the wave. Since one cycle requires λ / v seconds, in each second we have v/λ cycles. Thus the frequency, denoted by the Greek letter nu (ν), is $\nu = v/\lambda$ cycles per second. This relation between frequency, wavelength, and speed can be rewritten to show directly one important characteristic of a wave—the product of wavelength and frequency equals the speed:

$$\lambda\nu = v \tag{4-3}$$

The disturbance that creates a series of waves determines the frequency. Therefore, if the source oscillates more rapidly, the frequency of the resulting wave increases. Equation 4-3 says that the wavelength must correspondingly decrease to keep the product $\lambda\nu$ unchanged. This means that for a higher frequency one crest has less time to move away before the next crest is produced.

Unlike a water wave, a light wave needs no medium to exist. Light is an electromagnetic wave, which can be radiated by electrons in an atom or molecule. The electric and magnetic fields forming this wave exist even in the absence of matter, as we know because starlight comes to us through the vacuum of outer space. Light has an impressively high frequency and correspondingly a very short wavelength. A typical frequency is $\nu = 6 \times 10^{14}$ cycles per second. Several years ago by international convention it was decided to honor Heinrich Hertz' discovery in 1885 of radio waves with the term "hertz" (abbreviated "Hz") to replace "cycles per second." For monochromatic light having $\nu = 6 \times 10^{14}$ Hz, Eq. 4-3 predicts a wavelength in vacuum of

$$\lambda = \frac{v}{\nu} = \frac{3 \times 10^8}{6 \times 10^{14}} = 0.5 \times 10^{-6} = 5.0 \times 10^{-7} \text{ m} \tag{4-4}$$

As mentioned previously, when we speak of wavelength it is convenient to use one-billionth of a meter (10^{-9} m) as the unit of length. This is called the "nano-

TABLE 4-1 Conventional Prefixes
Indicating Various Multiples of 10,
Together with the Equivalent Value in
Powers of Ten Notation

pico	(p)	10^{-12}
nano	(n)	10^{-9}
micro	(μ)	10^{-6}
milli	(m)	10^{-3}
centi	(c)	10^{-2}
kilo	(k)	10^{+3}
mega	(M)	10^{+6}
giga	(G)	10^{+9}

meter,'' abbreviated "nm.'' The calculated wavelength in Eq. 4-4 can be expressed as $\lambda = 5.0 \times 10^{-7}$ m $= 5.0 \times 10^{+2} \times 10^{-9}$ m $= 500 \times 10^{-9}$ m, or 500 nm. The prefix "nano'' is one example of several prefixes that we will find useful (Table 4-1).

The preceding calculation of a typical wavelength for light shows that even though light has an incredibly high speed (v in the numerator of Eq. 4-4), the frequency with which it is launched by an atom is so much larger numerically (ν in the denominator) that the wavelength λ is extremely short.

The frequency of a light wave does not change when it moves through different materials, such as glass, plexiglass, or water. However, its speed does change, according to Eq. 4-2. The wavelength therefore adjusts to the new medium when light passes from one medium to another. The slower the speed, the shorter the wavelength. This is evident because on going into a medium of slower speed, during each cycle the disturbance does not travel as far as it did in the first medium. If λ_0 denotes the wavelength in a vacuum, then in a material of refractive index n the wavelength is λ_0/n.

■ **4-3 ELECTROMAGNETIC SPECTRUM**

Electromagnetic waves can exist at virtually any frequency. They are classified either according to frequency or, as shown in Fig. 4-5, according to the corresponding wavelength in a vacuum. The eye is sensitive to only a narrow region of the spectrum, spanning roughly an octave (factor of 2) in frequency that is called "light." Because of differences in sensitivity among individuals, there are no precise boundaries denoting the wavelength range of light. For simplicity the visible portion of the spectrum is said to include wavelengths from 400 to 700 nm. Under appropriate conditions some observers can detect electromagnetic waves within the broader region from 380 to 800 nm.

At longer wavelengths, just beyond the red end of the visible spectrum, is the *infrared* region ("infra" meaning frequencies *below* those of red light). Radiation in the infrared region was demonstrated in 1800 by William Herschel, the English astronomer noted for his systematic studies of the stars and discovery of the planet Uranus. Herschel placed a blackened thermometer just outside the red end of the spectrum of sunlight formed by a prism and found that the temperature rose. As this suggests, infrared electromagnetic waves, like light, are produced by atoms and molecules. Nowdays they can be detected by special photographic films and solid state infrared detectors.

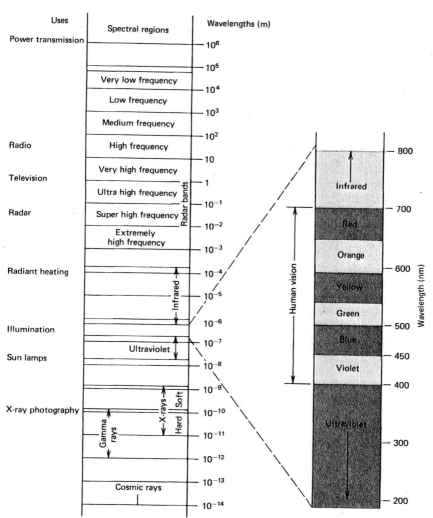

Fig. 4-5. Electromagnetic spectrum, showing the position of the visible region in relation to other regions. The boundaries between regions are indicated by wavelengths measured in vacuum.

Longer wavelengths between 100 μm and 10 cm fall in the *microwave* region. These waves can be produced by electrical circuits and are launched by appropriately shaped antennas. A common application is in long-distance communication and radar. Their ability to penetrate many objects that are visually opaque makes them useful for evenly cooking foods in microwave ovens.

Still longer waves carry broadcast signals for *television* and *radio*. Generally speaking, to radiate electromagnetic waves efficiently, the dimensions of the antenna must be about the same as the wavelength (or perhaps as small as one-quarter of the wavelength). Therefore it becomes impractical to broadcast a wave whose wavelength is much longer than a kilometer, although for some military communications this can be done.

Heinrich Hertz, working in Germany between 1885 and 1889, was the first to produce and detect electromagnetic waves, and he created them in the radio portion of the spectrum. When he produced a momentary electrical spark in the air gap between two oppositely charged metal needles, he observed that another spark appeared in a similar gap between a second set of needles positioned some distance away. As there were no electrical connections between one set and the other, Hertz reasoned that an electrical impulse must have traveled through space to cause the second spark. This experiment demonstrated that an electromagnetic wave carries energy from one place to another.

The existence of such waves had been predicted nearly 15 years earlier by the Scottish mathematical physicist James Clerk Maxwell. He had developed a theory to describe in a unified fashion the many phenomena in electricity and magnetism that had intrigued scientists during the preceding century. From this theory he derived an equation which showed that oscillating electric and magnetic disturbances in association should travel freely through space. The equation predicted that regardless of wavelength, these waves travel with the same speed. That this equation correctly describes the behavior of electromagnetic waves spanning such a wide range of wavelengths, as illustrated in Fig. 4-5, is truly a remarkable accomplishment.

Now we turn to consider the region of the electromagnetic spectrum with wavelengths shorter than those of the violet end of the visible spectrum. This is called the *ultraviolet* ("ultra" signifying frequencies greater than those in the violet). Ultraviolet radiation was discovered by J. W. Ritter in 1801 when he directed light from a prism onto a plate covered by crystals of light sensitive silver chloride, and found that the crystals that darkened first were positioned just outside the violet end of the spectrum; the crystals illuminated by violet darkened next, and finally those illuminated by blue. This change of silver chloride on exposure is the basis of the subsequent development of photography. Ritter's observation showed that electromagnetic radiation exists at wavelengths shorter than those of the violet light.

Beyond this region at still shorter wavelengths are the X-ray and then gamma-ray regions. In these regions of the spectrum, quantum effects (which will be discussed in Chapter 5) become dominant. Nevertheless, X rays and gamma rays are electromagnetic waves as are light and radio waves.

■ 4-4 INTERFERENCE OF TWO WAVES

The phenomenon of *interference* perhaps most graphically illustrates the wave aspect of light. This effect was first reported by Thomas Young, an English physician whose interests in science also led to important work in color vision, and in the theory of elasticity (Young's modulus). Young directed monochromatic light from a narrow portion of the spectrum straight toward two narrow, closely spaced slits, one just above the other, in an opaque screen. When the emerging light illuminates a white screen, the notion of raylike propagation led him to expect two elongated bright bands of light corresponding to the positions of the slits. Instead, there was a pattern of many bright spots ranging up and down the screen (Fig. 4-6). Especially remarkable was one spot that fell in the middle of the pattern at a position midway between the two bright slits, just where ray optics would predict a shadow! We shall now investigate how the wave nature of light provides a consistent explanation for this curious pattern.

First, consider what happens when two water waves of the same wavelength are superimposed in the same region of space (Fig. 4-7). If the waves arrive with crests coinciding (the waves are then said to have the "same phase"), the disturbance of one wave adds to the disturbance of the other. The net amplitude is the sum of the amplitudes of the two waves, and a stronger disturbance is observed. This is called *constructive interference.* On the other hand, if the waves arrive with the crest of one coinciding with the trough of the other (waves "out of phase"), the net amplitude is reduced, since one wave produces an upward movement as the other produces a downward one. This partial cancellation of effects—or complete cancellation if the original waves had the same amplitude—is called *destructive interference.* Whether constructive or destructive interference occurs at a particular location depends on the relative phase of the two arriving waves. The early Polynesian navigators learned this by experience, for there is evidence that even though they were out of sight of land they were able to find their way by observing the interference patterns from waves coming around nearby islands.

To understand Young's two slit experiments we need only figure out where light

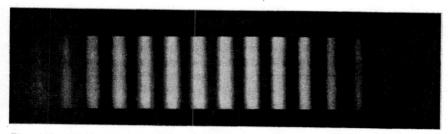

Fig. 4-6. Interference pattern produced when two beams of light emerging from closely spaced slits illuminate a screen. (From M. Cagnet, *Atlas of Optical Phenomena,* Springer-Verlag, 1962.)

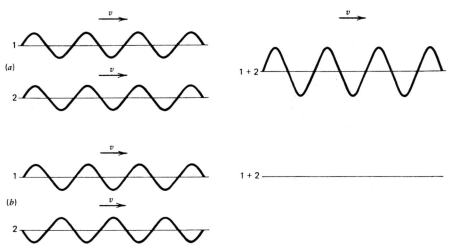

Fig. 4-7. (*a*) Constructive interference between two waves with the same wavelength and phase. (*b*) Destructive interference between two waves of the same amplitude and same wavelength that are out of phase.

from one slit arrives with the same phase as light from the other. At such locations constructive interference produces a large amplitude for the wave. This represents a place where light therefore has high intensity. Where light from one slit arrives out of phase with respect to light from the other, destructive interference creates a dark region. Figure 4-8 illustrates the arrangement for the two slit experiments. The illumination is arranged so that light emerges from each of the two slits with the same phase. Because of diffraction the light waves spread upward and downward behind the narrow slits as they travel toward the screen. The use of horizontal slits instead of round holes ensures that we need not be concerned with spreading in the horizontal direction. Of course at the midpoint of the screen where the two beams arrive after traveling identical distances from the holes, each has undergone the same number of cycles in transit so they arrive in phase. This explains the central bright region.

Somewhat higher on the screen we find a place where light from the lower slit has traveled a half wavelength further than light from the upper one; the extra half cycle puts the two out of phase and, consequently, we see a dark region. In similar fashion proceeding higher we come upon one bright region after another, separated by dark areas. The bright regions occur where the lower wave has traveled one, two, three, or any higher multiple of wavelengths further than the upper wave. The two interfere constructively at these locations. Owing to the slit arrangement, the pattern must be symmetrical up and down. Thus we can understand why a series of bright regions appears instead of only two as predicted by ray optics. The precise positions of these regions can be predicted as explained in Advanced

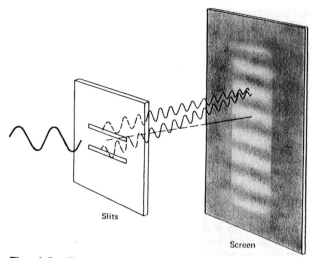

Fig. 4-8. Thomas Young's two-slit interference experiment, showing constructive interference when light emerging from the lower slit has one wavelength further to travel than light from the upper slit.

Topic 4B at the end of the book. Measuring where they occur provides a means of determining the wavelength of light.

Coherence of Light

It is impossible to produce an interference pattern by simply placing a common incandescent lamp in front of a mask containing two small slits. Even if the light were first made monochromatic by imposing a spectral filter in front of the lamp no pattern would be seen. The reason is that the source of light is comparatively large, and light arrives at the slits from a wide range of directions. Consequently the interference pattern produced by light from one small area of the lamp will be shifted on the viewing screen with respect to the pattern from another. The overall effect of superimposing the patterns from all the various areas produces an approximately uniform illumination at the screen. The light incident on the mask is said to be *spatially incoherent.* Incoherence means that the light reaching one point on the mask is quite independent of the light reaching another point, so there can be no systematic interference pattern. The spatial coherence is improved if a card-board containing a pinhole is placed between the lamp and mask. The pinhole then represents a small source (the size of the hole) from which the overlapping effect is greatly reduced. The pinhole serves to limit the area of the lamp that illuminates the mask. Because each area of the lamp emits light independently of the other areas, this limiting effect is what improves the coherence. (It was just this improvement in the experiment that allowed Young to observe the interference pattern!)

Interference patterns are particularly striking when produced by a laser. Unlike the common incandescent lamp that emits light in all regions of the visible spectrum, the laser emits all its power at one wavelength. It is an intense source of monochromatic light. As will be described in Section 7-6, the light is also spatially coherent across the beam, so the beam can be aimed right at the two slits in Young's experiment and need not first be passed through a pinhole. For this reason an interference pattern of impressive breadth, containing many bright and dark spots, can easily be created.

Interference From a Thin Film

Nature provides many splendid examples of interference when light reflects from upper and lower surfaces of a thin film such as an oil slick, the inner and outer surfaces of a soap bubble, or the more complicated structure of a peacock's feather. The essential point is that light coming to the eye from a particular direction is made up of superimposed beams that have traveled different path lengths from the source. If the original light from the source was white, the SPD (spectral power distribution) arriving at the eye will be enhanced at wavelengths for which the difference in path lengths produces constructive interference. The intensity in other spectral regions where destructive interference occurs will be reduced.

Let us consider an example where the resulting color can be predicted. In Fig. 4-9 the cross section of an oil film on a smooth surface is shown. This kind of film can be quite colorful as shown in Color Plate 12. If the thickness of the film is denoted by t, the extra distance traveled by light reflected from the lower surface instead of the upper surface of the film is twice the thickness, or $2t$. Light with this wavelength $\lambda = 2t$, or nearly this wavelength, goes through a complete cycle while traveling the extra distance. It therefore emerges from the film with the same phase as light reflected from the top of the film. To predict which wavelengths interfere constructively we must remember that the wavelength of light in the film differs from the wavelength in air. The wavelength is n times shorter in the film, where n is the refractive index of the film. So constructive interference is seen for light whose wavelength in air is $2nt$. For example, if the film thickness at one location is $t = 175$ nm, the beam passing straight through it and back has traveled a distance $2t = 350$ nm. Since for $n = 1.7$ this corresponds to a wavelength in air of $2nt = 590$ nm, incident light of wavelength $\lambda = 590$ nm (yellow in color) will interfere constructively and be reflected more strongly than light of other wavelengths. Therefore the surface of the oil at this location appears yellow.

A film can be viewed from different angles (Fig. 4-10). When looking in a direction further away from where we stand, the path length in the film is longer. The wavelength that exhibits constructive interference thus varies with film thickness *and* with viewing angle. This is why as our eye scans over an oil film many colors are seen. The colors of the spectrum repeat as we swing our gaze further away. The first repeat corresponds to a difference in path length of *two* wavelengths, the second to *three* wavelengths, as so forth. An oil film can be remarkably colorful.

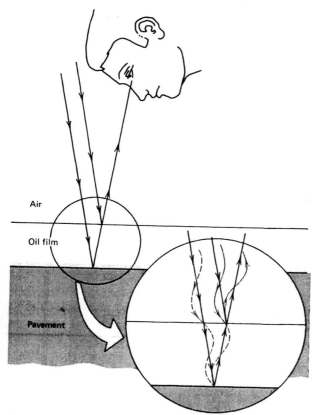

Fig. 4-9. Reflected light beams from upper and lower surfaces of an oil film that are superimposed as they travel toward an observer's eye.

Phase Change on Reflection

The preceding discussion neglected one important complication that must be considered when a surface reflects light: the relative values of the refractive indices. When light encounters a medium of *greater* index it suffers a reversal of its electric field on reflection, as though it had traveled an extra half-cycle. On reflection a crest becomes a trough, and vice versa. In the previous example of the oil film we supposed that the index of the oil film has a value somewhere between that of air and that of the underlying surface. Consequently the beam reflected by the upper surface of the oil changes by half a cycle on reflection, but so does the beam reflected at the underlying surface. As a result of both beams undergoing the same change, there was no net effect on the interference condition.

When light encounters a medium of *lower* refractive index it suffers no change

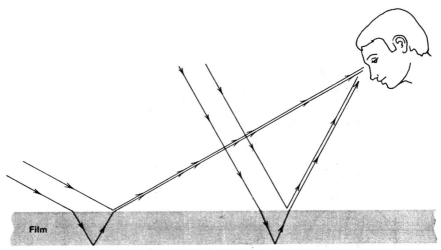

Fig. 4-10. The difference in path length for light reflected from the upper and lower surfaces of a film that travels toward the eye from the same location depends on our direction of view.

in phase on reflection. Also, light that is refracted at a surface and passes *into* the second medium suffers no phase change at the surface, regardless of whether the refractive index of the new medium is greater or less than the index of the original medium.

Color of a Soap Bubble

How do we know that such phase changes occur on reflection? From the remarkable fact that a very thin soap film appears *black* in reflected light. Because there is a negligible difference between the path lengths for light reflected from the front and back surfaces, this destructive interference must arise from converting crests into troughs on reflection from one of these surfaces. Detailed studies show that the phase change takes place at the back surface.

Color Plate 13 shows the colors seen on a soap film when illuminated by diffused sunlight. Gravity pulls the film so that it is thicker at the bottom than at the top. It is thin enough at the top for it to appear nearly black. However, since the actual thickness is a larger fraction of a spectral blue wavelength than red, the color is bluish.

Recall from Section 3-5 that the attributes of a color that we call hue and saturation have a more precise characterization when the color is represented on the chromaticity diagram. The coordinates uniquely identify the color's chromaticity. Benjamin Bayman and Bruce Eaton of the University of Minnesota have calculated the chromaticity coordinates for colors of a soap film of various thicknesses, as illustrated in Fig. 4-11. The point S represents direct sunlight illuminating the film.

For a film thickness of 10 μm the path difference tends to favor short wavelengths, so the color is blue. A thicker film of 200 μm allows wavelengths near 580 nm to interfere constructively, and this yields a saturated yellow. A somewhat thicker film favors still longer wavelengths, so the color changes to red. Between 200 and 250 μm there is still some constructive interference at very long wavelengths, but it occurs for very short wavelengths as well, since the difference in path lengths allows one additional short wavelength to be fit in. Correspondingly, the color is magenta. Tracing the curve around for still thicker films we see that the colors proceed from blue through the spectral colors to red, passing once again through the region of magenta for thickness from about 440 to 500 μm. The sequence of colors repeats many times with increasing thickness, as seen in Color Plate 13. When the film is a great many wavelengths thick, light at each of several wavelengths spaced across the spectrum interferes constructively, and the color is distinctly unsaturated. A very thick film has the hue of the source illuminating it (represented in Fig. 4-11 by the point S).

A soap bubble displays similar interference colors from light reflected by its inner and outer surfaces. Because its thickness is fairly uniform the various colors arise

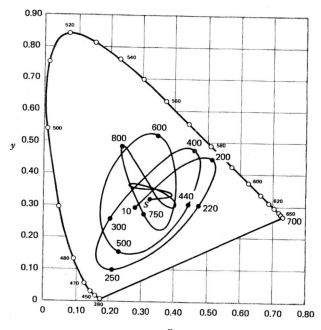

Fig. 4-11. Colors for various thicknesses of a soap film, illuminated by sunlight, at a 45° incidence, as described by their positions on the C.I.E. chromaticity diagram. The complicated curve shows how the chromaticity smoothly changes with thickness, the numbers indicating the thickness in micrometers. (From a calculation by B. Bayman and B. Eaton.)

from light entering the film at differing angles and traveling correspondingly different distances through the film before being reflected toward the eye from various places on its surface.

Interference Filters

Some filters rely on the effect of interference to produce monochromatic light. Coating a piece of glass with just the right thickness of a material having a different refractive index than that of the glass can produce constructive interference for light of a particular wavelength that travels straight through and light that is reflected twice, once at the surface between the material and glass and again at the surface between the material and air. For other wavelengths the interference is destructive. If several coatings of different materials are applied, the effect is made more selective, limiting even more the spectral region of transmitted wavelengths.

Nonreflecting Surfaces

In a high quality optical device such as a camera lens, which is composed of several individual lenses, reflection at the surface of each individual lens reduces the amount of light reaching the film and also produces extraneous bright spots ("flare"). To eliminate this, lenses are coated with a thin transparent layer that suppresses reflection. Such an antireflective coating is based primarily on the interference effect. If the light reflected from the surface of the coating interferes destructively with that reflected from the surface between the coating and the lens, destructive interference occurs. Transparent magnesium fluoride is often used to coat lenses for this purpose. Many nonreflecting glasses used to protect paintings and displays rely on destructive interference as well. A coating of magnesium or calcium fluoride, whose refractive index is between that of air and glass, is applied with a thickness of about a quarter wavelength in the center of the visible spectrum. The reflectance for light coming straight onto the surface is thereby reduced from the original value of about 4% to only 1%. Destructive interference at the ends of the spectrum is less effective, and this gives the glass a slightly purple hue. Even without interference, the coating would cause some reduction in the amount of reflected light because of the details of the laws of reflection.

Newton's Rings

Although Newton championed the ray theory of light, he also provided a detailed description of a phenomenon that we now recognize as an example of interference. If a piece of glass with a slightly curved bottom is placed on a flat piece, with overhead reflection from a point source, a series of concentric colored circles appears. The sequence of spectral colors arises from interference between light that is reflected at the curved surface and light reflected at the underlying flat surface (Color Plate 14). The phenomenon is known as *Newton's rings*. Its appearance provides a means of judging the flatness of a piece of glass or other transparent material. The same effect is often seen on photographic projection slides where glass covers protect the transparency. Because the transparency sandwiched between the glass is not perfectly flat, interference occurs between light reflected by the inner surface of the glass and the transparency. If the glass is sufficiently thin, it may even occur between the two surfaces of the glass itself. A distorted version of Newton's rings indicates the irregularity of the differences in path length for light.

■ 4-5 INTERFERENCE OF MANY WAVES

Iridescence

A brilliant, colorful display can be seen on the surfaces of some beetles, bird feathers, and butterfly wings as a consequence of interference between light waves reflected from close-lying layers of differing material. This effect is known as *iridescence*. Pearls also exhibit iridescence as the result of light being reflected from regularly spaced layers of calcium carbonate ($CaCO_3$) that are separated by organic films. Another example is seen in ancient glassware, caused by chemical changes that alter a thin outer layer of the glass.

The brightness of iridescent colors comes from the feature of constructive interference whereby the net amplitude of two waves superimposed with the same phase is the sum of their individual amplitudes, but the intensity increases as the square of the net amplitude. Because of this dependence on the square of the amplitude, two waves of equal intensity add constructively to produce *four* times the intensity of either, not twice as we might naively expect. That is, two waves of equal amplitude add to produce a wave of twice the amplitude of either, and squaring this net amplitude does indeed produce four times the intensity of either wave. Constructive interference produces unusually high spectral reflectance in certain directions, but this is compensated by especially low spectral reflectance in other directions of destructive interference. This must be so, because the total radiant power reflected into all directions can never exceed the total incident power.

A particularly striking example of color from interference is found in bird feathers. These include some of those on the peacock (Color Plate 15), the head of the male Mallard duck, and some hummingbirds and butterflies. Although the color of most feathers comes from selective absorption, these feathers have a covering of thin scales with an unusual structure on their surface. As seen by an electron microscope (Fig. 4-12), the surface of one type of scale has many parallel ribs supported by a delicate framework, and from each rib protrude numerous closely spaced horizontal ridges. Light reflected from these ridges interferes constructively in a particular direction to provide a bright color, while in other directions the interference is destructive.

One characteristic of iridescent colors is the shift of constructive interference to

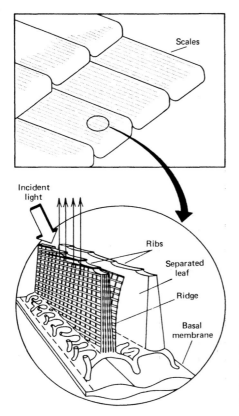

Fig. 4-12. (*a*) Scales on the surface of a butterfly wing; (*b*) a closeup of the surface showing parallel ribs and their many horizontal ridges from which light is reflected.

different wavelengths as the viewing angle changes, although the shift may be rather subtle in some cases (Color Plate 15). If you look closely, it is possible to see butterflies change colors as they dart from one place to another. It has been suggested that this rapid variation in color makes it difficult for predators to gauge their distance.

The silvery, pearl-like belly scales of the herring contain platelike crystals lying parallel to the scales. This "pearlescent" material was at one time extracted and injected into beads to simulate pearls. Recently a commercial process has been devised to achieve an equivalent effect by coating tiny specks of mica with thin layers of titanium dioxide. The high refractive index of titanium dioxide ("titanium white") reflects light strongly and provides a luster similar to that of natural pearl.

■ 4-6 DIFFRACTION

From earlier examples in this chapter it should be apparent that diffraction is a common phenomenon. It occurs both at edges and when light passes through a small hole (which is defined by its edge!). It causes the light streaks seen when with nearly closed eyes we squint at a bright lamp, and it produces the colored spectra arranged in the shape of a cross when you view a lamp through a thin curtain or other closely woven fabric. You can even see diffraction patterns produced *inside* your eye. If you squint with eyelids almost closed while looking at a brightly illuminated wall, you can see individual and perhaps groups of rings glide across your field of view, moving irregularly in various directions. The patterns are caused by diffraction around the edges of small particles, probably red blood cells, suspended in the fluid of the eye; as the particles float about, the patterns move with them.

How diffraction patterns are created was first explained by the seventeenth century Dutch astronomer and physicist Christiaan Huygens. His interest in optics led him to take up Robert Hooke's wave theory of light to explain how light propagates through space. He proposed that each point along a wave crest can be imagined as a source of disturbances (called "secondary waves") that move forward and spread out in spherical patterns. The portions of the secondary waves spreading parallel to the crest tend to cancel, whereas portions moving in the direction of travel of the original wave reinforce. With no obstruction present the total effect is to move the crest forward undeflected at the speed of light. But at the edge of an opaque object the spreading pattern of secondary waves is not fully cancelled, and an interference pattern from the secondary waves causes the light and dark fringes. Far to the side of the edge of the obstruction, a balance exists between secondary waves, so their net result is a uniform illumination. Huygens invented a geometrical construction (called Huygens construction) that can predict the details of diffraction patterns.

Airy's Disk

Diffraction is often an entertaining phenomenon, but in some cases it is undesirable. In an optical instrument, like a telescope, the presence of diffraction imposes a fundamental limitation on the quality of its image. This is also true for cameras. As light enters a camera, only the light that is intercepted by the lens is brought to a focus on the film. When photographing a point source of light (such as a star or distant streetlamp), we hope that the image will also be similarly sharp. But, in fact, diffraction from the edge of the lens causes some distortion of the image. The image on the film in this case is not a point but a central bright disk called "Airy's disk" with surrounding bright and dark rings of rapidly decreasing intensity, as shown in Fig. 4-13. This effect was first analyzed mathematically by the British astronomer Sir George Airy in 1834, and it bears his name. As in all examples of diffraction, the smaller the opening the larger is Airy's pattern. A telescope of large diameter produces a smaller pattern and therefore a sharper, more pointlike, image of a star.

This problem of diffraction by a circular hole can be mathematically developed in great detail. However the essence of the result is apparent from simple considerations. We are primarily interested in how large the pattern will be, for its size is a measure of the quality of an optical instrument. In Fig. 4-14 we show that the size of various features of the pattern (e.g., where the edge of Airy's disk lies and

Fig. 4-13. Enlargement from a photograph of a typical diffraction pattern of a distant point object produced by a lens. (From M. Cagnet, *et al.*, *Atlas of Optical Phenomena*, Springer-Verlag, 1962.)

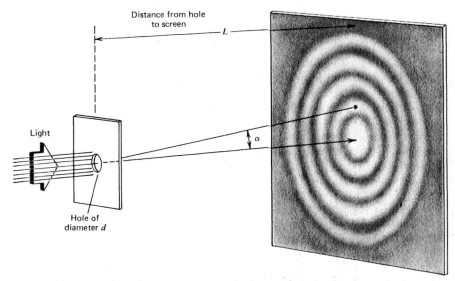

Fig. 4-14. Airy diffraction pattern formed by light of wavelength λ passing through a small hole of diameter d. The position of any point on the pattern is designated by the angle α with respect to the axis passing through the center of the hole. The angle α shown locates the center of the first bright ring outside of the central Airy disk.

where the first bright ring occurs) can be specified by the value of the corresponding angle from the axis, denoted by the Greek letter alpha (α). To predict the size of Airy's disk, only two quantities are relevant: the wavelength of light λ and the size of the hole (its diameter d). An appropriate combination of these two lengths should describe by how much diffraction spreads the light, because they are the only quantities that are needed to define the physical situation. Since an angle has no dimensions (it is not a length that is measured in meters or time that is measured in seconds), the appropriate combination is just the ratio λ/d. This is the simplest possible combination. There will also be a numerical factor that should multiply this ratio, but its value cannot be determined by these simple dimensionality arguments.

The combination λ/d does in fact prescribe the value for an angle, provided that we recognize the angle is expressed in *radians* and not degrees (one radian equals 57.3 degrees). The measurement of angles in radians is discussed in Advanced Topic 4A. The angle given by representative values of λ and d is generally very small, because λ is so much smaller than the diameter of a typical lens. The angle becomes larger if openings are made smaller, as we expect for a diffraction phenomenon. An exact calculation shows that the edge of Airy's disk lies at $\alpha = 1.22$ (λ/d). For an application of this formula consider a typical telescope having a diameter of $d = 12$ cm. The angular size of Airy's disk for $\lambda = 5 \times 10^{-7}$ cm is:

$$\alpha = 1.22\frac{\lambda}{d} = 1.22 \times \frac{5 \times 10^{-7}}{1.2 \times 10^{-1}} = 5 \times 10^{-6} \text{ radians}$$

The bright and dark rings surrounding this disk are irregularly spaced, as indicated in Table 4-2. These angles are so small that we have difficulty appreciating what such small numbers represent. For this example, Airy's disk is about the same size as the angular size of an ant viewed from a distance of 1 km.

In practice the size and shape of subjects of angular size less than (λ/d) cannot

TABLE 4-2 Angles Indicating the Directions where Light and Dark Rings Are Found in the Diffraction Pattern of Light from a Distant Point Source, When Focused by a Lens of Diameter d.[a]

Intensity	α (Radians)
First maximum (center of the pattern)	0
First minimum	$1.22(\lambda/d)$
Second maximum	$1.64(\lambda/d)$
Second minimum	$2.23(\lambda/d)$
Third maximum	$2.67(\lambda/d)$
Third minimum	$3.24(\lambda/d)$

[a]See Fig. 4-14 for the pertinent diagram. The lens diameter is assumed to be much larger than the wavelength λ of light.

Source: F. A. Jenkins and H. E. White, *Fundamentals of Optics* (McGraw-Hill, New York, 1950), p. 294.

be determined by a telescope. Smaller subjects are represented by their diffraction patterns. Increasing the diameter of a telescope decreases the size of the diffraction pattern and could reveal the shape of the subject, but another effect then limits the quality of the image for diameters above about 12 cm—atmospheric distortion. As will be explained in Chapter 14, this produces the "twinkling" of stars. Astronomical telescopes such as the 200 inch (500 cm) one at Mt. Palomar in California have larger diameters for another advantage—to capture and concentrate more light, so that fainter stars can be seen.

Rayleigh's Criterion for Resolution

Light directed through the center of a lens passes through it along an essentially straight line, as will be explained in Chapter 8. Then the place where the lens focuses the light from a point source must also lie on this line. As shown in Fig. 4-15, this notion allows us to determine the closest angular separation between two sources that still permits them to be discerned as distinct. If they are closer, their diffraction patterns overlap so much that they appear to be one source. This condition is called the *resolution* of the optical device. In the late eighteenth century Lord Rayleigh, noted for his work on the theory of sound, properties of fluids, scattering of light, and the discovery of the gas argon, suggested a criterion based on the angular size of Airy's disk; two point sources brought close together become unresolvable when the edge of the disk from one source just meets the center of

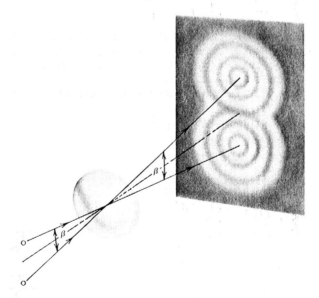

Fig. 4-15. The angular resolution of a lens is the smallest angle β for which images of two point sources are seen as distinct.

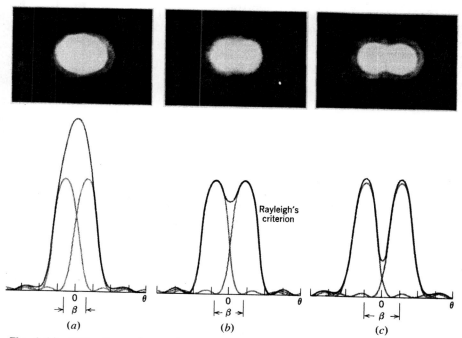

Fig. 4-16. Diffraction patterns of two distant point sources when photographed by a lens of 10 cm diameter are illustrated in these sketches. The corresponding intensity plot for each pattern and the total intensity are shown below, when the angular separation of the sources corresponds to the separation of their patterns by (*a*) one-half the radius of Airy's central disk; (*b*) the radius of the disk, corresponding to Rayleigh's criterion; and (*c*) twice the radius of the disk, where the two patterns can be clearly resolved. (From Haliday and Resnick, *Fundamentals of Physics,* Second Edition, Wiley, New York, 1981.)

the disk from the other. This is illustrated in Fig. 4-16. For this separation the angle denoted by the Greek letter beta with a subscript r (β_r) defined in Fig. 4-15 matches the radial size of Airy's disk. With the size of the disk as given in Table 4-1, Rayleigh's criterion for the resolution of a lens is:

$$\beta_r = 1.22 \left(\frac{\lambda}{d}\right) \text{ radians} \tag{4-5}$$

An interesting application of Rayleigh's criterion can be made to the human eye. In this case the pupil, not the lens diameter, determines the opening through which light is admitted. Under dim illumination the pupil opens to a diameter of about d = 5 mm, whereas in bright daylight it reduces the amount of admitted light by contracting to a diameter of about d = 2 mm. Diffraction is more significant in the latter case, and the corresponding angular resolution of the eye can be predicted from Eq. 4-5. The resolution for $\lambda = 500$ nm is:

$$\beta_r = 1.22 \left(\frac{500 \times 10^{-9}}{2 \times 10^{-3}} \right) = 3 \times 10^{-4} \text{ radians}$$

This is extremely small but nevertheless significant. Let us calculate the distance between the centers of the diffraction patterns appearing on the retina of the eye. We need only multiply the angle β_r by the distance between pupil and retina, approximately 2 cm. The result is a distance of 6 micrometers between the centers on a bright day. With a pupil size of 5 mm for cloudy conditions, the resolution is the equivalent of 2 micrometers. It should then be no surprise that the closest spacing of the cones, which serve as the photoreceptors at the center of the retina, is approximately this distance, about 2 to 5 micrometers center to center. This correspondence is an example of nature's economy in which no more receptors are provided than are consistent with the ultimate resolving quality of the eye's optical system. Projecting this angle β_r for a pupil size of 5 mm into the space in front of the eye corresponds to a 3 mm separation at a distance of 10 meters.

Diffraction Gratings

Nature provides many colorful effects made possible by diffraction. Whenever light encounters a small object or an edge, diffraction takes place. This concentrates light in particular directions according to wavelength, as can clearly be seen in the phenomenon of the sun's glory to be described in Section 14-3. Some of the most spectacular effects in the animal world rely on interference as well as diffraction. Portions of the surface of creatures such as certain beetles and wasps have a hard corrugation, with many sharp edges arranged in closely spaced rows. Each of these edges diffracts light into a wide range of directions. Color is a consequence of the fact that these edges have a fairly regular spacing (Fig. 4-17) with the distances between edges being only a few times greater than the wavelength of light. This structure of parallel edges is called a *diffraction grating*. The regularity causes interference to take place between the waves diffracted by the edges. Spectral colors can be seen in those directions where the interference is constructive. Examples of this effect are shown in Color Plate 16 for the wing cover of a ground beetle and a portion of the abdomen of a wasp. These colors change rapidly with the direction of view, so as these animals move there is a rapid play of color that serves to confuse a predator.

The colors of mother-of-pearl are attributed to a similar combination of diffraction and interference. The inner lining of the oyster shell has microscopic layers of crystals inclined at an angle, and the rows of edges of these crystals act as a diffraction grating. As for the beetle and wasp illustrated earlier, the best color is seen when the illumination comes from a single, small source of light such as the sun. If the source is extended, as with a common fluorescent lamp, the angles for constructive interference overlap for different wavelengths and the colors are unsaturated.

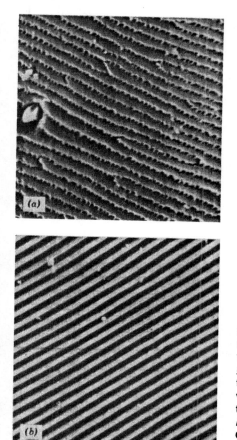

Fig. 4-17. Scanning electron micrographs of diffraction gratings on the (a) wing cases of the ground beetle, *Irida-gonum quadripunctum* and (b) stridulatory file on the abdomen of the mutillid wasp, *Mutilla europaea*. (From H. E. Hinten, "Natural Deception" in *Illusion in Nature and Art*, R. L. Gregory and E. H. Gombrich, Eds., Scribner's, New York, 1980.)

A carefully manufactured diffraction grating is a modern scientific tool for measuring the spectral power distribution of light. Commonly made by engraving closely spaced parallel lines on glass, a grating separates the directions traveled by light of different wavelengths in an accurately predictable way. The direction for a particular wavelength can be calculated from the interference condition for the known spacing between lines and the angle of the incoming beam of light. A grating composed of many lines provides more effective separation of wavelengths than just a few lines. This can be seen by considering diffraction from just two lines, because the pattern of light is similar to what is produced by the interference of light passed through two narrow slits, as shown in Fig. 4-8. A slight shift from the direction for optimum constructive interference has little effect for the waves from two adjacent edges, because the relative phase of the waves is only slightly changed. However, there is a correspondingly greater effect for the waves from edges that are 10 or 100 neighbors away, because of the greater difference in phase. Therefore the participation of many edges causes the intensity to decrease very sharply with slight departures from the optimum angle for constructive interference. This permits components of light having only slightly differing wavelengths to be resolved.

When the difference in path lengths for light diffracted from adjacent lines is one wavelength, the interference is said to be of *first order*. At a greater angle the path length is two wavelengths, and the interference is *second order*. With a carefully engraved grating, interference of much higher order can be seen. Even the abdomen of the wasp shown in Color Plate 16*b* displays several orders of interference as our eye scans across it. A similar colorful effect can be seen by viewing a video disk (Color Plate 17) or phonograph record at an angle, when it is illuminated by a single lamp some distance away. Note that the multilid wasp (Fig. 4-16*b*) has a more regular diffraction grating than the ground beetle (Fig. 4-16*a*), so that it produces a diffraction pattern exhibiting many more orders.

Opalescence

The opal provides another example of iridescent colors, where diffraction as well as interference plays an important role. The precious variety of stone is prized for the flashes of different colors seen in the depth of its body when rotated. It is particularly striking when viewed under a single, small white lamp. The patterns of color vary markedly with the direction of view, and this effect is called *opalescence* (Color Plate 18). This phenomenon is explained by the structure of the material. Opals are formed underground in trapped pools of water containing dissolved silica (SiO_2). In time, as water evaporates, the silica molecules attach to one another and form tiny spheres ranging in diameter from 150 to 300 nm. These spheres settle to the bottom where they nestle into tightly packed layers. Further loss of water causes the spheres to harden into a compact glasslike mass, with tiny voids between the spheres containing some of the remaining water. The body of the opal is divided into grains, each one of which has its layers of uniform spheres oriented in a particular direction.

The voids interrupt the fairly uniform refractive index of the medium and diffract light. Since they are so small, they cause a substantial portion of the incident light to spread out in all directions. The periodic arrangement of voids permits constructive interference between the diffracted waves in particular directions that depend on wavelength (Fig. 4-18). Grains containing different size spheres, and correspondingly different spacings between voids, produce differing colors in a given direction. Opals containing only small spheres give only short wavelength violet and blue colors, whereas those with larger spheres display colors of the full spectrum in various directions. As an opal is turned, the random orientations of the grains produces a changing pattern of color from various depths, accounting for its "fire."

In addition to color from diffraction and interference, more uniform colors throughout an opal may result from selective absorption of light by impurities within the material. These colors range from black through grays and unsaturated hues to a white or milky appearance, depending on the colorant.

The phenomenon of interference between the diffracted waves from periodically spaced bodies is not limited to visible electromagnetic radiation. It can occur whenever the distance between the bodies is comparable to the wavelength of the radiation. A counterpart of opalescence for X rays is commonly used by crystallographers for studying the arrangement of atoms in a crystal. The regular spacings between atoms in crystals such as diamond, copper, and silicon vary from one material to another but are typically 0.3 nm. When X rays with about this wavelength or shorter are directed into a crystal, they are diffracted by the atoms and interfere constructively in particular directions. As shown in Fig. 4-19, a film exposed to these diffracted waves reveals this pattern, which displays the symmetry of the crystalline arrangement of atoms. From measurements of the directions for constructive interference and the known value of the X-ray wavelength, the spacing between the atoms can be calculated. In this way the incredibly short distances between atoms are determined.

■ 4-7 LIGHT SCATTERING

Our previous discussion of diffraction revealed that wave aspects of light are clearly demonstrated when light meets objects whose size is comparable to wavelengths of the visible spectrum. Color effects in the atmosphere such as the glory and sun's corona are due to diffraction from such tiny water droplets (see Chapter 14). When light meets objects of a much larger size such as dust suspended in the air, reflection and refraction govern the dominant interaction. The coloration of light is predicted by the reflection and transmission curves for the material comprising the dust. Diffraction occurs at the edges, but it is a more subtle effect. On the other hand, for very small particles, those whose diameter is a small fraction of a wavelength, an extreme limit of diffraction is found. It is given the special name *Rayleigh scattering* after the Cambridge physicist, Lord Rayleigh. The simple term *scattering* is applied to the effect of particles regardless of their size.

Rayleigh Scattering

Small particles that scatter light have too little material to impose much selective absorption. Their primary effect is to deflect a portion of the incoming light in nearly all directions. The pattern of scattered light is independent of wavelength. The effectiveness for scattering light is less for smaller particles, which must be so, because a particle that is vanishingly small should produce no scattering at all. Futhermore, a given particle scatters short wavelengths more effectively than long.

Consequently blue light is scattered more effectively than red. This dependence on wavelength is very strong: the intensity of scattered radiation varies as the fourth power of the frequency (ν^4) or as λ^{-4} (Rayleigh's inverse fourth power law). Therefore, if white light with power spread evenly across the spectrum is incident, the intensity of scattered blue light ($\lambda = 450$ nm) is six times greater than the intensity of scattered red light ($\lambda = 700$ nm). Rayleigh scattering endows the sky with its blue color. As explained in more detail in Chapter 14, a portion of the nearly

Fig. 4-18. (*a*) Electron micrograph of the arrangement of silica spheres in a gem opal from Brazil, showing at the right a region where they are ordered and at the left a region where they are disordered and give no opalescence. (Courtesy of J. V. Sanders.) (*b*) How voids between the spheres diffract light that interferes constructively in a particular direction. Only two diffracted waves are shown for clarity.

white sunlight is scattered down to us, and this scattered light heavily emphasizes the blue.

A remarkably wide range of objects owe their color to Rayleigh scattering. Tiny fat particles in milk, for example, cause the scattered light to be bluish. For the same reason the transmitted beam, deficient in blue, appears reddish. Scattering is also responsible for the blue of cigarette smoke.

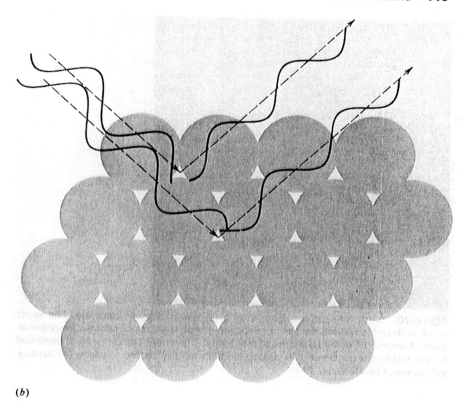

(b)

Imagine that we stand on any ordinary seaside pier, and watch the waves rolling in and striking against the iron columns of the pier. Large waves pay very little attention to the columns—they divide right and left and re-unite after passing each column, much as a regiment of soldiers would if a tree stood in their road; it is almost as though the columns had not been there. But the short waves and ripples find the columns of the pier a much more formidable obstacle. When the short waves impinge on the columns, they are reflected back and spread as new ripples in all directions. To use the technical term, they are "scattered." The obstacle provided by the iron columns hardly affects the long waves at all, but scatters the short ripples.

From Sir James Jeans, *The Stars in Their Courses*, (Cambridge University Press).

Fig. 4-19. X-ray diffraction pattern from a crystal of bismuth vanadate. The spots occur in directions where the film is strongly exposed because of constructive interference of waves diffracted by atoms of the crystal. The bright central circle is produced by the incident X-ray beam that passes through the film before reaching the crystal. (Courtesy of Gu Benyuan.)

In the biological world coloration is often due to a combination of scattering and selective absorption. Melanin is a common blackish-brown pigment that in high concentration appears nearly black. The bluish sheen of the hair of some orientals comes from Rayleigh scattering by these pigment particles. A similar effect is seen in feathers of the crow and bowerbird. Blue is also produced by scattering from some types of protein particles, as found in the scales of the skink displayed in Color Plate 19. Colors produced by scattering, unlike those due to interference, do not change with the angle of view, which allows the two effects to be easily distinguished.

The color of the iris of the human eye is also due to scattering. The iris has a layered arrangement with small particles of a type of melanin (specifically, eumelanin) distributed in varying amounts in each layer. Melanin is the same pigment responsible for giving skin its color: Negroes have melanin in high concentration, Caucasians in much lower concentration, and albinos none at all. When present in the iris in high concentration, it accounts for brown eyes; when present only in the inner layer of the iris the color is light brown. There are no blue or green pigments in the eye. Such colors arise from the structure of the iris and not from selective absorption. Blue is produced when the outer layer is devoid of pigment and the inner layer has melanin pigment dispersed as very fine particles. These particles create the blue by scattering. Green arises when the particles are somewhat larger, so that diffraction adds intermediate wavelengths to the scattered light. The albino eye is red because it lacks all pigment and only the color of the circulating blood is seen.

Melanin pigments in different forms and concentrations color hair brown, black, red, and blond. But hair turns white for structural reasons as well as from the absence of pigment: numerous small pockets of air form in each hair shaft and scatter light in a wide range of directions. Diffuse reflection from minute pockets of air is also responsible for the white of a pig's hair and of the polar bear's fur.

The chameleon provides a fascinating color effect. One layer of skin cells contain a yellow pigment carotenoid, a second layer includes tiny cells that scatter light and so is blue, and a third layer has the brown pigment melanin. The nervous system of this reptile controls processes that transfer the melanin through the light-scattering layer and into the yellow layer. This causes a progressive change from a bright, unsaturated green (produced by the overlap of the reflected yellow and scattered short wavelengths) through a darker, more saturated yellow-green and finally dark brown. With this process the chameleon can adjust its camouflage to blend with its changing surroundings.

Liquid Crystal Displays

Many electric watches, clocks, and hand calculators that are designed to consume only a small amount of power from a battery display their numerals by light that is scattered from a *liquid crystal.* A liquid crystal consists of densely packed long organic molecules in which the molecules are aligned in a particular direction. The term "crystal" denotes this regularity in orientation. On the other hand the term "liquid" indicates that the molecules can move relative to one another as in a common liquid, while keeping their alignment.

When an electrical voltage is imposed across an appropriate liquid crystal, the electric field causes the molecules to reorient and form an irregular pattern. This irregularity causes the incident light to be scattered, with a high percentage of it emerging. The size of the irregularity is sufficiently great to give the reflected light the same color as the incident light. If the metal electrodes that provide the voltage have the shape of a numeral, this number is seen in daylight as white scattered light. Because the display relies on scattering, the numeral cannot be seen in the dark. In another type of display the liquid crystal is disordered *unless* the electric field is imposed. Then the electric field produces a dark region, leaving the surrounding area white. The most common type of liquid crystal display depends on properties of polarized light, a topic that we shall discuss in the following section.

Labeling Tape

Scattering is also the origin of the white letters that appear when commercial labeling tape is embossed by a pressing device. The tape consists of vinyl chloride; when this material is deformed or stretched, its molecules are separated, creating tiny spaces throughout the region. These voids scatter light so effectively that they make the letters appear white when viewed under white light.

■ 4-8 POLARIZED LIGHT

The motion of water in a water wave on the sea is essentially vertical, whereas the wave with its crests and troughs travels horizontally. Because the motion is perpendicular to the path of the wave, it is said to be a *transverse* wave. Light too is a transverse wave: its electric and magnetic fields are directed perpendicular to its path (Fig. 4-20). For our purposes the magnetic field can be disregarded, because it is the electric field that interacts most strongly with atoms and molecules. The direction of the electric field is said to be the *polarization direction* of the wave.

A light beam with its electric field oriented in only one direction is said to be *completely plane polarized* (Fig. 4-21a). Most light encountered in nature is not of this type. Light coming directly from the sun is *unpolarized* (Fig. 4-21b). Components of the light are found with their electric fields in all possible orientations about the light path (but always perpendicular to it). Or light may be *partially polarized*, with a predominance of components polarized in a particular direction (Fig. 4-21c). This is often the case for light reflected from a surface.

Unpolarized light can be polarized by four methods: absorption, reflection, refraction, and scattering. The first is the most used in practical work. We shall examine each method in turn.

Polarizing by Absorption

For centuries the peculiar optical property of certain crystals such as tourmaline have proved fascinating. We know now that the molecules in the crystal are aligned so that they strongly absorb light when its electric field is in a particular direction, but weakly absorb light when it is polarized in the perpendicular direc-

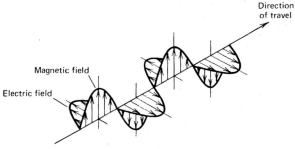

Fig. 4-20. Electromagnetic wave with electric field in the horizontal direction and magnetic field in the vertical direction. As the wave travels past a stationary observer, the electric field at his position would reverse direction periodically, but would always remain in the horizontal plane.

(a)

(b)

(c)

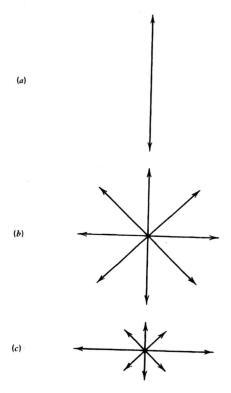

Fig. 4-21. Schematic representation for various types of polarized light, imagined to be traveling toward the reader: (*a*) completely polarized beam, with polarization direction vertical; (*b*) unpolarized beam; and (*c*) beam that is partially polarized in the horizontal direction.

tion. If the thickness of a tourmaline crystal is longer than about 2 mm, only the perpendicular component emerges. The orientation of its electric field is called the *transmission direction* of the crystal. However, since even light polarized in this direction suffers some selective absorption, the emerging beam is colored. Another device, the Nicol prism, consisting of two crystals of calcite cemented together in the proper orientation, is used today to polarize light without appreciable change in color. This is based on principles of refraction that will be discussed shortly.

The most common type of polarizer depending on absorption is now the Polaroid filter. Invented by Edwin Land in 1938, one type is a sheet of plastic in which innumerable tiny crystalline needles of herapathite are suspended. Herapathite's transmission direction coincides with the long axis of the needle. In manufacture, the plastic sheet is stretched as it hardens; this aligns all crystallites so that their transmission directions coincide. The Polaroid filter thus transmits only the component of light with its electric field parallel to this transmission direction and absorbs light polarized in the perpendicular direction (Fig. 4-22). Another type of polarizer is made by absorbing iodine or a dye in a stretched sheet of polyvinyl alcohol. The dye polarizer has the property of polarizing only certain regions of the visible spectrum and transmitting both components of light in the other regions.

The transmittance of a polarizing filter depends on the polarization condition of

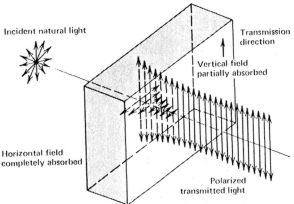

Fig. 4-22. A polarizing filter passes the component of light polarized in the transmission direction and absorbs the component of light that is polarized perpendicular to the transmission direction.

the incident light. When unpolarized or polarized light is passed through an ideal polarizer, it is possible to predict the transmittance. For simplicity we assume 100% transmittance of white light polarized in the transmission direction. The key concept is that completely polarized light can always be considered as the sum of two polarized beams traveling in the same direction but with polarization directions at right angles to each other. The two beams if added together reproduce the original one. A similar division can be made with unpolarized light. Consequently, with an incident unpolarized beam we may consider that half of the intensity corresponds to the component of the electric field aligned parallel to the transmission direction, and this component passes right through; the other half corresponds to the component at right angles to this component and so is completely absorbed. Consequently, for unpolarized light, the transmittance of a polarizing filter is 50%.

Two polarizers in sequence provide a filter whose transmittance can be continuously adjusted. The first polarizer reduces the intensity of the beam by 50%, and the emerging light is polarized. The closer this polarization direction lies to the transmission direction of the second polarizer (Fig. 4-23), the greater the intensity of the light that emerges. For the special case when the transmission direction of the second polarizer is placed 45° to that of the first, half of the incoming beam's intensity is associated with the component of its electric field along the transmission direction, and this half passes through; the other half of the intensity is absorbed. The transmittance is then 50% through the first polarizer and 50% through the second, for a net transmittance of 25%. By varying the angle ϕ between the transmission directions of the polarizers, the net transmittance can be varied from 0% when perpendicular ($\phi = 90°$) to 50% when parallel ($\phi = 0°$) (Fig. 4-24).

No existing polarizer is ideal. Common polarizers transmit only 40% of unpolarized incident light. The difference between 40 and 50% occurs because some light

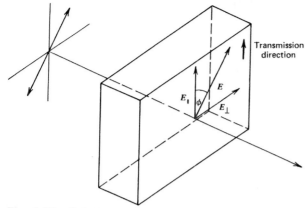

Fig. 4-23. Polarized light incident on a polarizing filter with an angle ϕ between the polarization direction and orientation of the transmission direction of the filter. The electric field E of the light can be considered to be the sum of a component $E_{\parallel}$ parallel to the transmission direction and component $E_{\perp}$ perpendicular to the transmission direction.

is absorbed in the polarizer and some is reflected at its surfaces. Futhermore, absorption of the perpendicular component is not complete. The net transmittance for crossed commercial Polaroid filters is not 0% but about 0.05%.

Polarized light afforded one of the first opportunities for three-dimensional movies. We judge the relative distances of subjects by their slightly different positions in the scenes viewed by our two eyes. Two separate 2-D pictures projected onto a screen through perpendicularly oriented polarizers allow the left and right eyes of viewers wearing special glasses to see the separate views and form a 3-D impression. If the original pictures were taken with cameras separated by the inter-pupil distance, and if they are projected in proper register, reasonably good depth perception is achieved (Fig. 4-25). The general principle used here is that the same beam of light (in this example, the light reflected from the screen) can convey two separate pictures, contained in perpendicularly polarized components. The two can be separately detected through suitably oriented polarizers worn by the viewer.

Example 4-2 Transmittance of a Polarizing Filter

When polarized light is directed into a polarizing filter, the transmittance can be predicted if the angle ϕ between the polarization and transmission directions is known (Fig. 4-23). The amplitude of the electric field E of the incident beam can be considered as the sum of a component $E_{\parallel}$, which lies parallel to the transmission direction of the polarizer, and a component $E_{\perp}$, which is at right angles to $E_{\parallel}$. Only the parallel component is transmitted, and the other is absorbed.

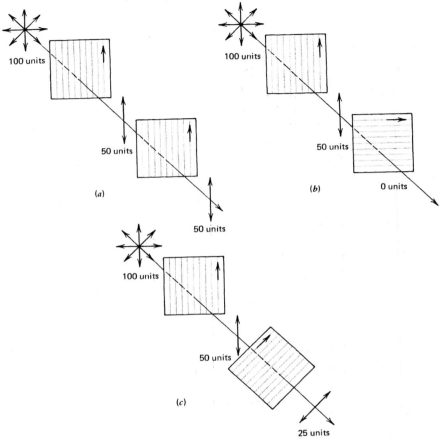

Fig. 4-24. Initially unpolarized light passing through two ideal polarizing filters whose transmission directions are at various angles.

The amplitude E_1 is the projection of E onto the transmission direction as given by the trigonometric cosine of the angle between them: $E_1 = E \cos \phi$. This predicts the amplitude of the emerging light wave. The intensity of light is proportional to the *square* of its amplitude and thus is proportional to $E^2 \cos^2 \phi$. The variation of intensity as the square of the cosine is known as *Malus' law*, named after Étienne Malus who in the early 1800s carried out pioneering studies of how light can be polarized. For example, when light is incident at an angle $\phi = 45°$, the emerging intensity is $\cos^2 45° = \frac{1}{2}$ of the incident intensity. The transmittance is $\frac{1}{2}$, or 50%. The transmittance for other angles can similarly be calculated by looking up the value of the cosine of the angle in a trigonometry table.

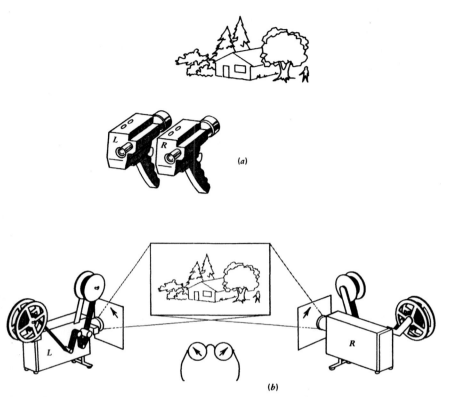

Fig. 4-25. Method for using polarized light to produce a three-dimensional perception when viewing a movie through appropriately polarized glasses: (*a*) filming; (*b*) projection and viewing.

Polarizing by Reflection

When light arrives at a surface the percentage of light reflected depends on the direction of its polarization. When the electric field is parallel to the surface, more is reflected than when it is not parallel. This means that specularly reflected light— from a road surface, the ocean, or ice—is partially polarized in the horizontal direction. People who wear polarizing sunglasses take advantage of this property. The transmission direction of these sunglasses is vertical; therefore, they selectively absorb the component of light that was specularly reflected toward them from any horizontal surface. This dramatically reduces "glare."

As an illustration of the importance of this effect, consider what happens when light meets a sheet of glass. The incident, reflected, and refracted beams are shown in Fig. 4-26. All three beams lie in a common plane oriented perpendicular

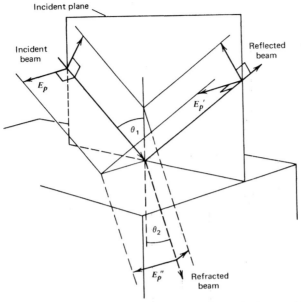

Fig. 4-26. Two components of polarization of a beam of light incident on a glass surface as they separately divide into reflected and refracted portions. The component of the incident beam parallel to the surface is denoted E_p, the component of the reflected beam E_p', and the component of the refracted beam E_p''. The corresponding components not parallel to the surface are shown but not labeled.

to the surface. This is called the *incident plane*. First, we direct our attention to the component of light whose polarization lies parallel to the surface of the glass. The reflectance of this component varies with the incident angle, which we shall denote by θ_1 (the Greek letter theta, followed by a subscript "1" to denote that it refers to an angle in the first medium). The reflected component is also polarized parallel to the surface. The solid curve in Fig. 4-27 shows how the reflectance for the component in this direction increases dramatically as θ_1 increases. When the incident beam meets the glass head-on ($\theta_1 = 0°$), only 4% of the intensity of this component is reflected, the rest being refracted. But the percentage increases to 100% when the incident beam just grazes the surface ($\theta_1 = 90°$). A photographer knows that the glare of this reflected light from a window can be eliminated effectively by using a properly oriented polarizing filter in front of the camera's lens (Fig. 4-28).

Now we turn our attention to the other component of the incident light, that does not lie parallel to the surface of the glass. The dashed curve in Fig. 4-27 shows that the reflectance of this component initially decreases as the beam is tilted away from the head-on case. At a particular angle θ_1 the reflectance becomes exactly zero: all of this component is refracted and none is reflected. This angle is called *Brewster's angle*. Étienne Malus discovered this effect in 1808, and only a few

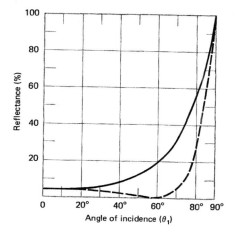

Fig. 4-27. Reflectance from the surface of a plate of glass (refractive index, $n = 1.52$) depends on the incident angle θ_1. The solid curve describes the reflectance for the component of incident light that is polarized parallel to the surface. The dashed curve describes the reflectance of the component that is not parallel to the surface.

years later the Scottish physicist David Brewster found from experiment that this occurs when the incident angle θ_1 and the refraction angle denoted by θ_2 add up to 90°: $\theta_1 + \theta_2 = 90°$. Since the law of refraction (to be discussed in detail in the next chapter) sets a fixed relationship between θ_1 and θ_2 that depends on the refractive indices of the two media, only one value of θ_1 can satisfy both this refraction requirement and the condition for zero reflectance. Brewster's angle thus varies with the type of glass: for $n = 1.5$ it is about 57°.

Figure 4-27 shows that for most angles the reflectance is greater for the component of incident light lying parallel to the surface of the glass. This feature bears out the claim that reflected light is at least partially polarized parallel to the surface.

(*a*)   (*b*)

Fig. 4-28. (*a*) Reflections in the window of a flower shop obscure our view of the interior. (*b*) When a polarizing filter is placed before the camera lens with its transmission axis horizontal to absorb the vertical component, which is parallel to the surface of the glass, the reflections are virtually eliminated.

At Brewster's angle the reflected light is completely polarized. At this angle the other component of the incident beam suffers no reflection at all. All of the other component is transmitted into the glass. This fact is exploited in building lasers, where even the 4% reflectance for head-on incidence is avoided by attaching the end windows onto the laser tube at Brewster's angle. Reflection from metals does not exhibit selective polarization since essentially all light that is incident on a metal is reflected.

Polarizing by Refraction

In 1669 Erasmus Bartholin, of the distinguished Scandinavian family noted for its achievements in medicine and science, discovered an interesting effect when light passes into a crystal of calcite in a particular direction. The refracted light divided into two rays that have differing refraction angles. This effect is called *birefringence*, or "double refraction." Subsequent experiments have shown that the two rays have different polarization directions. It has since been found that the speed of light in such birefringent materials differs for the two polarization directions. Indeed, the division of a single beam into two led Christiaan Huygens to the idea of describing light as a transverse wave 150 years before Étienne Malus provided the detailed explanation of birefringence. Calcite, commonly known as Iceland spar, has been known for its fascinating optical properties since the Middle Ages when sailors began to make regular voyages to Iceland. Quartz is another birefringent crystal. The Nicol prism is a polarizer that is made from two crystals of calcite, properly cut, and cemented together so that one of the two polarized waves emerges in the direction of the incident beam.

Stress Birefringence

Polarized light has important applications in the study of the strength of materials. When a block of plexiglass, or a similar material, is squeezed in a given direction, its molecules are forced closer together in the same direction. One consequence is that the electric field of a light beam polarized in the stressed direction interacts differently with the molecules than when the beam is oriented in the unstressed direction. Therefore the speed of light differs for components of a beam with the two polarizations. This effect is called *stress birefringence*. The advance of one component compared with the other changes the nature of a polarized beam, provided its polarization is not initially parallel to either of the special directions. Consequently patterns of stress are revealed if the plexiglass is placed between two polarizers whose transmission directions are perpendicular and a beam of light is directed through the sandwiched arrangement (Color Plate 20). Without stress the plexiglass is not birefringent, and no light emerges from the second polarizer. With stress the intensity of the emerging beam varies from one place to another over the plexiglass, depending on the distribution of stress. Such a device using crossed polarizers is called a polarimeter. By using plexiglass models it can reveal where stress concentrates at weak points in the design of bridge piers, tools, and machine parts. Since the alteration of the speed of light by stress birefringence may depend strongly on the wavelength, the effect with crossed polarizers can be quite colorful.

Polarizing by Scattering

In addition to absorption, reflection, and refraction the fourth means of polarizing light is through scattering. As described earlier, very small particles scatter light in virtually all directions. The scattered light is found to be completely polarized. The direction of polarization is determined by the direction of the electric field of the light before it was scattered. This will be explained in detail in Chapter 14, where we discuss light scattering in the sky. Our principal interest in this method of polarizing light lies in the fact that skylight is polarized. Bees, pigeons, and some insects take advantage of this, because their eyes are sensitive to the polarization direction of light. Even on overcast days they are able to detect the direction of the sun from the pattern of polarized skylight, and this serves as an aid to navigation.

SUMMARY

Light is a form of energy. Most of its properties can be explained by a simple wave theory, first proposed in 1665 by Robert Hooke. The wave phenomena of interference and diffraction were investigated in great detail during the nineteenth century, culminating in the complete theoretical formulation by James Clerk Maxwell, which explains a vast range of phenomena involving electromagnetic waves. Light waves, like water waves, are characterized by a wavelength λ, an amplitude A, and a speed v. The speed and wavelength are changed when light travels through different materials, such as air or glass. Diffraction patterns are observed when light passes through small holes or passes close to sharp edges. Interference colors are produced when light is reflected from both surfaces of thin layers such as soap films. Very striking interference effects are seen when reflection occurs at many layers, as in peacock feathers. Diffraction and interference both occur in devices like the diffraction grating, and in nature in the opal. X rays, similarly diffracted from the atoms in a crystal, produce characteristic diffraction patterns that reveal the crystal structure.

Scattering of light by very small particles, often in combination with selective absorption, accounts for many phenomena such as the bluish color of smoke, the differences in the colors of eyes, skin, and hair, and the properties of liquid crystals used in electronic displays. Because light is a transverse wave, the electric field may take on different directions perpendicular to the direction of travel of the light wave. Polarized light can be produced from initially unpolarized light by absorption, reflection, refraction, and scattering. Polarized light is important in photography, visual glare reduction, and in the analysis of stress patterns in models of buildings, bridges, and other pieces of architecture and machines. ■

QUESTIONS

4-1. Which form of electromagnetic radiation has a longer wavelength than light, infrared or ultraviolet?

4-2. When an electromagnetic wave passes into a medium of greater refractive index, how do its speed, frequency, and wavelength change?

4-3. Comparing two light waves, if one has twice the amplitude of the other, by how much is its intensity greater?

4-4. What is the frequency of a light wave with a wavelength in a vacuum of 589 nm emitted by a sodium arc lamp in a street light?

4-5. If two waves of equal wavelength and intensity interfere constructively, how much greater is the intensity than the intensity of each separate wave?

4-6. In Thomas Young's experiment on interference between monochromatic light emerging from two slits, how is the distance changed between the bright fringes produced on a screen if: (a) the screen is moved further from the slits; (b) the slits are moved closer together; or (c) the wavelength of the light is increased?

4-7. A very thin soap film appears black in reflected light. Explain why. If you look at the film from the other side and see the transmitted light, does it also look black? Explain.

4-8. What should be the thickness of an antireflection coating on a lens if the coating material has a refractive index midway between that of air and that of the lens?

4-9. What phenomenon causes the iridescence of a butterfly wing? How can you tell by looking at the color that it is not due to selective absorption by a pigment?

4-10. What name is given to the phenomenon when light passes through a small hole and spreads out on the other side?

4-11. One type of popular camera has a lens whose diameter is about 2.5 cm, and the lens is positioned about 7 cm from the film. What is the size of Airy's disk on the film if a picture of a distant point source of light is taken?

4-12. Could you see the diffraction pattern described in Question 11 if you looked at the developed negative while holding it 25 cm from your eye?

4-13. How does the pattern of light from a diffraction grating differ from the pattern from opal?

4-14. What condition must be satisfied by the size of a particle if its effect on light is an example of Rayleigh scattering? Why does the pattern of scattered light not depend on wavelength, unlike diffracted light?

4-15. Explain why light coming directly toward us from a lamp is unpolarized.

4-16. If a sheet polarizer is placed in a beam of completely plane polarized light so that the orientation of its transmission axis is halfway between the direction that transmits the maximum intensity and the direction for minimum intensity, what fraction of the incident intensity is transmitted?

4-17. At what angle should a flashlight beam be directed toward a window glass to have the greatest amount reflected?

4-18. When light is incident on a sheet of glass at Brewster's angle, describe the polarization of the reflected and refracted rays.

FURTHER READING

C. J. Campbell, C. J. Koester, M. C. Rittler, and R. B. Tackaberry, *Physiological Optics* (Harper and Row, Hagerstown, Maryland, 1974). Excellent discussion of optical phenomena produced by the wave aspects of light, as well as ray optics, vision, and the measurement of light and color.

W. Swindell, Ed., *Polarized Light* (Dowden, Hutchinson, and Ross, Stroudsberg, Pennsylvania, 1975). Collection of classic papers on the topic, beginning with Bartholimas. From "Benchmark Papers in Optics." (Distributed by Halsted (Wiley), New York.)

W. A. Shurcliff and S. S. Ballard, *Polarized Light* (Van Nostrand, Princeton, 1964).

H. Kubota, *"Interference Color"* in *Progress in Optics* Vol. 1 (North Holland, Amsterdam, 1961).

R. L. Gregory and E. H. Gombrich, *Illusion in Nature and Art* (Scribner's, New York, 1973). We particularly recommend the chapter by H. E. Hinton on "Natural Deception," which shows many examples of diffraction and interference phenomena in the animal world.

C. V. Boys, *Soap Bubbles: Their Colors and the Forces Which Mold Them* (Dover, New York, 1958).

F. J. Kahn, "The Molecular Physics of Liquid-Crystal Devices," *Physics Today,* Vol. 35 (May 1982), p. 66. A short description of the operation of liquid-crystal display devices.

W. A. Shurcliff, "Resource Letter PL-1 on Polarized Light," *American Journal of Physics,* Vol. 30 (1962), p. 227. An annotated list of books and review articles on polarized light, as well as a list of sources of materials for demonstrations.

From the *Scientific American*

J. H. Rush, "The Speed of Light" (August, 1955), p. 62.

P. Baumeister and G. Pincus, "Optical Interference Coatings" (December, 1970), p. 58.

J. Walker, "The Amateur Scientist: The Bright Colors in a Soap Film are a Lesson in Wave Interference" (September, 1978), p. 232.

H. F. Nijhout, "The Color Patterns of Butterflies and Moths" (November, 1981), p. 140.

N. Tinbergen, "Defense by Color" (October, 1957), p. 48.

K. H. Drexhage, "Monomolecular Layers and Light" (March, 1970), p. 108.

P. J. Darragh, A. J. Gaskin, and J. V. Sanders, "Opals" (April 1976), p. 84.

H. E. White and P. Levatin, "'Floaters' in the Eye" (June, 1962), p. 119.

V. K. La Mer and M. Kerker, "Light Scattered by Small Particles" (February, 1953), p. 69.

G. H. Heilmeier, "Liquid Crystal Display Devices" (April, 1970), p. 100.

■ CHAPTER 5 ■

LIGHT, MATTER, AND QUANTA

How does matter alter the spectral composition of light it transmits or reflects? There are diverse causes, but they have a common feature: the interaction of light with electrons. The study of this interaction at the turn of the century produced a crisis in physics when it was realized that portraying light as a simple wave and atoms as little balls failed to explain what was observed in several crucial experiments. A resolution of this dilemma was achieved by revolutionary notions introduced in the twentieth century that we label as *modern quantum theory*. In this chapter we investigate several ways that light interacts with matter and show how a variety of effects related to color can be understood only within the context of quantum theory. We are led to the notion that light has a dual nature depending on circumstance: in some cases its wave character predominates, in others it behaves more like a stream of particles, known as *photons*. It is unsettling to imagine something that is both a wave and a particle, yet numerous experimental tests have established the correctness of this view. These two aspects of light will become apparent as we proceed through this chapter. Atoms, molecules, liquids, and solids affect light differently. How this comes about can be understood from just a few basic principles of quantum theory. We can apply these to explain a rich variety of color phenomena.

▪ 5-1 CHANGING LIGHT'S DIRECTION

Whether we speak of a wave or a particle there is one thing they have in common—their direction of motion. We can draw a line showing this path, and if unaffected by an obstruction the line will be straight. This leads to the notion of a *light ray*. The ray traces the path taken by a narrow beam of light and shows how the path may be bent (refracted) on passing from one medium to another. Tracing these rays is the simplest way to show how matter affects the direction of light by refraction or reflection at a surface. *Ray optics* (or *geometrical optics*) is a procedure by which the path of light can be predicted according to the laws that govern these effects. In this section we review the basis for these laws.

Hero's Law of Reflection

Humans enjoy looking at themselves, so it is no wonder that when metallurgy evolved during the Copper Age one of the first implements to be developed was a mirror. A flat, polished copper surface serves admirably, although it imparts a red hue to the image, and when touched loses its shiny surface through oxidation. The modern version of the mirror is a coating of aluminum on one side of a glass plate, the glass serving to support and protect the reflecting surface of aluminum.

When a light ray is incident on any reflecting surface, the law of reflection governing the direction of the reflected ray is the same. It was discovered in the first century A.D. by Hero of the city of Alexandria, who conducted a series of careful measurements with mirrors. Hero announced that regardless of the angle at which light is incident on a flat surface, the reflected ray leaves at the same angle. It is

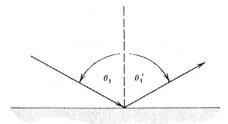

Fig. 5-1. Directions of incident and specularly reflected rays at a reflecting surface.

conventional to measure angles with respect to a line perpendicular to the surface, as shown in Fig. 5-1. We can denote the value of the incident angle by the symbol for the Greek letter theta: θ_1; and the value of the reflection angle by the same symbol with a prime added: θ_1'. Hero's law of reflection can then be expressed in mathematical shorthand as

$$\theta_1 = \theta_1' \qquad (5\text{-}1)$$

Notice that this is true regardless of the SPD of the incident beam of light. The equality of incident and reflection angles can be predicted by considering the details of how the electric field of the incident ray affects the electrons in a solid; but the calculation is highly technical and gives us no new insight. Instead we merely note that what happens to light is akin to the rebound of a ball when thrown against a smooth wall: the angle at which it rebounds matches the angle at which it was thrown. (The *percentage* of incident light that is reflected from a surface depends on the nature of the reflective material as we saw in Section 2-6).

Snell's Law of Refraction

The law governing refraction remained undiscovered for 15 centuries after the work of Hero, coming even after the invention of the telescope by the Dutchman Hans Lippershay in 1608. It was his compatriot the mathematician Willebrord Snell van Royen who led the way by patiently and systematically measuring the incident angle and corresponding refraction angle when a narrow beam of light was directed toward the flat surface of a transparent object. By 1621 Snell had tabulated the refraction angles for a variety of solids and liquids. René Descartes later found a mathematical formula to relate angles of incidence and refraction, and this subsequently became known as *Snell's law*. It is such a simple relationship that Snell's law has become the essential ingredient in designing optical systems for cameras, telescopes, or microscopes. With it, the path of a light ray can be predicted as it advances from one lens to another, no matter how complicated the arrangement.

We noted in Section 4-2 that the speed of light varies from one medium to another as indicated by the refractive index of the medium. A light ray impinging on a surface between media of differing refractive indices becomes divided so that

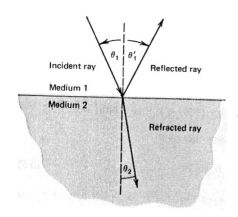

Fig. 5-2. Angles denoting the direction of incident, reflected, and refracted rays at a surface.

part of its energy is reflected and the rest refracted. This is evident in Color Plate 1 where a beam of light strikes the prism. As shown in Fig. 5-2, the directions of the incident, reflected, and refracted rays are measured by the angles they make with the line perpendicular to the surface. Willebrord Snell found that the angle of refraction θ_2 may be greater or less than the incident angle θ_1, depending on whether the new medium has a smaller or larger refractive index.

The credit for explaining this effect goes to Christiaan Huygens. In 1678 he suggested that this bending of the ray's path is due to differing values for the speed of light in the two media. He based his explanation on the wave nature of light. Figure 5-3 illustrates how the portion of a wave crest entering a medium with a lower speed is bypassed by the portion of the crest remaining in the faster medium. The key aspect of wave motion to keep in mind is that a wave moves at right angles to its crests. The figure shows that when the direction of the crests is rotated the direction of motion of the wave must similarly be altered. In this case, on enter-

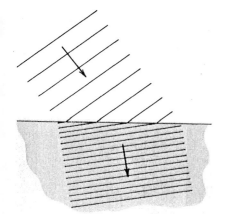

Fig. 5-3. Effect on the orientation of wavecrests and corresponding direction of travel when light passes into a medium in which it travels more slowly.

ing the slower medium the wave travels closer to the line which is perpendicular to the surface.

This is why in the ray picture of refraction, when going into a medium of greater refractive index the refraction angle θ_2 is smaller than the incident angle θ_1, and when going into a medium of smaller index θ_2 is greater than θ_1. It was Snell who discovered that every medium can be assigned a characteristic number to describe its refractive effect, and it was Huygens who interpreted this refractive index as giving the speed of light (Eq. 4-2).

The mathematical formula known as Snell's law is:

$$n_1 \sin \theta_1 = n_2 \sin \theta_2 \qquad (5\text{-}2)$$

where n_1 and n_2 are the refractive indices of the media in which θ_1 and θ_2 are measured. (Sin is an abbreviated form of sine.) This says that the product of n and $\sin \theta$ in one medium equals the product in the other medium. The symmetrical form of this equation tells us that if the direction of travel of the light ray is reversed so that it originates in medium "2" with incident angle θ_2, it would then be refracted at the angle θ_1 in medium "1."

Given the value of any angle, the value of the sine of that angle can be obtained from standard trigonometric tables. For convenience we show the same information in the curve in Fig. 5-4. It shows that the value of $\sin \theta$ increases from 0 to 1 as the angle θ increases from 0° to 90°. Since Eq. 5-2 says that the product of n

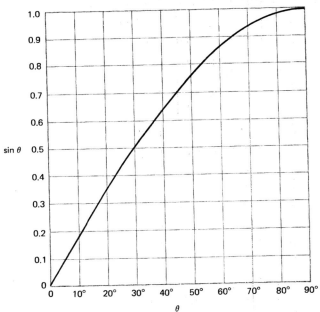

Fig. 5-4. How the sine of an angle depends on the angle.

TABLE 5-1 Refractive Index Values for Some Common Materials at a Wavelength of 550 nm

Vacuum	1.0
Air	1.0003
lucite	1.25
water	1.33
fused quartz	1.46
crown glass	1.52
amber	1.55
flint glass	1.62
diamond	2.42
silicon (in infrared)	3.4

and sin θ in medium "2" must match the product in medium "1," we see that for media with larger values of n the value of sin θ and consequently the value of θ itself must be smaller. That is why a medium with larger n has the direction of the ray closer to the perpendicular line.

Refractive Index

The speed of light in fact does not vary dramatically from one medium to another— only by a factor of 2 or 4 at most. We recall Eq. 4-2 to show how it is related to the refractive index:

$$n = c/v \qquad (5\text{-}3)$$

where c is the speed of light in a vacuum. Table 5-1 lists values of n for a number of common materials. For practical use, the refractive index for air can be taken as $n = 1$, the same as for vacuum. Snell's law can be used in ray optics without regard for its underlying physical significance. The most common application is in predicting the refraction angle θ_2 when n_1, n_2 and the incident angle θ_1 are known.

Example 5-1 Refraction at a Surface

A flashlight shines toward the surface of a pond, with its ray making an incident angle of $\theta_1 = 60°$. What is the refraction angle when the beam enters the water? The answer given by Eq. 5-2 is written explicitly by first dividing both sides of the equation by n_2. This gives

$$\sin \theta_2 = \left(\frac{n_1}{n_2}\right) \sin \theta_1 = \left(\frac{1}{1.33}\right) \sin 60° = (0.75) \times (0.87) = 0.65$$

Referring to the curve in Fig. 5-4, we see that the angle having this value for its sine is $\theta_2 = 41°$. This is the refraction angle in the pond.

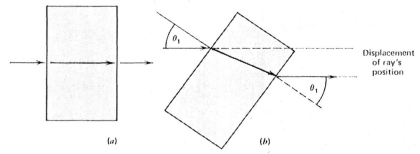

Fig. 5-5. (*a*) The position of a ray is unaffected if it comes to a surface head-on. (*b*) But if the ray is incident at an angle θ_1, refraction on entering the medium is exactly compensated by refraction in the opposite direction when the ray leaves this medium (for surfaces that are parallel), so that the ray is displaced to the side.

Snell's law predicts that light entering a plate of glass ($n = 1.5$) is refracted so that it travels more nearly perpendicular to the surface. When it is refracted again as it emerges from the other side of the glass, as shown in Fig. 5-5, the light emerges with its original direction. However the path is shifted to the side of the original one, by an amount that increases with greater angle of incidence.

■ 5-2 CHARACTERISTIC FREQUENCIES

Differences in refractive index of two adjacent media lead to the remarkable refractive phenomena illustrated in the preceding section. In turn, careful measurement of the incident and refraction angles provides a way to determine the values of the indices. When this is done for different wavelengths across the spectrum, the refractive index is generally found to vary with the wavelength. The effect is small for common transparent materials, amounting to a few percent at most. But this is enough to produce colorful effects, as Newton discovered in his experiments with a prism. The variation of refractive index with wavelength is called *dispersion*. There is a close correspondence between wavelengths showing a strong dispersion and wavelengths at which a medium selectively absorbs light. Dispersion and selective absorption give us insight into how matter interacts with light.

Phenomenon of Resonance

We begin with a simple experiment: Imagine three identical weights suspended from strings of differing length. Each weight, when moved to the side and released, swings back and forth with a pendulum motion having a frequency (number of swings each second) denoted by the Greek letter nu with subscript zero: ν_0. This so-called "natural frequency" is highest for the weight on the shortest string. Now if we push back and forth on the frame to which the strings are attached at a frequency ν that is very close to the natural frequency ν_0 of one of the weights, then that weight will begin to swing strongly even though the others barely move. The phenomenon—the very efficient response of an object to a driving force whose frequency is matched to the natural frequency—is called *resonance*. We intuitively take advantage of resonance when pushing someone on a swing. It is also responsible for the proverbial shattering of a glass by the powerful voice of an opera singer who sings the "right note."

Let us imagine a second experiment in which a weight is suspended from a spring. If we pull down (or push up) on the weight and then release it, the weight will oscillate up and down at a natural frequency that we can also designate by the symbol ν_0, although its value will differ from that of the previous experiment. The stiffer the spring and the lighter the weight the higher is the natural frequency ν_0. This arrangement is called an "oscillator," just as we can call the pendulum in the previous experiment an "oscillator."

Now suppose that we attach the top of the spring to an arrangement connected to a motor that moves the spring up and down. Suppose also that the frequency of this driving motion can be adjusted by turning a knob controlling the speed of the motor. For each frequency of the driving motor we measure the amplitude of the weight's oscillation. We shall find that this response depends on the driving frequency as indicated schematically in Fig. 5-6. The response is sharply peaked where ν equals ν_0, which is the condition of resonance. For very low driving frequencies the response approaches a small but constant value, while for very high driving frequencies it falls steadily toward zero.

The sharpness of the peak is different for different oscillators. If the weight and spring are immersed in oil, the oscillation dies out quickly when you stop driving it. The oscillation is said to be "strongly damped," and the peak shown in Fig. 5-6 will be low and broad. On the other hand, if the weight is suspended in air, we know that it continues to oscillate for a long time after you stop driving it. The oscillation is weakly damped, and the peak in its response curve will be high and narrow.

Atoms (and molecules as well) are also oscillators. The small, positively charged nucleus of the atom is surrounded by a cloud of negatively charged electrons. If one of the electrons is displaced slightly from its orbit and then released, it will respond to electrical forces within the atom and oscillate at its natural frequency ν_0. Since the electron is so light and the "spring" forces are very strong, the resonant frequency of an electron is extremely high—typically around 10^{15} Hz—which is in the same range as the frequency of light.

Classical View of Selective Absorption

Because an electromagnetic wave such as light has an electric field, it can perturb an electron: the electric field pushes the negatively charged electron in one direction while pulling the positive nucleus in the other. Thus, we can understand the selective absorption of particular wavelengths by pigments and dyes as a manifestation of resonance. The atoms and molecules in the pigment are driven by all the frequencies of the electromagnetic waves present in the incident light. But they

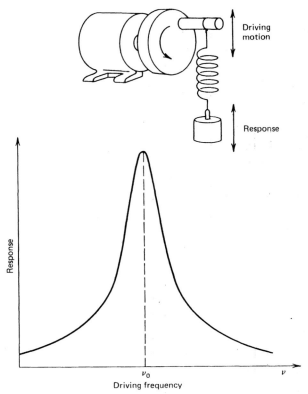

Fig. 5-6. Response of an oscillator whose natural frequency is ν_0 to a driving force of frequency ν.

respond most strongly to frequencies close to their natural frequencies and therefore absorb light in those spectral regions. The SPD (spectral power distribution) of light reflected from a pigmented surface illuminated by white light is thus determined by the resonant frequencies of the pigment.

In a leaf, for example, each chlorophyll molecule has numerous resonance frequencies in the red and blue spectral regions. (The light that it absorbs is utilized in photosynthesis.) Consequently the reflected light will be deficient in red and blue components and will be predominantly green. As a young leaf develops it produces more chlorophyll, which allows greater absorption in the red and blue, so that the growing leaf becomes a progressively darker and more saturated green.

Similarly the electrons of neon atoms present in the gas within a neon lamp oscillate at their natural frequencies, when struck by free-flying electrons that comprise the electrical current flowing in the lamp. Before these atoms can lose this extra energy by other means, they give it up by radiating electromagnetic waves at the same frequencies. Since many of these natural frequencies are in the red portion of the spectrum, a neon lamp is red.

Dispersion

Now we return to Fig. 5-6 and consider what will happen when a monochromatic light wave passes through a region of space containing a large number of identical atoms. If the frequency ν of the light matches the natural frequency ν_0 of the atoms, then the light will be absorbed. But if ν_0 is in the ultraviolet spectral region while the frequency of the light ν is in the visible, the light will be far below the resonant frequency and there will be no absorption. As indicated in the figure, however, the atoms will still respond somewhat to the light even though they do not absorb it. As the light wave progresses through the region containing the atoms, there will be a continual exchange of energy back and forth between the light wave and the atoms. This interaction reduces the speed of the light wave.

According to Eq. 5-3 the slower the speed the greater the refractive index of the medium. It is reasonable that the value of n increases with the amplitude of the atomic response (the height of the curve in Fig. 5-6 at the frequency ν of the light wave). It also increases with the number of atoms in the volume. The increase with density explains why light travels faster in hot air than in cold air: heating causes the air to expand, reducing the number of molecules present in a given volume and thus reducing the refractive index n. As we shall see in Chapter 14, this fact is at the center of all mirage phenomena.

Now we consider the relation between n and ν. Returning again to Fig. 5-6, suppose that ν_0 is in the ultraviolet, as is the case for most optical materials such as glass. Then for visible light the frequency ν will be to the left of ν_0, with the blue closest to the peak and the red farthest from it. Since the refractive index increases with increasing response, while the response increases as we go from red at the left to blue further to the right, we see that the refractive index should be greater for blue light than for red light. This is *dispersion*—the variation of refractive index with wavelength. In our example with the refractive index greater in the blue than red we have what is called *normal dispersion*. Examples of normal dispersion are shown in Fig. 5-7. The opposite case, which would occur if ν_0 were at longer wavelengths in the visible portion of the spectrum, is called *anomalous dispersion*. Anomalous dispersion will be discussed further in Chapter 11 when we deal with optical effects in paintings.

In Chapter 8 we shall see that the operation of optical elements such as prisms and lenses depends on the bending of light at the surface due to refraction. Since optical glass has its resonance frequencies in the ultraviolet and therefore exhibits normal dispersion, blue light will be refracted more strongly than red. This accounts for the formation of a spectrum by a prism. Dispersion also produces the colorful display of a diamond. Table 5-1 shows that diamond has a remarkably large refractive index; it also has a correspondingly pronounced dispersion. This explains the striking spread of the spectrum when refracted by the faces of a dia-

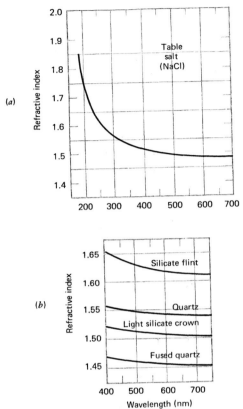

Fig. 5-7. (*a*) The variation of the refractive index with wavelength across a wide spectral range for common table salt. (*b*) Dispersion across a narrower range for quartz and several types of glass used in optical instruments.

mond. However, dispersion is undesirable in certain situations, such as in optical systems found in cameras, and therefore must be corrected by adding compensating elements to the lens, as will be discussed in Section 8-5.

■ 5-3 QUANTUM EFFECTS

The preceding section presented the classical view of how light interacts with matter, as it was accepted at the end of the nineteenth century. It was extremely successful in explaining many phenomena related to light and color, but spectroscopists who studied the wavelengths at which an atom absorbs light were unable to explain their results with models based on the classical idea of oscillators. Some radically new notions were needed. The quantum theory in physics, advanced over a revolutionary 30-year period beginning in 1900, provided them. Many of its concepts are so novel when first encountered that they may seem unbelievable. This is simply because we have no direct experience with processes at the atomic level, and so our intuition is of little help. Even when we observe consequences of quantum effects, such as fluorescence, we do not recognize them for what they are.

The first topic we address is the nature of light itself. We shall see that the wave aspect considered in the previous chapter is a simplified view and goes only partway toward explaining many phenomena that we encounter daily. We shall then discuss how quantum effects influence the interaction between light and matter at the atomic level.

Photons: "Particles" of Light

During the last decade of the nineteenth century several experiments involving the study of light revealed phenomena that had no ready explanation in the wave theory, which had been so successful since the experiments of Thomas Young. As Albert Einstein was later to say, "a profound change in our views of the nature and constitution of light is indispensable." Indeed, Einstein himself is commonly given credit for introducing the startling idea that light is a collection of particles, now called *photons*. But quantum theory does not permit (without resorting to its specialized language) an exact specification of the "size" of a photon, nor exactly where it might be found at a particular instant.

We have no experience with individual photons, because in common situations we encounter so many that their individual character is swamped by their overall effect. For instance a common 100 watt lamp emits about 10^{20} photons each second, a truly uncountable number. Each photon has wavelike properties with a particular frequency ν_p. Its wavelength adjusts to the medium in which it travels, as we described in Eq. 4-1.

Each photon represents an exact quantity of energy, contained in its electric and magnetic fields. The amount of energy is determined by its frequency. This is why the frequency of an electromagnetic wave remains unchanged when it passes from one medium to another, while its wavelength changes. The fact that a photon's energy is determined by its frequency was proposed by Einstein in 1905 to explain what is known as the *photoelectric effect*. He sought to explain how electromagnetic waves affect the electrons in a metal plate. If the plate is mounted within an evacuated chamber, it is possible to detect electrons that escape from the plate into the vacuum and strike a collector whenever light of an appropriate wavelength strikes the plate. In fact, the energy of the liberated electrons is greater for blue light than red, even when the two lights have equal intensity! This is shown in Fig. 5-8. By our classical ideas that "intensity" is the number of watts incident on each square meter of the plate, we would not expect the energy of the electrons to change if the blue light and red light have the same intensity. Another remarkable feature was discovered when the light's intensity was diminished by first passing it through a neutral density filter. The number of liberated electrons was reduced, but

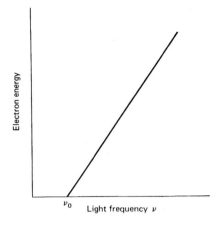

Electron energy

ν_0 Light frequency ν

Fig. 5-8. Energy of electrons emitted from a metal surface increases with the frequency ν of the incident light. The threshold frequency ν_0 for the photoelectric effect depends on the metal used.

the energy of each electron remained unchanged! If every electron were equally affected, we should not expect that result.

Energy of a Photon

There is one logical explanation for this curious behavior: that the liberation of an electron is an "all or nothing" affair. This happens if a beam of monochromatic light is composed of many individual particles of the same energy. If an electron absorbs radiation it must absorb a photon; therefore it acquires exactly that amount of energy. Measurements of the photoelectric effect showed that the energy of the liberated electron increases linearly with increasing photon frequency. Assuming that each electron absorbed only one photon, Einstein deduced that the energy E_p of a photon must be proportional to its frequency ν_p:

$$E_p = h\nu_p \qquad (5\text{-}4)$$

The proportionality constant h in this expression is called Planck's constant. It is named in honor of Max Planck, who first proposed that the intensity of light can only change in discrete steps. To Einstein goes the credit for recognizing that this discreteness is a fundamental characteristic of light itself. Planck's constant is a very small number: 6×10^{-34} joules per hertz. Even if ν_p is large, say about 6×10^{14} hertz for light, the energy of a photon is incredibly small: $E_p = 6 \times 10^{-34} \times 6 \times 10^{14} = 3.6 \times 10^{-19}$ joules. If you compare this figure with the 100 joules per second of electrical power drawn by a common 100 watt lamp, you can appreciate the fantastically large number of photons it gives off each second.

We can understand why inserting a neutral density filter in the light beam leaves the electron's energy unchanged but diminishes the number of electrons. The filter leaves the photons' frequency unchanged and simply absorbs a certain fraction of them. Fewer photons striking the metal plate produce a corresponding decrease in the number of liberated electrons. Thus we conclude that the power in a mono-chromatic beam of light increases in proportion to the rate at which photons are emitted by the source. A light source having a broad spectral power distribution such as a 100 watt lamp emits photons having a similarly large range of frequencies.

Thus light has a particle–wave duality. In certain situations the particle aspect is most evident; in others, the wave aspect. Generally speaking the higher the frequency the more apparent is the particle aspect, because each photon has a correspondingly greater energy.

The photoelectric effect has a practical application in the "electric eye" used in doorways to sense when someone approaches. A beam of light directed across the path strikes a metallic surface sealed within a small vacuum tube. When the beam is interrupted by the person's body, the accompanying decrease in photoelectric current from the tube triggers a mechanical device for opening the door.

■ 5-4 EXCITING ATOMS

Inspired by Einstein's earlier work on the particle–wave duality of radiation, Louis de Broglie, while still a graduate student in France, extended the idea to material particles in 1924. His notion has fundamental importance when applied to particles of atomic size or smaller. De Broglie suggested that particles also have wavelike properties. How might we convince ourselves that an electron behaves as a wave? By observing the same type of interference phenomena that convinced us that light is a wave! Interference patterns can in fact be observed when a stream of electrons is directed toward a crystalline solid. The electrons that are diffracted from the individual atoms are found to interfere constructively in certain directions, producing a strong electron beam; they interfere destructively in other directions so that no electrons are found. This is proof of the wave aspect of electrons. An example is shown in Fig. 5-9.

This wave aspect of an electron imposes an important condition on the type of orbit it may have when bound to an atomic nucleus. A simplified view is given in Fig. 5-10. As an electron makes a complete revolution, the wave must join smoothly onto itself. This means that the path of the orbit must be an integral number of wavelengths—λ, 2λ, 3λ, etc.—in circumference. Furthermore the speed of the electron (which is related to λ) must have a fixed relationship to the radius of the orbit in a way that is determined by the strength of the attractive force of the nucleus. To satisfy both requirements only certain orbits are possible for an electron. The circumference cannot be arbitrary in size but has only certain values in a quantitative sense. We say that the circumference is "quantized." Different orbits are distinguished by the number of wavelengths that fit in. The orbits are correspondingly organized in a sequence of shells about the nucleus. According to quantum theory only two electrons can be found in any one orbit. (Actually the "spin" of one electron must point in the opposite direction to the "spin" of the other, but to explain this "pairing" in detail would take us too far afield). The larger

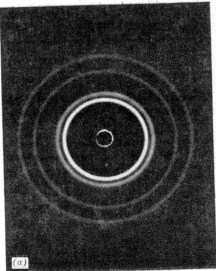 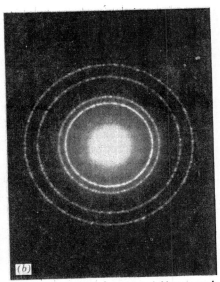

Fig. 5-9. Direct comparison of the diffraction patterns produced when (*a*) X rays and (*b*) electrons of appropriate energy are diffracted by the same aluminum foil. The similarity between the patterns is evidence that the electrons have a wave property and that their wavelength in this instance is the same as the wavelength of the X rays. (From the Physical Science Study Committee film *Matter Waves*. Educational Development Center.)

the shell the more orbits that can be fit in. Thus two electrons fill the innermost, or first shell as in the helium atom; the next shell holds eight electrons as in neon; and so on. The electrons in any shell having its full complement are all paired, and this arrangement is particularly stable.

Quantized Energies

Because of the wavelength quantization condition for each orbit, the energies permitted for an atomic electron are correspondingly quantized: each orbit has a specific energy. The energy of an electron is determined by the orbit it occupies. When

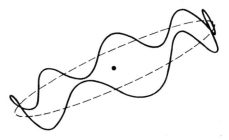

Fig. 5-10. Portrayal of the wave describing an electron in its orbit about the nucleus in an atom. The circumference of the orbit (dashed curve) must be of the proper size to accommodate an·integral number of wavelengths.

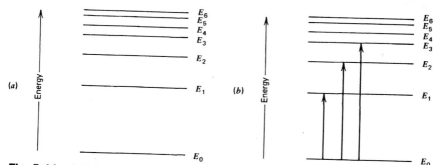

Fig. 5-11. (a) Quantized atomic energies arranged in order of increasing energy for a hypothetical atom. (b) Some possible transitions to excited states for an atom that initially is in the ground state.

all the electrons of an atom occupy the lowest energy orbits, the atom has its lowest total energy. This is said to be the *ground state* of the atom. Should an electron absorb energy from a light wave and move into an orbit of higher energy, the total energy of the atom would be correspondingly higher. Because these energies are quantized, the sequence of possible values can be displayed graphically as rungs on a ladder (Fig. 5-11a). The bottom rung represents the energy of the ground state. Higher rungs represent what are called *excited states,* with the electron in orbits of successively higher energy. The positions of the rungs are characteristic of the given type of atom.

Quantum Theory of Selective Absorption

This quantization of allowed energies in an atom is only one of the many surprising predictions that come from the quantum theory. These notions were developed around 1920, notably by physicists Niels Bohr, Erwin Schrödinger, and Werner Heisenberg. Quantum theory provides a conceptual framework whereby we can understand why an atom absorbs light of only certain frequencies and not others. This is because the energy of a photon absorbed by an atom must be just the proper value to excite an electron from one state to another. This requirement merely expresses the law of conservation of energy: that energy may be converted from one form to another but is never destroyed. Therefore, if a photon is absorbed by an atom, its energy must appear in some other form. In this case it corresponds to the extra energy the electron must have to go into a higher energy state. A few of these possible transitions from one state to another are illustrated in Fig. 5-11b. The energy of a photon, represented by the length of an arrow in the figure, must just span the distance between two states. Should a photon not have the correct energy to connect two states, it will pass by the atom without being absorbed.

The requirement for energy conservation can be stated succinctly by an equation. When an electron is excited from a state of energy E_0 to another of energy, say, E_1 the energy increase must match the energy E_p of the photon that was absorbed:

$$E_1 - E_0 = E_p = h\nu_p \qquad (5\text{-}5)$$

This is the condition in the quantum theory that replaced the notion of a "resonance" in the classical theory of the preceding section. Because the differences between the energies of the states vary from one element to another, they form a signature for each type of atom. If white light is passed through a gas containing atoms of a given element, the pattern of wavelengths across the spectrum where light is absorbed can be used to identify the element. Such an absorption spectrum for a hot gas of mercury atoms is shown in Fig. 5-12. This was recorded by passing the light emerging from the gas through a spectrograph, which separated its spectral components so that they could be registered on film. Wherever light had been absorbed by the mercury atoms we see a dark line across the developed print. The appearance of this spectrum suggests the name that has been applied to it: a *line spectrum*. This property of providing a signature makes atomic absorption spectroscopy a highly sensitive means of detecting small concentrations of contaminants, such as pollutants in air and water, as well as lead in human blood.

Once an electron is excited to a higher energy state, it can subsequently drop back down to one of the lower states and in the process emit a photon. The photon carries away the energy given up by the electron. Consequently photons can be emitted at any frequency that corresponds to the possible downward transitions between states on the ladder diagram in Fig. 5-11. The spectral power distribution from excited atoms is therefore a line spectrum, just as the absorption spectrum is. Color Plate 21 gives examples of the emission spectra of mercury, sodium, and hydrogen.

Because of the stability of an atomic shell containing its full complement of paired electrons, the energy required to excite one electron can usually be supplied only by photons from the ultraviolet, or in extreme cases the X-ray regions of the spectrum. As a consequence, these filled shells have no direct influence on the color of materials. Selective absorption in the visible usually comes from exciting an unpaired electron from an incomplete shell, which in most cases is the outermost shell. These outer electrons are also responsible for the chemistry of atoms. That is, they determine how an atom will combine with other atoms to form

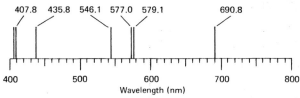

Fig. 5-12. Absorption spectrum for mercury atoms. The vertical lines indicate wavelengths where photons are absorbed because their energy corresponds to the difference in energies between two electronic energy levels.

molecules. Consequently the light absorptive properties of an atom may be strongly affected by the chemical environment in which it resides.

Color Identifies Atoms

Atoms produce characteristic colors when heated sufficiently to excite their electrons into higher states. This property serves as the basis for the *flame test* used by analytical chemists to identify elements in a sample. When sodium is heated in the flame of a bunsen burner, its single outermost electron is promoted to the first excited state—actually either of two excited states lying close together in energy. When an electron in the higher of these two states drops back into the ground state, it emits light at a wavelength of 589.1 nm. An electron that starts from the lower of the two excited states emits at a wavelength of 589.6 nm. Photons of these energies produce the characteristic, nearly saturated yellow-orange color that distinguishes sodium (Color Plate 21).

These lines of the spectrum are called the "D" lines and are so much stronger than other lines that a sodium lamp is sufficiently monochromatic to produce excellent diffraction and interference effects when its light is passed through narrow slits. However the slits must be reasonably close together since the light is not fully coherent. For mercury, the prominent lines are in the green and violet regions of the visible spectrum (Color Plate 21), and for neon in the red. Some of these lines arise from transitions from one excited state to another excited state, instead of to the ground state. ∎

5-5 EXCITING MOLECULES

The structure of a molecule permits new types of excitations that are not possible with a single atom. These include not only different shapes for electron orbits but also vibrations and rotations of the molecule itself. The water molecule, for example, consists of an oxygen atom to which two hydrogen atoms are attached. The positions of the three atoms form an angle that is slightly larger than a right angle (Fig. 5-13a). The nature of the electron bonding holding the atoms together is such that the single electron from each of the hydrogen atoms preferentially stays close to the oxygen atom, which lacked two electrons in its outermost shell. Consequently the hydrogen ends of the molecule are left with a net positive charge whereas the oxygen has acquired a net negative charge.

This "electric polarization" endows water with the ability to interact strongly with an electromagnetic wave (Fig. 5-13b). The electric field pulls the hydrogens in one direction while pulling the oxygen in the opposite direction. This causes the molecule to vibrate with the electric field. By the classical physics of the nineteenth century any amplitude of vibration is possible, corresponding to the amount of energy absorbed. But according to the quantum theory, only certain definite excitation energies are allowed. If the energy necessary to excite this vibratory motion is matched by the energy of a photon, absorption is possible. This is an example of *molecular excitation*. It takes place at frequencies that are characteristic of the particular molecular properties. Since the heavy masses of the nuclei makes vibratory motion sluggish, absorption occurs at comparatively low frequencies, most importantly in the infrared portion of the spectrum.

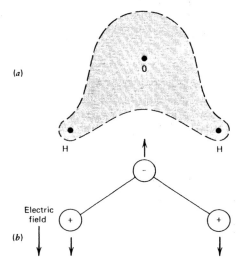

Fig. 5-13. (*a*) Cloud of electrons orbiting around the oxygen atom and two hydrogen atoms of the water molecule. (*b*) The net negative electric charge on the oxygen and the positive charge on each hydrogen causes them to be pulled in opposite directions by the electric field of the electromagnetic wave.

Absorption Bands

In addition to the vibratory motion we must consider the fact that the molecule can be set into rotation. This also corresponds to a molecular excitation, but it takes rather little energy to cause rotation. As a consequence of rotational excitation, each frequency that can excite a vibrational mode is surrounded by a family of closely spaced frequencies representing the additional excitations of differing rotational states. Since these individual frequencies often cannot be resolved, the absorption spectrum appears as a sequence of bands, spaced across the infrared portion of the spectrum as shown in Fig. 5-14. For the water molecule the absorption bands extend into the long wavelength region of the visible spectrum. For this reason when water molecules are concentrated together—as in pure water and ice—we see a pale blue color. It is caused by absorption of photons

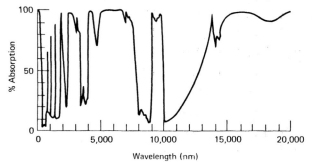

Fig. 5-14. Regions of the spectrum at infrared and shorter wavelengths in which a water molecule can absorb photons. These regions are indicated by the percent of sunlight that is absorbed on passing through the atmosphere.

that excite vibrations and rotations of the individual molecules. Figure 5-14 also shows absorption in the ultraviolet, which comes from electronic excitations.

Color from Organic Molecules

Modern dyes are made from organic molecules, that is, molecules containing one or more carbon atoms. Selective absorption by such molecules is explained by the nature of the excited states of the electrons. The theory accounting for these orbits was advanced by the chemists Robert Mulliken, Linus Pauling, and many others in the 1930s. They explained that an electron state is not confined to a single atom but extends across several atoms that comprise the molecule. Excited states of such "molecular orbitals" involve subtle changes that require correspondingly low excitation energies. The orbits may be sufficiently extended that absorption takes place in the visible portion of the electromagnetic spectrum and is not limited to the ultraviolet.

Many organic molecules have a characteristic structure in which carbon atoms are joined by a system of alternating single and double bonds. In a single bond between two atoms, a pair of electrons is shared between the partners, and for a double bond, two pairs. Benzene is an example of such a molecule, consisting of six carbon atoms arranged in a ring. Each has a hydrogen atom protruding outward, as shown in Fig. 5-15a. There are two ways that the double bonds can be assigned, and both arrangements have the same electronic energy. Since neither arrangement is favored over the other, the chemist says that the molecule "resonates" between the two possible states. Another way to visualize the situation is to imagine all of the carbon atoms to be joined by single bonds with the remaining electrons spread over the entire ring (Fig. 5-15b). For a small molecule like benzene the lowest excited state for such an extended orbit is accessible only by

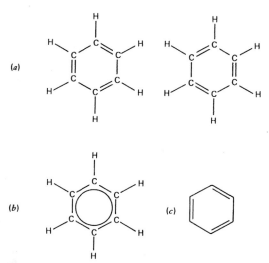

Fig. 5-15. (a) Two energetically equivalent arrangements of double bonds in the benzene molecule are said to be in resonance with each other. (b) The combination of two arrangements is represented by an extended orbit for the electrons. (c) Another symbol for benzene, equivalent to (a) but without the C and H atoms shown explicitly.

absorption of a photon from the ultraviolet region of the spectrum; therefore benzene is colorless.

There are three principal ways by which the structure of an organic molecule leads to colorful effects. The first is by the addition of one or more groups of atoms that serve as *electron acceptors* or *electron donors*. Both donors and acceptors extend the spatial extent of the electron orbits and lower the energy of the first excited state. An acceptor and donor are particularly effective when attached to opposite ends of the molecule. Figure 5-16a shows the *p*-nitrophenolate ion that in water solution has a net negative charge and pronounced yellow color. This ion consists of a benzene ring to which an oxygen donor is attached on the right and a NO_2 acceptor group on the left. In solution the unbonded electron that was originally localized around the oxygen is transferred to the NO_2 group, and to help fill the outermost shell of oxygen, which lacks two electrons to be completed, a double bond is formed with the benzene ring. Similarly a double bond forms with the nitrogen of the NO_2 group. Thus the electron states of the benzene ring extend onto these groups, with a corresponding reduction in the energy of the excited state.

The presence of donors and acceptors tends to lower the excitation energy of a resonating molecule so that absorption may occur in the visible portion of the spectrum. However, another factor determines how strong that absorption will be. We noted earlier when discussing absorption by a water molecule that the polarization of its charge endowed it with the ability for strong absorption. A "polar" molecule is one whose center of negative electronic charge is shifted relative to the center of positive nuclear charge. A benzene ring by its symmetry is nonpolar, but the addition of electron acceptor or donor groups that themselves are polar causes the entire molecule to become polarized. This encour-

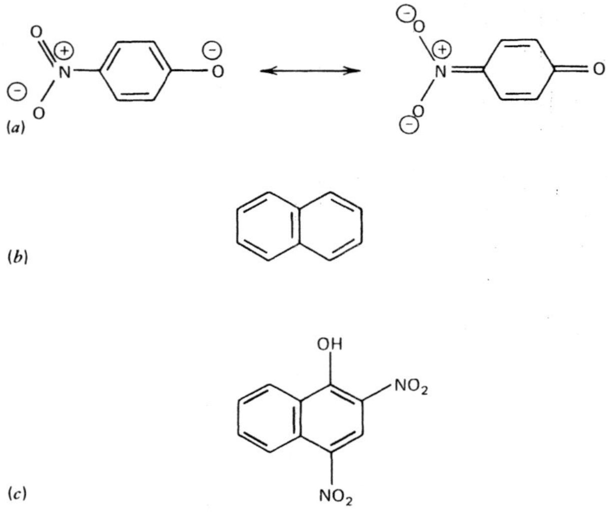

Fig. 5-16. (*a*) Change of the polarization of the *p*-nitrophenolate ion when it goes into solution. (*b*) Colorless napthalene. (*c*) The dye Martius yellow. In interpreting these schematic diagrams, we understand the carbon atom to be located at every junction. There is a hydrogen atom extending from each junction where there are three (but not four) bonds and there is no other side group.

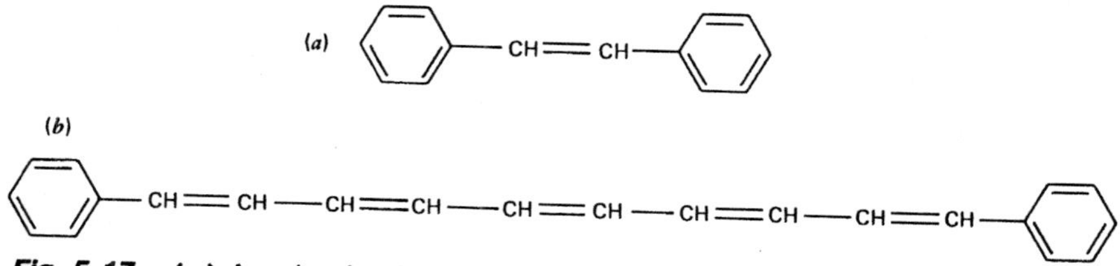

Fig. 5-17. (a) A molecule with few conjugated bonds (1,2 diphenylethene) absorbs at short wavelengths, in this case most strongly at 319 nm and therefore is colorless; (b) one with many conjugated bonds (1,10-diphenyl-1,3,5,7,9-decapentane) absorbs at longer wavelengths (most strongly at 424 nm) and is orange.

ages strong absorption, because the charge imbalance allows the electric field of light to couple more strongly to the molecule.

For example, Fig. 5-16b shows the molecule known as napthalene, which is a combination of two benzene rings. This geometry supports more extended orbits for the electrons than a single ring; nevertheless the peak absorption still lies in the ultraviolet, with no appreciable absorption in the visible. The addition of two NO_2 groups, which serve as acceptors, and an OH group, which serves as a donor, creates a molecule that is polar and strongly absorbs in the blue. This molecule is shown in Fig. 5-16c and is the dye called Martius yellow.

The second type of structure providing selective absorption in the visible is a long string of carbon atoms joined by alternating single and double bonds. This sequence is said to consist of *conjugated bonds*. For example, the diphenylpolyene series of molecules shown in Fig. 5-17 has various numbers of carbon atoms in the chain of conjugated bonds that link the rings found at each end. Molecules with more conjugated bonds display absorption at longer wavelengths.

Finally, the third structure of importance is an extensive system of rings joined by conjugated bonds and stabilized by a transition element such as iron, manganese, or copper. Hemoglobin of red blood cells is an example. It consists of a protein called globin that is bound to an iron-bearing group called heme, which provides the color. Examples of similar colorants found in nature are discussed in the following section.

Biological Colorants

Some well-known colors in nature result from selective absorption due to the excitation of electronic states that extend over a large number of atoms. One example we have just mentioned is the red color of blood. Another is the green color of plants, provided by the absorptive properties of molecules of *chlorophyll*. This molecule is the major pigment of most higher plants and algae; its structure is shown in Fig. 5-18. The nitrogen atoms found at the apex of each of four pentagonal rings are bonded to a single magnesium atom at the center. Each pentagon is also bridged to its neighbors by conjugated bonds. This network of pentagons and conjugated bonds is also found as the central structural element of heme in hemoglobin (where iron replaces magnesium) and in vitamin B_{12} (where cobalt replaces magnesium). In addition to the chlorophyll molecules shown in Fig. 5-18 there are also two other forms of chlorophyll. In a plant most of these forms serve as antennas to absorb photons, and they transfer much of this energy to a relatively few chlorophyll molecules that are specialized to carry out a sequence of chemical reactions.

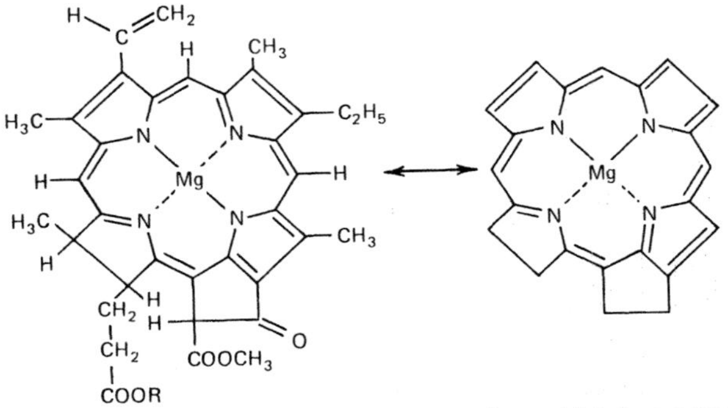

Fig. 5-18. The structure of chlorophyll-*a*, which has a magnesium atom at its center. The pattern of alternating single and double bonds is in resonance with the one sketched at the right without its side groups explicitly shown.

Because these reactions depend on the absorption of light they are known as *photo-chemical reactions*. The specific reactions involving chlorophyll are called *photosynthesis*. Two processes occur simultaneously in photosynthesis, a chemically reducing one in which carbon dioxide (CO_2) is converted into a carbohydrate such as sugar ($C_{12}H_{22}O_{11}$) and an oxidizing process in which oxygen is produced from water. About nine photons are needed for the assimilation of one CO_2 molecule.

Members of another family of pigments called *carotenoids* are universally present in organisms that carry out photosynthesis. These pigments protect chlorophyll from harmful reactions involving oxygen by absorbing some of the excess energy that chlorophyll acquires. Carotenoids with more than nine alternating pairs of single and double bonds, as shown in Fig. 5-19 for the molecule beta-carotene, are particularly effective, because the excited electronic states are spread over such a long distance that their excitation energies are low. The color of carotenoids can be seen in the absence of chlorophyll, as in the carrot or in flowers such as tulips and dandelions. In autumn, when the chlorophyll in green leaves breaks down after a sharp frost, the effect of carotenoid pigments becomes apparent with the display of various red, orange, and yellow colors. Carotenoids also give color to marine life such as salmon as well as crustaceans including lobsters and shrimp. Flamingos acquire their pink color from eating carotenoid-rich crustaceans. These pigments, which contain vitamin A, are also a basic element of visual pigments in the retina of the eye.

On the other hand, the colors of red cabbage, rhubarb, beetroot, red roses, and red wine are due to the pigment *anthrocyanin*, as is the purplish color of autumn leaves. Still another pigment, *phycoerythrin*, is found in certain marine algae. It absorbs light having wavelengths less than about 570 nm, from the yellow through violet, to provide energy for the biochemical reactions that keep the algae alive. These algae are responsible for the coloration for which the Red Sea was named.

The brownish-black pigment of soil is composed of long-chain molecules (polymers) that are folded back and forth and have many bonds joining close-lying segments of the molecules. *Humic acid* is the principal colorant of dark soil, coming from decayed organic matter. Other colors of soils, such as the reds and ochers in the deserts of Arizona and New Mexico, come from minerals, not from carbon-based pigments.

A pigment extremely important to human life is *melanin*. The elementary molecular unit

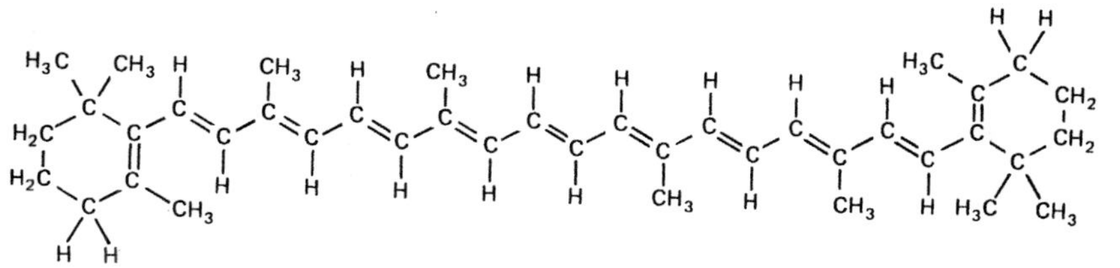

Fig. 5-19. The carotenoid pigment beta-carotene is formed with a remarkable number of conjugate bonds.

of most forms of melanin has the *endoquinone* structure shown in Fig. 5-20. The molecule itself is unstable, but if a quiltlike pattern of these molecules is assembled, with additional units stacked above and below to form a tiny particle, an extremely stable arrangement is created that is virtually indestructible. Some fossils a million years old have been found containing melanin. This pigment colors the dark ink that octopus and squid release as a smoke screen when they are disturbed.

Melanin is also the pigment that gives skin its color. Melanin is produced in various forms depending on genetic factors. The black and brown pigments are designated as eumelanin, and the red and yellow pigments as pheomelanin. Melanin strongly absorbs light, and its absorption varies inversely with wavelength. Consequently it is more effective in absorbing high energy photons that are harmful to skin tissue. The outermost layer of skin or epidermis is composed of two regions. The surface layer which is about 25 μm thick, contains dead cells, some of which contain melanin. Just below the surface layer is a layer of comparable thickness that consists of living cells. This layer contains *mela-nocytes* that produce melanin and store it for injection into the other cells. The dermis lies underneath and consists mainly of connective tissue and blood vessels. If exposed to ultraviolet radiation, molecular bonds in the dermis are disrupted and substances are released that digest proteins and dilate the capillaries. The increased amount of blood is evident in the color of hemoglobin that can be seen as the skin becomes red. Inflammation and pain develop in about two hours, the familiar condition of "sunburn".

The destructive effect of ultraviolet radiation is greatly reduced by the absorptive prop-

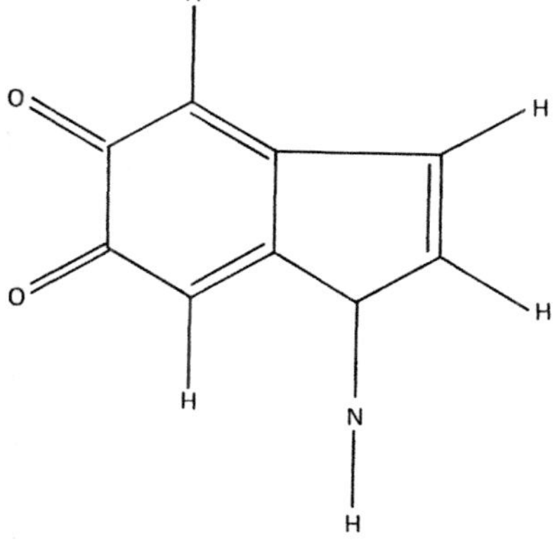

Fig. 5-20. The endoquinone structure is characterized by a benzene unit to which two oxygens are attached and a five-sided unit containing nitrogen and hydrogen.

erty of melanin in the epidermis. Only 10% of the radiant energy in the region of 300 nm, at the extreme limit of the solar SPD reaching earth, penetrates the 50 μm to the dermis. Nevertheless the skin has a protective mechanism; upon exposure to unusual amounts of ultraviolet it produces more rapid growth of skin cells and faster release of melanin by the melanocytes. This rapid cell division may triple the thickness of the layer of dead cells, and the corresponding increase in melanin over a period of days is seen as tanning. The increased absorption reduces transmission through the epidermis by an additional factor of ten.

Liquids

Molecules retain their identity when they condense to form a liquid. Yet the buffeting that each receives from the movement of its neighbors causes slight distortions of its shape that perturb the electron orbits. As a result the sharp line spectrum of the molecule is broadened in a liquid. A similar effect, called "pressure broadening" occurs in a gas at high pressure when atoms or molecules frequently strike each other with great force. The consequence in a liquid is typified by the absorption spectrum for the dye shown in Fig. 2-18. This is called a *continuous spectrum* to distinguish it from a line spectrum. A continuous spectrum has some absorption to a greater or lesser degree over the entire visible spectrum. The amount of broadening varies with the nature of the substance.

There is a distinctly different effect from the *chemical change* that takes place when certain materials dissolve in a liquid. Some molecules break into two fragments of opposite charge when dissolved in water, thereby changing the electron orbits. For example, copper chloride ($CuCl_2$) appears yellow in powder form. But when dissolved in water, the copper ends up deficient of two electrons that were transferred to the two chlorine atoms to complete their outer shells, and the color of the solution is green.

5-6 COLOR IN SOLIDS

Atoms comprising a solid are also bound together by electrical forces. In common materials the binding is strongest if the atoms are arranged in a particular geometry appropriate for the elements involved. In some cases one or more electrons may be shared between an atom and its neighbors ("covalently" bound together). Good examples of this are diamond and the semiconductors germanium and silicon. In other solids—such as common table salt—an electron may be completely transferred to a neighbor (ionic binding). But the binding in most nonmetallic solids lies somewhere between ionic and covalent. Metals are quite different: the electrons of the outermost shell are free to move throughout the crystal. This property endows metals with their electrically conducting quality and high reflectivity. These electrons also help to bind the atoms together.

Gems Colored by Impurities

It is remarkable that the color of many insulating solids comes from the effect of small concentrations of impurities lodged within the host structure. This is true for many gems. In most cases these impurity atoms are examples of the so-called *transition elements*, such as iron, cobalt, chromium, and copper. These atoms are unusual because one of the inner shells of electrons is only partially filled. It is possible to excite one of these unpaired electrons to a higher state with a photon of the visible spectrum. Still the chemistry of these transition elements is largely determined by electrons of the outermost shell, leaving the inner electrons to provide their distinct colors through more subtle effects of their environment.

Both ruby and emerald derive their dramatically different colors from trace amounts of the same element—chromium. The basic material of ruby is an oxide of aluminum known as alumina, with the formula Al_2O_3. When pure it is the colorless gem sapphire, but if a few percent of the aluminum atoms are replaced by chromium atoms the characteristic ruby-red color appears (Color Plate 22). The interactions between the electrons of chromium and its host medium perturb the allowed atomic states: their energies are shifted upwards or downwards in varying amounts. Therefore the allowed energies of the excited states form bands on the energy scale instead of sharp rungs of a ladder.

The placement of the center of each band and its breadth depend on the strength of the electrical forces exerted by the neighboring atoms. In ruby each chromium atom is surrounded by six oxygen atoms. Three excited states of chromium are responsible for visible effects, and they are depicted in Fig. 5-21. Not all of these excited states are accessible from the ground state. "Selection rules" from quantum theory discourage a direct transition from the ground state (labeled "$4A_2$") to the first excited state (2E). But both of the next higher states ($4T_2$ and $4T_1$) can readily be entered. These absorption processes correspond to absorption bands falling in the green-yellow and the violet portions of the spectrum. This allows the red portion of the spectrum to be transmitted as well as some blue, endowing ruby with its deep red color containing a slight contribution of blue.

The red color may be made all the stronger by a second effect of the selection rules. They preferentially encourage transitions from the two excited 4T states to the 2E state instead of directly down to the ground state. The corresponding photons for these rela-

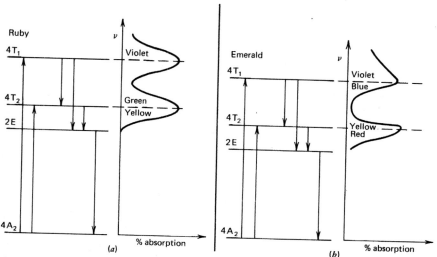

Fig. 5-21. Excited state energies of the chromium impurity in (*a*) ruby and (*b*) emerald, represented by sharply defined values. To the right of each diagram are shown the actual bands of the absorption spectrum that give each gem its characteristic color.

tively small differences in energy are in the infrared and therefore are not seen. But the subsequent transition from the 2E state down to the ground state corresponds to a photon lying squarely in the red portion of the spectrum. This emission of red photons is seen in good quality gems as an increase in the saturation of the color. However the effect is quenched in many natural rubies by the presence of iron impurities. (This emission line is also utilized in the ruby laser—see Section 7-6.)

Emerald also derives its color from chromium impurities. The host material in this case is beryllium aluminum silicate ($Be_3Al_2Si_6O_{18}$). As in ruby each chromium is surrounded by six oxygen atoms, but the influence of its environment is somewhat weaker because of a slightly different arrangement of the atoms. This slightly reduces the energy differences between the excited states of chromium, as depicted in Fig. 5-21b. The violet absorption band is correspondingly shifted down into the blue, and the green-yellow band into the red. This leaves a high transmission in the green, which accounts for the color of emerald (Color Plate 23).

Other gems similarly acquire their colors from impurities. Jade and aquamarine depend on small concentrations of the transition element iron for their colors. In many crystals and compounds the color is due to transition elements occurring in substantial concentrations. This is true for most pigments in paints. The red garnets are colored by iron; and turquoise and malachite owe their colors to copper.

Blue sapphire provides an example of a slightly different color process. Blue sapphire is composed of the same host material as ruby, but both iron and titanium are present in modest quantities. In this compound, iron transfers two electrons to its neighbors, so the ion has a net positive charge. Titanium in sapphire also has a net positive charge because it gives up four electrons to its neighbors. An excited state can be formed when an electron is liberated from the iron and travels to the titanium to reside, leaving both with net charges of three. The excitation of such a *charge transfer* can be accomplished by a wide range of photon energies, resulting in a broad absorption band extending from the red into the green. This gives blue sapphire its deep blue color.

It may seem surprising, but *defects* in a crystal also can produce color. A missing atom or other structural defect in the arrangement of atoms may create new electron states centered at the defect. The corresponding ladder of allowed energies determines the absorption spectrum. Another type of defect can be produced by natural radiation from cosmic rays or radioactive materials in the ground. An electron may absorb sufficient energy to become liberated from its atom, leaving the atom with a net charge and an unpaired remaining electron. The absence of an electron at such an atom is called a "hole." Defects of this kind are called *color centers*. Quartz may become colored in different ways by hole color centers. The crystal is composed of basic units of silicon dioxide (SiO_2). Amethyst is a form of quartz containing iron as an impurity. Exposure to X rays may eject an electron from an oxygen atom that subsequently becomes trapped at a neighboring iron atom. The resulting hole color center at the oxygen ion has a set of excited states producing absorption throughout most of the visible spectrum except for the highest frequencies. The color of amethyst is correspondingly violet. Ultramarine blue, obtained from the semiprecious gemstone lapus lazuli, also derives its color from a color center.

Smoky quartz as well derives its color from a hole color center, but here aluminum impurities trap the liberated electron. Some familiar objects have similar sources of color. Old glass bottles containing iron and perhaps manganese turn purple when exposed over many years to the ultraviolet rays of sunlight. Some color centers may disappear when heated because the defect is repaired, although others remain quite stable.

High Reflectance of Metals

Metals have distinctly different properties from the electrically insulating materials we have just discussed. The nature of their atomic electron orbits is such that electrons from their outermost shells are free to travel throughout the material. Because of the extreme spread of the electrons' states, the energy differences between successive excited levels is very small. The energies are essentially continuously spread over a broad band. The entire population of electrons responds very strongly to light, and the resulting very high refractive index prevents light waves from entering the metal. This is what makes a metallic surface reflecting.

There is a band of higher excited states in a metal that may be excited by photons of sufficiently high frequency. The threshold for such absorption varies from metal to metal, falling in the green for copper and in the ultraviolet for silver. This is what makes the color of silver differ from that of gold or copper. The spectral reflectance curves for representative metals are shown in Fig. 5-22. Both copper and gold have lower reflectance in the blue end of the visible spectrum than silver, with copper appearing less saturated than gold because of its somewhat higher reflectance in the green.

Coloration of Semiconductors

In addition to their importance in the electronics industry, semiconductors in powdered form are very popular pigments. Semiconductors differ from metals in that the broad band of excited states is separated from the ground state by a gap of forbidden energies. Because of the gap there is a minimum frequency for a photon to excite an electron. Cadmium sulfide (CdS) has a gap corresponding to a wavelength of 478 nm; therefore

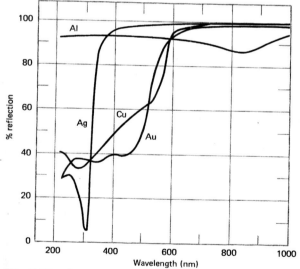

Fig. 5-22. Reflectance curves for clean surfaces of aluminum (Al), silver (Ag), copper (Cu), and Gold (Au). (From Wyszecki and Stiles, *Color Science*, Wiley, New York, 1967, after Haas, 1955.)

only blue and violet light can be absorbed and it appears yellow ("cadmium yellow"). By contrast cadmium selenide (CdSe) has a gap corresponding to 775 nm, and the crystal appears black. If some cadmium selenide is added to cadmium sulfide when formed at high temperatures where both are liquid, the cooled solid has a gap that is shifted from the blue toward longer wavelengths, depending upon the relative proportions. Attractive colors ranging from orange through red are the result. These compounds when powdered are sold as pigments under the name "cadmium red." Cinnabar, or mercury sulfide (HgS), is another semiconductor and it has a gap of 590 nm, thus allowing transmission of longer wavelengths. This red material makes up the pigment vermilion.

The famous semiconductors silicon and germanium have much narrower energy gaps that correspond to infrared wavelengths. Therefore they respond to visible light and are opaque. Diamond, formed with the same crystalline structure of atoms as silicon and germanium but made from carbon instead, has a gap of 229 nm and so is transparent through the visible spectrum and into the ultraviolet.

■ 5-7 QUANTUM EFFECTS IN EMISSION

When an atom acquires energy on absorbing a photon, it can subsequently lose that excess energy in many ways. Often it is transferred to a neighboring atom when the two atoms bump against each other, and the energy can then be converted into the energy of motion of the neighboring atoms. This represents a way of heating materials. It is also possible for the excited atom to give up its excess energy by emitting one or more photons. Several of these processes are explored in this section.

Fluorescence

Certain colors in modern posters appear so bright as to seem self-luminous. We may conclude that the abundant light that comes from the color cannot be explained by normal reflection. The pigment is probably an example of a *fluorescent* material, and it may very well emit more light at a particular wavelength than it receives at that wavelength. It does not emit more energy than the *total* amount it receives, because that is physically impossible. Fluorescence occurs when a substance absorbs electromagnetic energy at one wavelength and emits it at

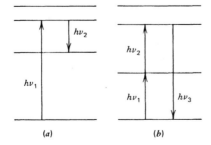

Fig. 5-23. (*a*) Typical sequence for fluorescence in which a photon of energy $h\nu_1$ is absorbed and a photon of lower energy $h\nu_2$ is emitted. (*b*) A less common possibility involves the absorption of two photons in succession, of energy $h\nu_1$ and $h\nu_2$, and then the subsequent emission of a photon with greater energy $h\nu_3$.

TABLE 5-2 Characteristics of Some Fluorescent Materials[a]

Phosphor	Excitation Range (nm)	Sensitivity Peak (nm)	Emitted Range (nm)	Emitted Peak (nm)	Color
Barium silicate	180–280	220	310–400	346	Black
Cadmium borate	220–360	250	520–750	615	Pink
Calcium halophosphate	180–320	250	350–750	580	White
Calcium silicate	220–300	254	500–720	610	Orange
Calcium tungstate	220–300	272	310–700	440	Blue
Strontium halophosphate	180–300	230	400–700	500	Cyan
Zinc silicate	220–296	254	460–640	525	Green

[a]All the materials absorb strongly in the ultraviolet at 253.7 nm, which coincides with a strong emission line for a mercury arc lamp, a common type of UV source.

another. Usually one high energy photon is absorbed, and the excited atom very quickly emits a photon of lower energy as one of its electrons drops to a lower energy orbit (Fig. 5-23*a*). For example, yellow fluorescence is emitted from a solution of rhodamine dye when it is illuminated by green laser light, as shown in Color Plate 24. In exceptional cases two or more lower energy photons are absorbed in succession, and the electron is excited to a sufficiently high energy level that it can subsequently emit a photon with a higher energy than provided by any one of the absorbed photons (Fig. 5-23*b*).

An intriguing display occurs when certain fluorescent materials are illuminated solely by an ultraviolet lamp (a so-called "black" light) because then no illumination is apparent. The material appears to emit light spontaneously! Various minerals produce a wide range of fluorescent colors under ultraviolet illumination, because of the differing energies of the excited states of their constituent atoms. An example is shown in Color Plate 25, and Table 5-2 gives some properties of several fluorescent materials.

If you should illuminate a room with an ultraviolet lamp after all other lamps have been turned off, you would see several common objects giving off a bluish-white light. They include white stationery and white shirts. Paper does this because it contains a fluorescent dye to enhance the amount of light that it gives off in the blue end of the spectrum to compensate for the natural color of paper, which tends to be an unsaturated yellow. Thus the color becomes white under incandescent room lighting and is supposedly more attractive for the user. White shirts may initially contain some fluorescent dye for the same reason, but in addition many modern detergents also contain a fluorescent dye to make clothing "whiter" after washing. This, of course, has no bearing on how clean are the emerging clothes. Eyeglasses and false teeth may also be fluorescent under ultraviolet because of minerals they contain.

Fluorescent colors that may be produced by light are commonly used for safety where a person or object must be seen. Orange is a favorite color for its excellent visibility against most backgrounds. Although pigments are more durable than dyes, many fluorescents fade within a few days when exposed to direct sunlight. Many organic compounds found in oil paintings of previous centuries fluoresce under ultraviolet illumination. Some waxes fluoresce white, while shellac is orange. Modern synthetic varnishes do not fluoresce and in this way can be distinguished from older materials.

Phosphorescence

If selection rules discourage an excited electron from dropping into a particular state of lower energy, it may still do so, although the process takes longer. Ordinarily an electron remains in its excited state for about 10^{-9} seconds. But an unfavorable selection rule may lengthen this time considerably. If there is a much longer delay and the electron emits a photon in the visible, having been originally excited by a photon of differing wavelength, the process is called *phosphorescence*. The distinction between phosphorescence and fluorescence is a matter of degree. Somewhat arbitrarily the division is set so that a time interval exceeding 1 microsecond is said to characterize a phosphorescent material. Some of these atoms store their excess energy for many minutes or even hours, and then glow in the dark.

The name "phosphorescence" was originally applied to light given off by the reactive element phosphorous and chemically similar substances when left exposed to air. They spontaneously combine with oxygen in a slow reaction and in the process emit light. The modern technical meaning of the word, however, is what we have given earlier.

Television relies on phosphors (phosphorescent materials) coated on the face of the screen of the picture tube to provide the light that we see. They are excited by absorption of the energy provided by a beam of electrons that travels inside the tube in the receiver from the rear toward the front and strikes the phosphor. In a black-and-white receiver, the phosphor emits a bluish-white light (which under typical conditions we usually perceive as white). However, in a color receiver, three separate phosphors emitting the additive primaries red, green, and blue are arranged in a pattern over the screen (Color Plate 4). Since the light from the red and the blue phosphors is confined to a very narrow spectral band, the colors are strongly saturated. Green appears less saturated (Chapter 13).

Before the health hazard of radioactivity was appreciated, the numerals of watch dials were painted with a pigment containing radioactive radium and a phosphor. The steady disintegration of radium nuclei provides X rays, which were absorbed by the phosphor and served as a continuous source of energy. This enabled the numerals to emit light day and night.

■ 5-8 NONRADIATIVE EXCITATIONS

Atoms can be excited in many ways other than by absorbing a photon. We mentioned that the element phosphorous spontaneously combines with oxygen when exposed to air. There is a transfer of energy to the phosphorous electrons during this chemical reaction which excites them to sufficiently high energy states that they can subsequently emit light when dropping into a lower state. This is an example of what is termed *chemiluminescence*, the emission of light as a result of a chemical reaction.

A related effect is *bioluminescence*, when light is produced by chemical reactions associated with biological activity. Bioluminescence occurs in a variety of life forms and is more common in marine organisms than in terrestrial or freshwater life. Examples include certain bacteria, jellyfish, clams, fungi, worms, ants, and fireflies. There is considerable diversity in how light is produced. Most processes involve the reaction of a protein with oxygen, catalyzed by an enzyme. The protein varies from one organism to another, but all are grouped under the generic name *luciferin*. The enzymes are known as the *luciferase*. Both words stem from the Latin *lucifer* meaning light-bearing. The various chemical steps leading to bioluminescence are yet to be explained in detail, but in some higher organisms the process is known to be activated by the nervous system.

The firefly is best understood. Its light organ is located near the end of the abdomen. Within it luciferin is combined with other atomic groups in a series of processes in which oxygen is converted into carbon dioxide. The sequence culminates when the luciferin is split off from the rest, leaving it in an excited state. The excess energy is released as a photon. The peak in the emission spectrum lies between 550 and 600 nm depending on the type of luciferase. This flash produced by the simultaneous emission of many photons serves to attract mates, and females also use it to attract males of other species, which they then devour.

Certain bacteria also produce light when stimulated by motion. This is why the breaking sea or a passing boat generate the greenish light seen in some bodies of water such as Phosphorescent Bay in Puerto Rico. Some fish have a symbiotic relationship with bacteria. The "flashlight fish" takes advantage of the light created by bacteria lodged beneath each eye. Certain other fish produce their own bioluminescence, which serves as identification. However the biological advantage—if any—of bioluminescence in some other organisms such as fungi remains a mystery.

Triboluminescence is the emission of light when one hard object is sharply struck against another. This contact, when atom scrapes against atom, excites electrons and disrupts electrical bonds. Light is then created when the electrons find their way to lower states. Triboluminescence is not to be confused with the glow of small particles that may be broken off by the impact. Such "sparks" are seen as a result of their high temperature. Light given off by hot objects is called *thermoluminescence*, or *incandescence*, which will be considered in detail in Chapter 7.

Another form of thermoluminescence is a basis for dating ancient ceramic objects. Quartz and other constituents of clay are continually irradiated by naturally occurring radioactive elements (e.g., uranium and thorium) and by cosmic rays. This produces defects in the material where electrons may be trapped. Heating pottery to 500° C releases the trapped electrons, which can then migrate back to their original atoms, where on returning to an atomic orbit they then emit a photon. The intensity of thermoluminescence is therefore a measure of the duration of irradiation since the time when the pottery had been previously fired.

Fig. 5-24. The Aurora over Alaska. (Photograph by Gus Lamprecht.)

Excitation is also possible by other means. The passage of an electrical current *(electroluminescence)* is one. The impact of high energy particles is another. The *aurora borealis* (Fig. 5-24) and its southern counterpart the *aurora australis* arise when a stream of high energy particles from the sun enters the earth's upper atmosphere and literally shatters some of the molecules of the air. This leaves their atoms in excited states, and the light subsequently given off is characteristic of the atoms. Although the oxygen molecule, a major constitutent of our atmosphere, has no emission in the visible, the oxygen atom can emit photons in either the red or green portions of the spectrum. Other atoms contribute to light at other wavelengths.

SUMMARY

The influence of material objects on the propagation of light is readily apparent at a boundary between different media where it leads to the effects of reflection and refraction. Two fundamental laws describe the changes in direction of travel when light rays reach such a boundary: Hero's law of reflection and Snell's law of refraction. Snell's law is particularly instructive because the refraction angle depends on the relative speed with which light travels in the two media and is therefore determined by the refractive indices of the media. Studies of refraction consequently give information on how strongly a particular medium interacts with light. In the classical view of physics the refractive index depends on frequency because each material has characteristic resonant frequencies at which its electrons respond most strongly to the driving forces of a light wave. Resonance also accounts roughly for the phenomena of selective absorption and dispersion. But the classical wave theory of light fails to explain some of the subtle aspects of the interaction of light with matter. These aspects can only be understood with the quantum theory. In this theory light has a dual nature depending on the phenomenon under study, in some cases exhibiting the wave properties of the classical theory and in others exhibiting particle properties. The photoelectric effect is an example that only quantum theory can explain: individual particles of light—photons—are absorbed one at a time, each causing a single electron to be emitted. A similar particle–wave duality exists for the electrons in atoms, molecules, and crystals and accounts for the occurrence of a precisely fixed set of allowed energies for the excited states of any material. The interaction of light with matter having these quantized energies accounts completely for selective absorption, luminescence, fluorescence, and other related effects. The arrangement of atoms in molecules and solids and the electron bonds that hold them together determine the allowed excited states that account for the characteristic color of dyes, pigments, gems, and phosphors. ■

QUESTIONS

5-1. When monochromatic light passes from air into another material, which of the following does not change, regardless of the material: its speed, frequency, wavelength?

5-2. When monochromatic light is specularly reflected by a mirror, does the reflection angle depend on the nature of the material? What determines the reflection angle?

5-3. If light passes into a medium where it travels more slowly, in what direction is it refracted?

5-4. What does the term *dispersion* signify?

5-5. How is the selective absorption of light by a blue pigment explained by the classical theory based on the notion of resonances?

5-6. Explain how the photoelectric effect gives evidence that the photon has a fixed amount of energy determined by its frequency.

5-7. If a photon has a wavelength in vacuum that is half as long as the wavelength of another photon, how are their energies related?

5-8. A microwave oven relies on electromagnetic waves whose wavelengths are about 10 cm. What is the energy of a single photon in the oven?

5-9. What are meant by the terms *quantum state*, *excited state*, and *ground state* for electrons in an atom?

5-10. Is it possible to have fluorescence following absorption of a photon having a frequency of 8×10^{14} Hz in which two photons are subsequently emitted, one with 6×10^{14} Hz and the other with 3×10^{14} Hz?

5-11. What is the difference between fluorescence and phosphorescence?

5-12. Why are the absorption peaks in the spectrum of a liquid generally not as sharp as those of a gas?

5-13. Give three ways in which the structure of an organic molecule leads to absorption in the visible portion of the spectrum.

5-14. Give an example of an electron-donating side group on an organic molecule and also an electron-accepting group.

5-15. Why is napthalene colorless while Martius yellow absorbs in the blue?

5-16. Draw the molecular structure near the iron molecule in hemoglobin that gives blood its color.

5-17. What is the process by which skin tans when exposed to sunlight?

5-18. What is the difference between the optical properties of a metallic conductor and a semiconductor?

5-19. Why are metals highly reflecting?

5-20. What materials form the pigments cadmium yellow and cadmium red? Why do these pigments have different colors?

5-21. What is responsible for the color of ruby and emerald? Why are they differently colored?

FURTHER READING

R. L. M. Allen, *Color Chemistry* (Appleton-Century-Crofts, New York, 1971).

T. B. Brill, *Light. Its Interaction with Art and Antiquities* (Plenum Press, New York, 1980). Detailed presentation of the coloring effects of pigments and the effects of molecular structure on the absorptive properties of organic colorants.

R. K. Clayton, *Light and Living Matter,* Vol. 1: *The Physical Part* and Vol. 2: *The Biological Part* (McGraw-Hill, New York, 1971). A concise primer on the electronic properties and chemistry of molecules participating in photosynthesis, photochemistry of vision, photoregulation of daily and seasonal cycles in organisms, photobiology, and bioluminescence. For the reader with a year of college chemistry (paperback).

D. L. Fox, *Animals, Biochromes, and Structured Colors* (Cambridge University Press, 1967).

H. M. Fox and G. Vevers, *The Nature of Animal Colors* (Sidgwick, London, 1960).

P. J. Herring, Ed., *Bioluminescence in Action* (Academic Press New York, 1978). Update of Harvey's classic work on this topic.

R. H. Eather, *Majestic Lights: The Aurora in Science, History, and the Arts* (American Geophysical Union, Washington, D.C., 1980).

P. Carruthers, ''Resource Letter QSL-1 on Quantum and Statistical Aspects of Light,'' *American Journal of Physics, 31,* 1 (1963). An extensive guide to books, review articles, original research reports, and films on the nature of photons, the interaction of photons and matter, and the coherence properties of light waves.

From the *Scientific American*

V. F. Weisskopf, "How Light Interacts with Matter" (September, 1968), p. 60.
S. B. Hendricks, "How Light Interacts with Living Matter" (September, 1968), p. 174.
A. Javan, "The Optical Properties of Materials" (September, 1967), p. 238.
K. Nassau, "The Causes of Color" (October, 1980), p. 124.
H. M. Fox, "Blood Pigments" (March 1950), p. 20.
S. Clevenger, "Flower Pigments" (June, 1964), p. 84.
K. V. Thimann, "Autumn Colors" (October, 1950), p. 40.
L. J. Milne and M. J. Milne, "How Animals Change Color" (March, 1952), p. 64.
R. J. Wurtman, "The Effects of Light on the Human Body" (July, 1975), p. 68.
W. D. McElroy and H. H. Seliger, "Biological Luminescence" (December, 1962), p. 76.
J. E. McCosker, "Flashlight Fishes" (March, 1976), p. 106.
H. F. Ivey, "Electroluminescence" (August, 1957), p. 40.

CHAPTER 6

MEASURING LIGHT

Light is a form of energy associated with the electric and magnetic fields that comprise electromagnetic waves. As described in Chapter 4, such waves have an enormous range of wavelengths. Waves between 400 and 700 nm differ from others only because of their ability to produce a response in the photoreceptors of the human eye. When there is such a physiological response, we say that the stimulus is "light." Electromagnetic waves with wavelengths shorter than 400 nm produce no such response and are classified as ultraviolet radiation (UV); waves longer than 700 nm are also ineffective in producing "light" and comprise infrared radiation. It then should be no surprise that radiation in the visible portion of the spectrum produces a physiological response whose strength depends on wavelength. Moreover, the range of wavelengths to which the eye is most sensitive corresponds to the maximum in the spectral power distribution (SPD) of our most common source of light—the sun!

In daylight vision, the eye is more sensitive to light in the yellow-green portion of the spectrum than in the blue or red. The "amount of light" coming from a source evidently depends on the shape of its SPD and the spectral sensitivity of the eye. The methods used to measure radiant power constitute a branch of physics known as *radiometry*. The science dealing with the measurement of light as opposed to radiant power is called *photometry*. Because we have as yet no means of electrically tapping the communication channels coming directly from photoreceptors of the retina, it is not possible to use the eye itself as an instrument for measuring amounts of light. Instead we must rely on the judgment of observers and the relationship of their perceptions to physical measurements of the corresponding SPD. Photometry is ultimately based on these judgments. Since the psychology of perception is involved, it was natural to give the name *psychophysics* to this branch of science dealing with the measurement of perception. Historically, the first serious psychophysical studies of vision were developed in the nineteenth century, and the names of Fraunhofer, Maxwell, Helmholtz, and Fechner are associated with this effort.

■ 6-1 RELATIVE SENSITIVITY CURVE

The central problem of photometry is how to relate a particular physical stimulus to the perceptual response that it produces. We already met this problem in Chapter 3, where we found that the perceptual responses of hue and saturation could be related to the physical stimulus (the SPD) through the C.I.E. chromaticity diagram, which is based on the results of psychophysical color matching experiments. In photometry, we seek to relate the amount of *power* in a beam of light (measured in watts) with its ability to evoke a perception of brightness, which can be loosely called the "amount of light," without reference to chromaticity.

The problem actually has two parts. First, we need to establish a procedure by which the relation between the amount of power and the amount of light can be determined. Second, we need a unit that can be used to quantify the amount of light. The procedure for relating the power in a beam to the amount of light evidently must take account of the SPD of the light. The method is closely related to color matching experiments that formed the basis for the chromaticity diagram described in Chapter 3, as will be explained in detail in the following section. The relative sensitivity of the eye for an average person must be established for monochromatic light at each wavelength across the spectrum. To do this a person with normal vision compares the brightness of monochromatic light having the wavelength of 555 nm, where the eye is most sensitive, with the brightness of a second monochromatic light at another wavelength of interest. To achieve a match in brightness we must reduce the power in the 555 nm beam to a certain fraction of the power in the second beam. The value of this fraction is taken as the measure of the observer's sensitivity to light at that wavelength compared with light at 555 nm. When this is done for a large number of wavelengths across the spectrum, a smooth curve can be drawn through the data to represent that observer's *relative sensitivity* at any wavelength.

After researchers obtained a consistent set of measurements of the sensitivity curves for a number of observers, an international committee known as the Commission Internationale de L'Eclairage (International Commission on Illumination) in 1924 adopted a representative curve. It is officially known as the *relative sensitivity curve for the C.I.E. Standard Observer*. Each wavelength λ has a corresponding value for the Standard Observer's sensitivity, and this is represented by the symbol V_λ. The value of V_λ is unity at the wavelength of 555 nm where the eye is most sensitive in daylight, and the values of V_λ decrease for shorter and longer wavelengths, becoming zero at the ends of the visible portion of the spectrum. The relative sensitivity curve for the standard observer is shown in Fig. 6-1. Accurate numerical values are given in Table 6-1.

There is as yet no complete theory for the physical and chemical aspects of the eye's photoreceptors that accounts for the sensitivity curve V_λ. It can only be deter-

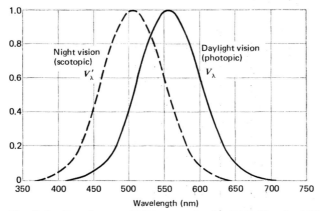

Fig. 6-1. Relative sensitivity curves for the C.I.E. Standard Observer for daylgiht vision (solid curve) and night vision (dashed curve).

TABLE 6-1 Values of the Relative Sensitivity Curve of the C.I.E. Standard Observer[a]

Wavelength λ(nm)	Photopic V_λ	Scotopic V'_λ
380	0.0000	0.0006
390	0001	0022
400	0004	0093
410	0012	0348
420	0040	0966
430	0116	1998
440	0230	3281
450	0380	4550
460	0600	5670
470	0910	6760
480	1390	7930
490	2080	9040
500	3230	9820
510	5030	9970
520	7100	9350
530	8620	8110
540	9540	6500
550	9950	4810
560	9950	3288
570	9520	2076
580	8700	1212
590	7570	0655
600	6310	0312
610	5030	0159
620	3810	0074
630	2650	0033
640	1750	0015
650	1070	0007
660	0610	0003
670	0320	0001
680	0170	0001
690	0082	0000
700	0041	0000
710	0021	0000
720	0010	0000
730	0005	0000
740	0003	0000
750	0001	0000
760	0001	0000
770	0000	0000
780	0000	0000

[a]To convert to spectral luminous efficacy in lumens per watt, multiply V_λ by 683 and V'_λ by 1700.

mined from psychophysical measurements. The resulting values of V_λ are only as accurate as the experimental data and can safely be applied only to situations that are comparable to those in the psychophysical measurements.

Figure 6-1 illustrates one reason for this word of caution. It was discovered that the sensitivity curve depends on whether the light is quite bright or rather dim. The eye itself adapts to the average brightness, and in the process its relative sensitivity changes. Thus we must distinguish between bright-adapted vision *(photopic vision)* and dark-adapted vision *(scotopic vision)*, and there are two corresponding curves V_λ and V'_λ. The peak for scotopic vision is shifted 43 nm toward shorter wavelengths, but there is little change in the shape of the curve. The physiological basis for the shift when adapting to dark conditions is the switch from cones to rods as the principal active receptors in the retina of the eye (to be discussed in Chapter 10). In addition, for sufficiently low levels of illumination when vision is exclusively by rods (scotopic vision), all perception of hue is lost and colors appear as grays of various brightness.

We emphasize that the curves in Fig. 6-1 express the relative sensitivity as determined by a brightness match; they do *not* indicate *how much* brighter one monochromatic light is then another having the same power. As an example of the use of these curves, with daylight levels of illumination the photopic curve shows that a monochromatic orange light at 625 nm has $V_\lambda = 0.5$; thus we conclude that we would need twice as much power from this light source than from a yellow-green light at 550 nm where $V_\lambda = 1.0$ for the two to appear equally bright. Similarly, to achieve equal brightness for a monochromatic light at 631 nm, where $V_\lambda = 0.25$, with the brightness of the 550 nm light, we would need four times more power. The relative sensitivity of our perception of light at other wavelengths similarly can be read from the curve.

Purkinje Effects

The shift of the sensitivity curve toward shorter wavelengths as the eye becomes adapted to darker surroundings produces unusual visual effects for an attentive observer. For example, red objects, which reflect light primarily at long wavelengths, become darker because it is in this spectral region where sensitivity is lost most dramatically. Blue objects become relatively brighter because sensitivity is enhanced for short wavelengths. This phenomenon was discovered by the Czechoslovakian physiologist Johannes Purkinje in the early nineteenth century. Another consequence of the Purkinje effect is an increase in brightness of green objects compared with the surroundings, especially evident at twilight.

Illumination engineers take advantage of our separate systems of rods and cones when designing lighting for such places as the control bridge of a ship. At night if a saturated red light illuminates the instruments, only the cones respond to the light. They adapt to the level of illumination and their sensitivity is correspondingly reduced. But the rods are not activated and so retain their full sensitivity when needed for vision outdoors.

■ 6-2 PSYCHOPHYSICAL MEASUREMENTS

Our human visual processes cannot accurately assign a numerical value to the brightness of a light that we perceive. Judgments vary too much from day to day for an acceptable determination of the sensitivity curve. Much better reproducibility is achieved when the eye is used as a "null" instrument, to indicate when two lights are identical. Then an observer need not form a judgment on "how much" brighter one light is than another. The procedure of matching brightness was first carried out by the nineteenth century German optician Joseph von Fraunhofer. From his early work several types of such matching experiments have been developed to determine the numerical values for the relative sensitivity curve V_λ. Two were found to give the most consistent results: the so-called *cascade method* and the *flicker method.*

In the *cascade method* the observer views a screen on which a sample of monochromatic light at one wavelength and a reference monochromatic light at another wavelength are projected side-by-side. The brightness of the reference is varied until the subject judges that it matches the sample. The observer's relative sensitivity to the sample is defined as the ratio of the power in the reference beam to the power in the sample when they match. This condition for matching brightness can be measured with satisfactory precision if the two wavelengths are nearly the same but not if they differ greatly. The precision can be maintained if wavelengths are compared in cascade, using the sample of one match as the reference for the next and proceeding in sequence from the center of the spectrum toward one end and then back toward the other. The procedure is subject to cumulative errors, but nevertheless most researchers believe it gives the most reliable results. This rather tedious procedure must be repeated for a number of observers under both photopic and scotopic conditions.

The *flicker method* is known as "flicker photometry" and has been used to compare monochromatic lights of quite different wavelengths. For this the observer views a screen on which the sample and reference lights are alternately presented, and what is seen depends on the differences in the wavelength and the frequency of alternation. As the frequency is increased the rather unpleasant alternation effect attributed to differing colors disappears, and yet a flicker is still perceived. By adjusting the brightness of the reference this flicker can also be made to disappear, and by definition the two lights are then said to have the same brightness. At very high frequency there is a perceived fusion of the two colors regardless of relative brightness. Consequently, there is only a rather delicate intermediate range of frequency in which the flicker method works. This method has an advantage over the cascade method in that observers with little training can participate, and they can rapidly compare colors of greatly differing wavelength. With sufficient experience, researchers learned how to control conditions so that the cascade and flicker determinations of V_λ produced similar curves, thus giving us confidence in the accuracy of the tabulated values.

■ 6-3 THE LUMEN

There is no absolute way that a number can be assigned to the amount of light from a source: a method has not yet been found to indicate *how much* brighter one source is than another, because we have no quantitative method of measuring the observer's perception of brightness. Nevertheless it has been found useful to define the "amount of light" from a source in such a way that we can determine whether it meets or exceeds a desired level. In this way we can predict whether a particular source is adequate for illuminating a desk, a room, or a parking lot. For such practical purposes the "amount of light" has been defined in such a way that for a given shape of the SPD it is proportional to the power that is in the beam. In other words, if the power at every wavelength is increased by a certain percentage, the "amount of light" increases by the same percentage. Therefore knowing the "amount of light" coming from each of two sources tells us which appears brighter, and it tells us by how much the power output of the other must be increased to make them equally bright.

When the "amount of light" coming from a source is given a numerical value, this number specifies how much light there is in terms of a unit amount called the *lumen* (abbreviated "lm"). Historically the lumen was first specified in reference to the light from a "standard candle": the lumen was the amount of light falling on a surface having an area of one square meter located at a distance of one meter from the candle. The original standard candle was made from the oil of the sperm whale by following a specified procedure in an attempt to ensure reproducibility from one candle to another. As technology made possible brighter and better controlled sources, the candle was replaced in succession by gas lamps, incandescent lamps, and most recently a hot oven held at the melting point of platinum (2042 K or 3216 degrees fahrenheit).

At a meeting in Paris of the General Conference on Weights and Measures in 1979 the definition of the lumen was dramatically changed. No longer is it defined by the light from a source such as a hot oven emitting radiation throughout the visible spectrum. Instead it was defined by reference to a monochromatic light. To paraphrase the conference's definition: "The lumen is the amount of light of monochromatic radiation whose frequency is 540×10^{12} hertz and whose power is $\frac{1}{683}$ watt." Stating the definition in terms of the frequency instead of the wavelength of the radiation makes it unnecessary to specify the refractive index of the medium through which light passes. For light in air this definition is equivalent to saying that "One watt of monochromatic radiation whose wavelength is 555 nm provides an amount of light of 683 lumens." The value of "683 lm/W" was chosen for this definition so that the lumen corresponds with an accuracy of a few percent to the lumen as previously defined in terms of the light from a standard candle. The numerical value is a purely arbitrary choice. It was selected for historical reasons, so that published tables for photometric quantities based on the former standard candle remain valid. The value 683 lm/W is commonly denoted by the symbol K_m.

Spectral Luminous Efficacy

Because the sensitivity curve V_λ decreases on going away from 555 nm toward either end of the spectrum, one watt of radiant power at any other wavelength will provide less light than at 555 nm. Multiplying the amount of power at the particular wavelength λ by the quantity $K_\lambda = K_m V_\lambda = 683 V_\lambda$ gives the number of lumens at that wavelength. The value of K_λ at each wavelength is called the *spectral luminous efficacy* of the Standard Observer. The largest value for the spectral luminous efficacy is simply $K_\lambda = K_m = 683$ lumens per watt for 555 nm. Radiation at other wavelengths has a lower spectral luminous efficacy as dictated by the curve V_λ. The corresponding values are shown by the curve in Fig. 6-2. This applies of course to photopic vision.

A different curve applies to scotopic vision because V'_λ for the dark-adapted eye

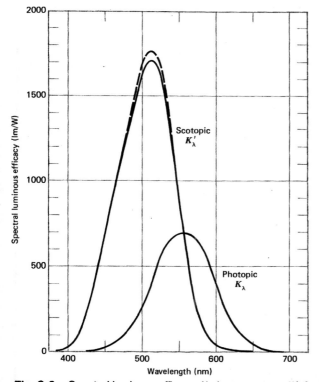

Fig. 6-2. Spectral luminous efficacy (in lumens per watt) for light-adapted (photopic) vision and dark-adapted (scotopic) vision. The dashed curve for scotopic vision is the former definition and the solid curve is the new one.

differs from V_λ. Formerly the maximum spectral luminous efficacy for scotopic vision was defined as $K'_m = 1754$ lumens per watt (dashed curve in Fig. 6-2). However now it is defined as $K'_m = 1700$ lm/W (solid curve in Fig. 6-2), because then K'_λ and K_λ have exactly the same value at 555 nm. Thus the same definition for the lumen applies to both photopic and scotopic vision. The slight adjustment of the scotopic curve is less than the inaccuracy in the psychophysical measurements that determined the original curve, so previously determined photometric values need not be corrected to be consistent with the new definition.

▪ 6-4 RADIOMETRIC AND PHOTOMETRIC DESCRIPTIONS

There are many applications in which the numerical prediction, specification, or measurement of light quantities are important. They include photography, the theater, advertising, and interior design, to name but a few examples. In most cases the radiation is not monochromatic but is spread over the spectrum. Then the total amount of light in a beam is calculated by adding together the contributions from the various wavelengths. This is known as Abney's rule. It is generally accepted in photometry, because it has been verified in many psychophysical studies where lights of differing SPDs but the same "total amount" are found to be equally bright. The procedure is to multiply the value of the SPD at each wavelength by the spectral luminous efficacy $K_\lambda = 683 V_\lambda$ for that wavelength and add together the results. In this way every radiometric (physical) quantity has a photometric (perceptual) counterpart. All photometric terms have "luminous" as an adjective in one form or other. To illustrate this, Table 6-2 summarizes the four most important radiometric quantities and their corresponding photometric quantities. Although we shall

TABLE 6-2 Summary of Radiometric and Photometric Quantities[a]

Radiometric Quantities		Photometric Quantities	
Name Unit (and Abbreviation)	Symbol	Name Unit (and Abbreviation)	Symbol
Radiant flux watt (W)	ϕ_e	Luminous flux lumen (lm)	ϕ_v
Radiant intensity watt/steradian (W/sr)	I_e	Luminous intensity lumen/steradian or candela (lm/sr or cd)	I_v
Radiance watt/steradian·meter² (W/sr·m²)	L_e	Luminance lumen/steradian·meter² or candela/meter² (lm/sr·m² or cd/m²)	L_v
Irradiance watt/meter² (W/m²)	E_e	Illuminance lumen/meter² (lm/m²)	E_v

[a]Symbols for the radiometric quantities have the subscript *e*, and photometric quantities have the subscript *v*.

explain each term in later sections, it is apparent that the form of the different terms distinguishes the physical quantity from its perceptual counterpart.

Radiant and Luminus Flux

First we introduce the concept of *radiant flux* (denoted by the symbol ϕ_e, where the subscript *e* reminds us that this quantity refers to energy). Since "flux" means "the rate of flow of a fluid or energy across or through a surface," *radiant flux* has been defined as the rate at which radiant energy is transferred from one place to another. It is a term in radiometry that signifies the same thing as radiant power; therefore, radiant flux is expressed as the number of watts involved. For example, a common 100 W incandescent lamp consumes electrical energy at a rate of 100 W. If 80% of this were converted into radiant power, the emitted radiant flux would be 80 W. *Luminous flux* (denoted by ϕ_v where the subscript *v* indicates that this quantity refers to a "visible" property) is the photometric equivalent of radiant flux. Luminous flux is therefore measured in lumens.

With these notions of radiant and luminous flux firmly in mind we can restate the definition of the lumen in its exact form:

> The *lumen* is the luminous flux of a monochromatic radiation whose frequency is 540×10^{12} hertz and whose radiant flux is $\frac{1}{683}$ watt.

Thus if the SPD of a source is known, its luminous flux can be calculated from this definition and the sensitivity curve of the Standard Observer. For each wavelength λ we compute the luminous flux by multiplying the radiant flux ϕ_e at that wavelength by the spectral luminous efficacy $K_\lambda = 683 V_\lambda$ appropriate for that wavelength. The result is the luminous flux ϕ_v at that wavelength:

$$\text{luminous flux} = (\text{spectral luminous efficacy}) \times (\text{radiant flux}) \qquad (6\text{-}1)$$
$$\phi_v = 683 V_\lambda \phi_e$$

To include contributions to the total luminous flux from wavelengths across the spectrum, the calculation in Eq. 6-1 should be done for a series of wavelengths at evenly spaced intervals, with ϕ_e representing the radiant flux found in the interval. Then these individual contributions should be added together. Thus the total luminous flux can be found if the SPD is known. A calculation based on the SPD of a common 100 watt incandescent lamp shows that its luminous flux is about 1700 lumens.

Example 6-1 Luminous Flux from a Line Spectrum

The principal emission of light by a mercury discharge lamp is in the form of a line spectrum. It emits radiant flux at wavelengths of 360, 408, 436, 546, and 578 nm. Let us suppose that the SPD has been measured, and we now proceed to calculate the total luminous flux that it emits.

We first set up a table, dealing with one wavelength at a time, and calculate the contribution to the luminous flux at each wavelength. The measured radiant flux at each wavelength is entered in column (2) of the table shown below. Then we consult Table 6-1 for the values of the sensitivity curve of the eye and enter them in column (3). When the values in column (3) are each multiplied by 683, we obtain the spectral luminous efficacy in column (4) giving the luminous flux (in lumens) if 1 watt of radiant flux at that wavelength is emitted. Then, multiplying each entry in column (4) by the actual amount of radiant flux in column (2) gives the corresponding contribution of luminous flux, entered in column (5). Adding these contributions we find that the total luminous flux is about 93,000 lumens. Notice that a large fraction of the radiant flux from this mercury lamp is emitted in the ultraviolet portion of the spectrum where the eye's sensitivity is negligible.

(1) Wavelength (nm)	(2) Radiant Flux ϕ_e (watt)	(3) Sensitivity V_λ	(4) $683V_\lambda$ (lm/W)	(5) Luminous Flux ϕ_v (lumen)
360	78	0	0.0	0
408	30	0.001	0.7	20
436	60	0.019	13.0	773
546	71	0.978	668.0	47,400
578	74	0.886	605.0	44,800
				Total: 92,993 lumens

Luminous Efficacy

The ratio of the luminous flux in a beam of radiation to the radiant flux is called the *luminous efficacy* of the radiation. This is determined solely by the shape of the SPD and not by the overall amount of power. Thus the luminous flux in a beam is the radiant flux multiplied by the luminous efficacy for the particular SPD. The luminous efficacy is denoted by the symbol K:

$$\text{luminous flux} = (\text{luminous efficacy}) \times (\text{radiant flux}) \qquad (6\text{-}2)$$
$$\phi_v = K\phi_e$$

Luminous efficacy specifies the total luminous flux for each watt of total radiant flux from a source. It is a measure of the overall effect. By comparison, the spectral luminous efficacy K_λ introduced earlier can be considered the luminous efficacy for monochromatic light at the wavelength λ. Both the spectral luminous efficacy and the luminous efficacy are expressed in the unit of lumens per watt.

For the best possible luminous efficacy in radiant power we should use monochromatic light at 555 nm, because V_λ is greatest for this wavelength. This light has a luminous efficacy of $K = 683$ lm/W. By comparison, white light with a flat SPD, indicating equal amounts of power in each 10 nm wavelength interval in the range 400–700 nm has $K = 187$ lm/W. And our example of a 100 W incandescent lamp, mentioned above, has an SPD with so much radiant flux in the infrared portion of the spectrum that its luminous efficacy is only 17 lm/W. Examples for other common light sources may be found in Table 7-2.

Example 6-2 Calculating Luminous Efficacy

What is the luminous efficacy for the mercury arc lamp described in Example 6-1? The total radiant flux ϕ_e is obtained by adding the contributions from column (2). The corresponding total luminous flux ϕ_v is the sum of the amounts in column (5). Therefore the luminous efficacy according to Eq. 6-2 is $K = \phi_v/\phi_e = 93,000/313 = 297$ lm/W. This is remarkably high, because so much power is concentrated in the center of the visible spectrum, where the eye is sensitive. This is one reason why mercury lamps are used for illuminating large outdoor spaces such as parking lots.

■ 6-5 PHOTOMETRIC QUANTITIES DIFFER FROM BRIGHTNESS

If we double the luminous flux given off by a lamp does its brightness double? The answer is no! The psychophysical methods involving the cascade or flicker techniques merely indicate how to make different lamps *equally* bright. A lamp that provides more lumens than another will appear brighter, but a photometric description does not tell us how much brighter.

Other psychophysical experiments have been devised to provide quantitative judgments concerning the relative brightness of different lamps or illuminated surfaces. The results were mentioned in Chapter 3 in regard to Munsell's numerical scale for the possible Value of a color. The psychologist's notion of brightness is found to increase approximately as the *logarithm* of the luminous flux entering the eye. The vertical scale for brightness (Fig. 6-3) is arbitrarily chosen to range from zero representing black to 10 representing dazzling white. The shape of this logarithmic curve has the interesting feature that each doubling of the luminous flux causes the perceived brightness to increase arithmetically by a fixed amount (1.5 units according to the graph) rather than to double! Thus at low levels of luminous flux the curve shows that the brightness of a surface increases rapidly as the flux increases; but brightness is much less sensitive to further increase in flux at high levels.

The advantage in using photometric quantities based on the lumen as the unit of measure of luminous flux is that it accounts for the contributions from various portions of the SPD to the total amount of light. Two equal surfaces of differing spectral characteristics but emitting the same number of lumens will appear equally bright. But if one surface emits more lumens than the other, their relative brightnesses can be gauged in qualitative terms by reference to a curve such as that of Fig. 6-3. ■

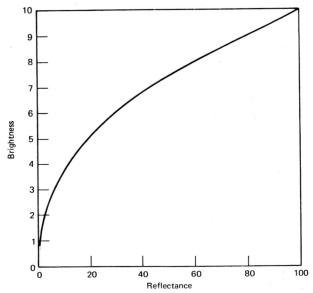

Fig. 6-3. Relationship between the perceived brightness of light diffusely reflected from a surface and its reflectance, when illuminated by an extremely bright source.

6-6 CHARACTERIZING LIGHT SOURCES

Looking around we see many different sources of light, and they are described in various ways. Some sources generate their own light and are said to be "self-luminous" such as a candle or the sun. Others such as the moon merely reflect light that was produced somewhere else. In either case photometric concepts can be applied to describe in precise terms the nature of the emitted light.

Total Radiant and Luminous Flux

The *total radiant flux* from a source is the total power radiated by it in all directions. The corresponding photometric quantity is called the *total luminous flux*. Take a candle as an example. Typically the total radiant flux amounts to 15 watts, spread throughout the electromagnetic spectrum. Most of the power appears in the infrared but a little is found in the visible portion of the spectrum. How to calculate its total luminous flux is shown in Example 6-3.

Example 6-3 Output of a Candle

Suppose that the SPD of a particular candle has been measured by using suitable instruments. Some variation in the radiant flux is found for candles prepared in different ways, so this example should be considered as illustrative and not representative of every candle. The data for the total radiant flux obtained within consecutive 20 nm regions of the visible portion of the spectrum are entered in column (2) of the table below. Each number gives the radiant flux within the 20 nm region centered at the wavelength indicated in column (1). The respective values of the spectral luminous efficacy corresponding to these central wavelengths may be obtained from Table 6-1 (or from Fig. 6-1 with less accuracy), and they are entered in column (3). The value of the luminous flux in each region is then calculated by multiplying the corresponding radiant flux by the spectral luminous efficacy. The result is entered in column (4). Then the total luminous flux is obtained by adding together all of the contributions in column (4). The value is approximately 10 lumens for this candle. As the values for V_λ are known to an accuracy of only 2%, there is no purpose in stating the result with greater precision.

(1)	(2)	(3)	(4)
		Spectral Luminous	
	Radiant Flux in	Efficacy	
Wavelength	20 nm interval	$K_\lambda = 683 V_\lambda$	Luminous Flux
(nm)	(mW)	(lm/W)	in 20 nm interval (lm)
400	4	0	0.00
420	8	2	0.00
440	16	15	0.00
460	27	40	0.01
480	41	94	0.04
500	73	220	0.16
520	112	484	0.53
540	171	651	1.06
560	232	679	1.56
580	318	592	1.89
600	425	431	1.83
620	554	260	1.43
640	707	119	0.84
660	884	41	0.36
680	1086	11	0.12
700	1285	2	0.03
		Total luminous flux =	9.91 lm

A candle has a relatively low luminous efficacy, because its flame has a comparatively low temperature. Consequently most of its radiant flux is found in the infrared portion of the spectrum. To illustrate this point, we note that the total radiant flux in the visible is given by the total of all the contributions listed in column (2), which is 59 mW. Since by comparison a measurement of the total radiant flux across the entire spectrum yields 15 W, the amount in the visible represents only 0.4% of the total. The luminous efficacy (Eq. 6-2) consequently has the small value of $K = 10$ lm/15 W $= 0.7$ lm/W. This is less than 5% of the luminous efficacy of a typical incandescent lamp, which operates at a much higher temperature.

Radiant and Luminous Intensity

Total radiant flux and total luminous flux are measures of total source output and are rather gross indications of the strength of a source. They do not indicate how the output varies with direction, and that detail is often important to know. For example, we may be

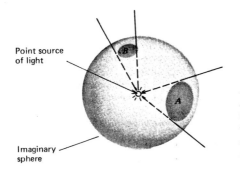

Point source of light

Imaginary sphere

Fig. 6-4. An angular range can be designated by the fraction of a sphere's area that it intercepts. Angular range *A* intercepts more of the area than *B* and is correspondingly larger.

more concerned with knowing how much light an automobile headlight directs forward than in knowing the total luminous flux. Thus we are concerned with how much power is concentrated within a narrow angular range in a particular direction. This is known as the *radiant intensity* (I_e) of the source in that direction. Its photometric counterpart, the *luminous intensity* (I_v), indicates how many lumens are concentrated within the same angular range.

How do we measure an "angular range"? A given angular range, or span of directions from a source, can be assigned a numerical value by comparison with the *total* angular range about a point. Imagine that the source is placed at the center of a large sphere, then a given angular range from the source can be indicated by the fraction of the sphere's area the range includes (Fig. 6-4). In solid geometry such an angular range is called a *solid angle* and by convention it is measured in the unit called the *steradian* (abbreviated "sr"). Also by convention the size of 1 steradian is chosen so that there is a total of exactly 4π steradians in total angular range about a point. Therefore an angular range that includes only half of the area of the imagined sphere is a solid angle of 2π steradians. And an angular range including one-quarter of the sphere is a solid angle of π steradians. Therefore the angular range encompassed by only 1 steradian is about $\frac{1}{13}$ of the total range about a point.

To determine the radiant intensity in a given direction from a source, we would measure the radiant flux emerging within a small angular range in the given direction and divide that number by the size of the angular range:

$$\text{intensity} = \frac{\text{flux emitted}}{\text{angular range}} \qquad (6\text{-}3)$$

Radiant intensity is thus indicated in the unit of watt per steradian (W/sr). Correspondingly the luminous intensity has the unit of lumen per steradian (lm/sr).

If this is done for an automobile headlight, we would find about 25 W of radiant flux concentrated within an angular range of only 0.01 steradian. Consequently the radiant intensity is $I_e = 25/0.01 = 2500$ W/sr. The intensity is very high, but only within a narrow range of directions straight ahead of the headlight. A reflector at the back of the lamp and a specially shaped glass lens in front serve to concentrate the flux. For an incandescent lamp such as this, the luminous efficacy K is typically about 20 lm/W, so the luminous intensity of the headlight is about $I_e = KI_e = 20 \times 2500 = 50,000$ lm/sr. Figure 6-5 shows how reflectors for other types of lamps provide various patterns of luminous intensity to direct light either fairly uniformly over an area or to concentrate it toward a specific spot.

An alternative unit is available for indicating the value of luminous intensity that avoids mention of the steradian altogether. It is called the *candela* (abbreviated "cd"). It is equal to 1 lumen per steradian. Thus the headlight in our example could alternatively be said to have a luminous intensity of 50,000 candela. Formerly the catchy term "candlepower" (abbreviated "cp") was used for the unit of measuring luminous intensity. Its value for practical purposes is essentially the same as one candela. But use of the term "candlepower" is now discouraged, because the technical word "power" should not be applied

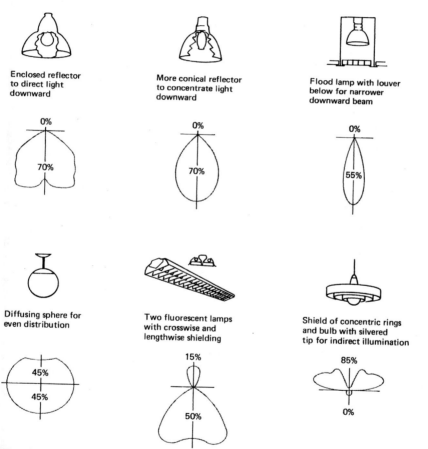

Enclosed reflector
to direct light
downward

More conical reflector
to concentrate light
downward

Flood lamp with louver
below for narrower
downward beam

Diffusing sphere for
even distribution

Two fluorescent lamps
with crosswise and
lengthwise shielding

Shield of concentric rings
and bulb with silvered
tip for indirect illumination

Fig. 6-5. Some common lighting fixtures and their intensity profiles, which give the luminous intensity in various directions. The percentages indicate the proportion of luminous flux directed upward and downward, the difference between the total and 100% representing absorption by the fixture.

TABLE 6-3 Luminous Intensity of Various Light Sources, in the Direction of Greatest Intensity[a]

Source	Luminous Intensity (Candela)
Light-emitting diode (LED)	0.005
Candle	1
100 watt incandescent lamp	150
Automobile headlamp (high intensity)	100,000
Lighthouse	300,000
Flashtube (peak value)	1,000,000

[a]The unit of one candela is identical to one lumen per steradian and is approximately equal to the formerly popular non-S.I. unit of one candlepower (to within 2%).

to a photometric quantity. Table 6-3 lists the luminous intensity of a few representative sources.

Example 6-4 Luminous Intensity of a "Point" Source

If a bare incandescent lamp hanging by a wire from the ceiling provides a total radiant flux of 200 W, what is its luminous intensity? We can easily obtain the answer within 10% accuracy if we neglect the small amount of light absorbed by the base of the lamp where the filament is mounted. We also assume that the emitted flux spreads evenly throughout the total angular range of $4\pi \approx 13$ sr surrounding the lamp. Then by Eq. 6-3 the radiant intensity is the total radiant flux divided by the angular range into which it goes: $I_e = (200 \text{ W})/(13 \text{ sr}) = 15$ W/sr. From the radiant intensity we can calculate the luminous intensity by using Eq. 6-2 and a typical value of $K = 20$ lm/W for the luminous efficacy of a high-power incandescent lamp. Then the luminous intensity is found to be $I_v = KI_e = 20 \times 15 = 300$ lm/sr. Equivalently we can say that the luminous intensity of this lamp is 300 cd.

The preceding example illustrates the fact that a simple relationship exists between the total luminous flux ϕ_v from a point source and its luminous intensity I_v provided that light is emitted uniformly in all directions:

$$I_v = \frac{\phi_v}{4\pi} \tag{6-4}$$

This equation expresses what we noted before: to find the intensity we divide the amount of emitted flux by the size of the angular range in which the flux is emitted. The same equation holds for the corresponding radiometric quantities if we replace the subscript v by e: the radiant intensity I_e is given similarly by the total radiant flux ϕ_e divided by 4π.

Radiance and Luminance

The concepts of radiant and luminous intensity discussed in the preceding paragraphs apply to *point sources*. In practice that means a source whose dimensions are less than one-tenth of the distance between the source and the subject being illuminated. The shape of the source and its size are then unimportant. But when we look at the surface of a desk or wall close-up, the source must be considered an *extended source*. To provide a useful characterization of the light coming from an extended source, it is necessary to take account of the amount of area that provides the light. Indeed, the actual area of the surface is less important than the *perceived* area. Figure 6-6 illustrates what is meant by the word "perceived": If the actual surface is tipped away from our line of sight, its perceived area is considerably smaller than the actual area. The perceived area suggests how concentrated the luminous flux is when it leaves the surface, and it is this perceived area that is taken into account when we judge the brightness of the surface.

For this reason the *luminance* of a surface is defined as the luminous flux spreading outward per steradian in a given direction *per square meter of perceived surface area:*

$$\text{luminance} = \frac{\text{lumens per steradian emitted}}{\text{perceived area of the source}} \tag{6-5}$$

The luminance of a surface has the unit of lumen per steradian per square meter (lm/sr·m^2), or candela per square meter (cd/m^2). The radiometric counterpart to luminance is *radiance*, which is the radiant flux per steradian per square meter of perceived surface area.

Notice that the luminance of a surface does not depend on how large the surface is: luminance describes a property of each small region of the surface. The luminance of an extended source such as a fluorescent lamp may very well vary from one place to another across the glass bulb. The luminance of a given region of the surface may also vary with direction, so that it is greater in some directions than others.

It is the luminance of a surface that determines its brightness. A subject having a greater luminance than another one appears brighter. Increasing the luminance of a reflecting surface by increasing the luminous flux falling on it (or in the case of a self-luminous surface by increasing the luminous flux it emits) makes it appear brighter. We recall, however, that brightness is a perception that we have no objective way of measuring; therefore it is not possible to say "by how much" the brightness increases when luminance increases by a given amount. There is only a rough, qualitative relationship between the two, as characterized in Fig. 6-3.

Luminance is a particularly useful photometric quantity when dealing with a diffusely

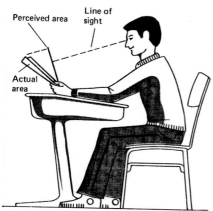

Fig. 6-6. Perceived area of an extended source, such as a sheet of paper, is less than the actual area if it is inclined to the line of sight.

Incident light

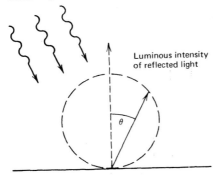

Luminous intensity
of reflected light

Fig. 6-7. The luminous intensity from a small region of a diffuse reflector varies as the cosine of the angle θ (Lambert's law).

reflecting surface such as fresh snow, because measurements show that such a surface has the same luminance in every direction. Therefore a diffusely reflecting surface appears equally bright regardless of the direction from which it is viewed. Surfaces that have some specular reflection are generally brighter in certain directions than in others.

The fact that a diffusely reflecting surface has the same luminance in every direction leads us to an interesting conclusion. As we look at a flat, diffusely reflecting surface at a greater angle from the perpendicular direction, a given apparent area of the surface corresponds to a greater actual area (Fig. 6-6). Therefore each region of the actual surface must provide a correspondingly smaller luminous intensity at greater angles so as to exactly compensate for the smaller perceived area which is proportional to the cosine of the angle θ. The luminous intensity from a region of surface that is so small as to be considered a point source thus must be proportional to the cosine of the angle from the

TABLE 6-4 Typical Values for the Luminance of Various Objects

Sources	Luminance (cd/m^2)
The star Sirius	15×10^9
Our sun	1.6×10^9
Carbon arc lamp	1×10^9
Xenon lamp	650×10^6
Mercury arc (1000 W)	300×10^6
Mercury vapor (high pressure)	30×10^6
Incandescent filament (500 W)	11×10^6
Sodium vapor lamp (high pressure)	6×10^6
Incandescent lamp (bulb frosted on inside—500 W)	300×10^3
Gas mantle	34×10^3
Snow in sunlight (maximum)	30×10^3
Fluorescent lamp	10×10^3
Candle flame	5×10^3
Moon (maximum)	5×10^3
Overcast sky (maximum)	5×10^3
Clear blue sky (maximum)	4×10^3

Source. Adapted from A. Stimson, *Photometry and Radiometry for Engineers,* Wiley, New York, 1974.

perpendicular (Fig. 6-7). This property of a diffusely reflecting surface was first noted by Johann Lambert in 1760 and is called *Lambert's cosine law*.

The fact that a surface obeying Lambert's law has the same luminance in every direction will be proved in Example 6-6. Furthermore it is interesting that an extended source appears to have the same brightness regardless of how far away it is, whereas a point source becomes dimmer with increasing distance. This fact will be shown in Example 6-7.

The luminance of familiar objects spans an impressive range: the sun has a luminance of 1.6×10^9 cd/m²; snow in sunlight, 3×10^4 cd/m²; a fluorescent lamp, 1×10^4 cd/m²; and snow in moonlight, 3×10^{-1} cd/m². Other examples are given in Table 6-4.

6-7 IRRADIANCE AND ILLUMINANCE

We often want to know how much light should be used to illuminate a room, a desk, or perhaps a hallway. More precisely, we need to specify how *concentrated* should be the light falling on the surface of interest.

Illuminance (E_v) is the luminous flux that is incident from all directions onto one square meter of the surface area:

$$E_v = \frac{\text{incident luminous flux}}{\text{surface area receiving it}}$$

(6-6)

Consequently the unit for expressing illuminance is the lumen per square meter (lm/m²). Sometimes this is called the "lux." The radiometric counterpart is called *irradiance* (E_e), which is the radiant flux falling onto one square meter of surface. The irradiance of sunlight on the ground varies with location, season, and time of day, but on a clear day at noontime it is typically 1000 watts per square meter. The luminous efficacy K of direct sunlight being about 100 lm/W, we deduce from the value of the irradiance that the illuminance on the ground is typically 100,000 lm/m². Commonly encountered levels of illuminance are included in Table 6-5. Summer sunlight provides a million times greater illuminance than moonlight. In normal offices, classrooms, and kitchens we would be comfortable with an illuminance in the range of 500 to 1000 lm/m². Higher levels are preferred for fine work when sewing and operating machine tools.

TABLE 6-5 Common Levels of Illuminance in Various Settings

	Lumens per Square Meter (lm/m² or lux)
Sunlight alone (maximum)	102,000
Television stage	25,000
Skylight alone (maximum)	16,000
Dull day	1,000
Merchandise display indoors	1,000
Recommended for reading	500
Public areas in buildings	300
Moonlight	0.4
Starlight	0.002

Note. To obtain the approximate values in footcandles, divide each number by 10.8.

To achieve a maximum illuminance for a given irradiance, we should concentrate the radiant energy in the yellowish green portion of the spectrum where the sensitivity curve V_λ of the eye exhibits its largest values. Designing light sources with this feature to conserve energy however is unacceptable in most applications, because much color information is thereby lost, and people do not react favorably to green illumination.

Other Units for Illuminance

Several other terms have been popular at various times for describing illuminance. An early English unit was the "foot-candle" (ft−cd), once defined as the illuminance on a surface placed a distance of one foot from the standard candle. From geometry we can deduce that a curved surface with an area of 1 square foot when placed 1 foot from the candle intercepts an angular range amounting to a solid angle of 1 steradian. It therefore receives a luminous flux of 1 lm. An illuminance of 1 foot-candle is therefore equal to 1 lm/ft². The preferred S.I. quantity of 1 lm/m² is consequently equal to 0.093 foot-candles (roughly, 1 foot-candle = 11 lm/m²).

Psychologists use a special unit called the "Troland" (formerly called "photon") for the unit of illuminance on the retina of the eye. It is the illuminance produced when viewing an extended source having a luminance of 1 cd/m² if the "apparent" area of the pupil of the eye is 1 mm². As we shall explain in Example 6-7, the illuminance on the retina from an extended source is independent of its distance (and so is its brightness). By "apparent" area in the definition of the Troland we mean that peripheral areas of the pupil admitting light off-center should be counted as less effective by an appropriate amount to produce the same retinal response as if light entered near the center of the pupil. This corrects for the Stiles–Crawford effect whereby the sensitivity of cones is decreased if light impinges on them from an angle. Light entering the eye from near the edge of the fully dilated pupil is typically one-tenth as effective as the same amount entering from the center of the pupil.

6-8 RELATING SOURCE OUTPUT TO ILLUMINANCE

We know from experience that when a source of light is moved away from a surface, the illuminance on the surface decreases. Figure 6-8 shows that a given luminous flux from the source is spread over a greater area when the surface is further away, and thus by Eq. 6-6 the illuminance is less. In fact, the illuminance in a given direction from a point source decreases as the square of the distance S from the source:

$$E_v = \frac{I_v}{S^2} \tag{6-7}$$

The origin of this expression is explained in Example 6-5. Because of this inverse square law, the illuminance weakens dramatically with distance. Doubling the distance reduces the illuminance by a factor of 4.

Example 6-5 Illuminance from a Point Source

Suppose that we wish to predict the illuminance on a desk when a single incandescent lamp is overhead. If the bulb is more than 2 m above it can be considered a point source. We let S represent the distance from the desk, and I_v is the intensity. To find

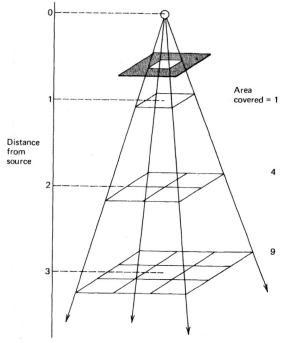

Fig. 6-8. As a surface is moved farther from a source, less light falls on a given area. The same luminous flux that covers an area of 1 m² at a distance of 1 m is spread over an area of 4 m² at a distance of 2 m, and is further spread over an area of 9 m² at a distance of 3 m. The illuminance on the surface decreases as the square of the distance from the source.

the illuminance we first calculate the luminous flux that falls onto a small area of the desk, say an area of 0.1 m². This area intercepts a certain angular range with respect to the lamp that is measured by the solid angle represented by this 0.1 m² area at a distance of $S = 2$ m. A sphere of radius S has an area of $4\pi S^2$; therefore, this 0.1 m² area is $0.1/4\pi S^2$ of the sphere's area, and it represents a solid angle of $(0.1/4\pi S^2) \times 4\pi$ steradians, or $0.1/S^2$ steradians. The luminous flux within this angular range is $(0.1/S^2)I_v$.

To deduce the illuminance we take this value for the flux and divide it by the area of the surface over which it is spread (Eq. 6-6). Thus the illuminance is $(0.1/S^2)(I_v/0.1) = I_v/S^2$. This is how Eq. 6-7 given above was deduced. As a practical example, suppose that the lamp emits 1700 lumens and has an intensity of $I_v = 150$ candela. Then the illuminance on a desk at a distance of 2 m is by Eq. 6-7: $150/2^2 = 150/4 = 37$ lumens per square meter. Comparing this value with those in Table 6-5 shows that this illuminance is not sufficient for comfortable reading. The lamp should be brought much closer to the desk. At one-third the distance ($\frac{2}{3}$ meter) the illuminance would be $37 \times (3)^2 = 333$ lm/m², a level that is acceptable for reading books with large type.

Example 6-6 Diffusely Reflecting Surface

There is a simple relationship between the illuminance falling on a flat surface and the luminance of the surface if the surface reflects diffusely. We consider a region of the surface that is so small that it may be considered a point source. Then we can let I_0 represent the luminous intensity of this region emitted in the direction perpendicular to the surface. Because the surface is a diffuse reflector, the luminous intensity I_v at any angle θ from the perpendicular decreases as the cosine of the angle: $I_v = I_0\cos\theta$ (Lambert's law). The total luminous flux emitted from this region is obtained from this expression by summing the contributions from all angles, ranging from $\theta = 0°$ (perpendicular to the surface) to $\theta = 90°$ (parallel to the surface). The total solid angle is 2π steradians, and the average value of $\cos\theta$ can easily be shown to equal $\frac{1}{2}$. Therefore the total luminous flux can then be expressed as $\theta_v = \pi I_0$ from this region.

If the surface absorbs none of the incident luminous flux, the incident luminous flux $E_v A$, where E_v is the illuminance and A is the actual area of the region, must match the luminous flux emitted by the region: $E_v A = \phi_v$. In a more realistic case the surface absorbs some of the light and the reflectance R is not 100%. Then this relation becomes: $RE_v A = \phi_v$. This allows us to relate the luminous intensity I_0 to the illuminance E_v by the formula:

$$I_0 = \frac{\phi_v}{\pi} = \frac{RE_v A}{\pi} \qquad (6\text{-}8)$$

Finally we note that the luminance L_v of the surface is the luminous flux per steradian per square meter of perceived area, and the perceived area for the small region of interest for a direction θ is $A\cos\theta$. Thus for any angle we have

$$L_v = \frac{I_v}{A\cos\theta} \qquad (6\text{-}9)$$

But the cosine in the denominator of this expression is compensated by the cosine in Lambert's law when I_v is replaced by

$$L_v = \frac{I_0\cos\theta}{A\cos\theta} = \frac{RE_v}{\pi} \qquad (6\text{-}10)$$

Therefore the luminance is independent of direction and can be calculated from this expression once the illuminance E_v on the surface is known.

Example 6-7 Brightness of Point and Extended Sources

The concepts of the preceding section explain why the brightness of a point source depends on its distance from us but why the brightness of an extended source does not. First we consider the point source. The key fact to appreciate is that the optical limitations of the eye (diffraction or lens imperfections) spread the luminous flux admitted by the pupil over a small region of the retina whose size is *independent* of the distance from the source. The illuminance on the retina—which determines the perceived brightness of the source—is thus proportional to the luminous flux entering the eye, or equivalently it is proportional to the illuminance on the pupil. But the latter was shown in Eq. 6-8 to decrease as the square of the distance from a point source. Consequently, the brightness decreases as the distance increases.

The case of the extended source is different, because now the image formed on the retina has a size that depends on the distance S from the source; and indeed the concepts of ray optics to be discussed in Chapter 8 tell us that the image dimension is proportional to S^{-1}. Consequently the area of the portion of the image corresponding to a small area of the source varies as the square of the distance, or as S^{-2}. The luminous flux coming from this small area of the source and entering the pupil also varies as S^{-2} according to Eq. 6-8, so the illuminance on the retina—which is the incident luminous flux divided by the area it illuminates—is independent of S. Thus the brightness of an extended source is independent of its distance. A person does not appear darker on walking away from us.

6-9   A REVIEW

Because so many new terms have been introduced in this chapter, it may help to fix them in mind by briefly reviewing them here. Radiant flux (in watts) has a photometric counterpart called the luminous flux (in lumens). A light source can be characterized by its total radiant flux (in watts) and total luminous flux (in lumens). The latter depends on details of the SPD and not simply on the total power that is radiated. A point source of light furthermore has the directional features of its radiated power specified by the radiant intensity in the various directions of interest (in watts per steradian) and by the corresponding luminous intensity in those directions (in lumens per steradian, or in candela). Perceptually, we judge the brightness of an extended source by the apparent area of the source. Therefore the radiometric quantity called radiance is introduced to indicate the radiant flux per steradian per square meter of perceived area. Radiance therefore has the unit of watt per steradian per square meter. Its photometric counterpart is luminance, which has the unit of lumen per steradian per square meter. When surfaces are illuminated with radiant energy we describe how concentrated it is by irradiance (in watts per square meter), and the corresponding photometric quantity is illuminance (in lumens per square meter).

SUMMARY

Photometry is the science dealing with the measurement of light. It affords a quantitative means of characterizing lights of various colors and intensities so that the condition under which they appear equally bright can be predicted. One watt of radiant flux at 555 nm wavelength is defined to have a luminous flux of 683 lumens. Any other light has proportionately more (or less) luminous flux according to how much the radiant flux of this monochromatic reference must be increased (or decreased) to achieve a brightness match, under identical conditions. The sensitivity curve of the C.I.E. Standard Observer describes what fraction of the radiant flux of this reference is required for a brightness match to one watt of monochromatic radiation at any other wavelength. Thus the sensitivity curve together with the measured SPD of any light permits a calculation of the corresponding total luminous flux of the light. By this procedure each radiometric quantity has a corresponding photometric counterpart. Luminous flux, luminous intensity, luminance, and illuminance are the relevant characteristics of light to consider when dealing with perceptual aspects related to brightness. The brightness of a large object is determined by its luminance; and the brightness of a tiny object by its luminous intensity. When deciding what type of reading lamp to buy, we are more concerned with the illuminance it provides on the magazine or book than by its total luminous flux.

QUESTIONS

6-1. Explain the difference between power (measured in watts) and energy (measured in joules).

6-2. What is the unit of measure for light? How is this unit defined?

6-3. How many lumens are emitted by the "standard candle" of the kind used as a standard light source early in this century?

6-4. If a laser produces 2 m W of red light at a wavelength of 630 nm, how many lumens of light is this?

6-5. Suppose that a special type of lamp emits light at three wavelengths: 2 W at 450 nm; 1 W at 550 nm; and 3 W at 650 nm. At which wavelength is the greatest luminous flux? What is the total luminous flux from the lamp? What is the total radiant flux?

6-6. If the illuminance on a desk is 1000 lumens per square meter (lm/m^2), what is the total luminous flux on the desk if it has an area of 2 m^2?

6-7. Suppose that a small incandescent lamp (a point source) provides an illuminance of 300 lm/m^2 on the floor 2 m directly underneath, what would be the illuminance if the lamp were raised to a height of 4 m above the floor?

6-8. What is meant by the luminous intensity of a point source of light?

6-9. If we were to measure the luminous intensity of a point source of light at various distances, how would the luminous intensity vary with distance?

6-10. Our eye judges brightness of a surface according to its luminance. What units are used to indicate the value of luminance?

6-11. If the luminance of a fluorescent lamp is doubled, is it correct to say that its brightness doubles? Why?

6-12. When an incandescent lamp has a luminance of 200 candela per square meter and a fluorescent lamp has 100 candela per square meter, which appears brighter? If one or the other is to illuminate a desk, can you determine which would provide the greatest illuminance? (Explain why or why not.)

6-13. How is the illuminance of a room affected if a yellow bulb in the lamp rated at 1700 lm is replaced by a bulb of the same rating but of a green color?

6-14. If a white spotlight for the stage provides an illuminance of 2000 lm/m^2, what is the illuminance if a filter having a transmittance of 30% is inserted in the beam?

FURTHER READING

R. S. Hunter, *The Measurement of Appearance* (Wiley-Interscience, New York, 1975). A practical discussion of all aspects of the appearance of objects and what aspects can be measured. Covers not only basic principles but also how specimens should be prepared and what instruments can best be used.

A. Stimson, *Photometry and Radiometry for Engineers* (Wiley, New York, 1974). A more specialized explanation of methods for measuring radiometric and photometric quantities.

G. Wyszecki and W. S. Stiles, *Color Science, Concepts, and Methods—Quantitative Data and Formulas* (Wiley, New York, 1967). Standard methods for measuring color and radiometric and photometric quantities. Extensive sets of tables, graphs, and references to the literature.

I.E.S. Lighting Handbook—1981 Applications Volume; and *I.E.S. Lighting Handbook—1981 Reference Volume* (Illuminating Engineering Society, New York, 1981). How to design lighting installations for a wide variety of spaces, including residences and public rooms.

F. E. Nicodemus, "Optical Resource Letter on Radiometry," *American Journal of Physics,* Vol. 38 (1970), p. 43. An annotated list of books and articles on radiometry and relevant background topics.

W. R. Blevin and B. Steiner, "Redefinition of the Candela and the Lumen," in *Metrologia,* Vol. 11 (1975), p. 97.

P. Giacomo, "News from the BIPM," in *Metrologia,* Vol. 16 (1980), p. 55. This article and the preceding one explain the new definition for the lumen.

Various manufacturers of lighting fixtures provide information and workbooks for estimating the lighting level that can be obtained with different lamps.

From the *Scientific American*

A. V. Astin, "Standards of Measurement" (June, 1968), p. 50.

J. M. Fitch, "Control of the Luminous Environment" (September, 1968), p. 190.

CHAPTER 7

PRODUCING LIGHT

Energy in the form of electromagnetic radiation is continually exchanged between every object and its surroundings. As described in Chapter 5, an atom (or molecule) can radiate energy if it is initially in an excited state. An atom can be excited by many different means. We have seen examples where energy is acquired by absorbing radiation or through chemical reactions. The radiant energy subsequently emitted by an atom has an SPD (spectral power distribution) that is characteristic of the differences between the energies of its allowed states. In this chapter we shall discuss a variety of important light sources whose SPDs display such atomic characteristics.

However, we first focus on another means of excitation: the transfer of energy when one atom (or molecule) makes contact with another. This occurs whether the atoms are free to move about as part of a gas, or have more restricted movement as in a liquid, or simply vibrate back and forth as in a solid. Atoms in these materials are continually in motion, and associated with this is an energy of motion, or *kinetic energy*. Each atom is continually jostled by its neighbors, and in the process from time to time some of their kinetic energy is transferred to it.

The temperature of the material is a measure of the average kinetic energy. The higher the temperature, the more kinetic energy each molecule has on the average. The increase in thermal agitation with increasing temperature brings atoms together more frequently and with greater speeds. Their collisions serve both to transfer energy and to disturb the atomic (or molecular) quantum states, thereby shaping the SPD of emitted radiation. Correspondingly, the characteristics of the emitted radiation are strongly influenced by the temperature. Consequently the radiation was once called "temperature radiation" but is now better known as *thermal radiation*. An object that is sufficiently hot to produce visible radiation is said to be *incandescent*. Examples of incandescent sources of light are the common table lamp and automobile headlights.

Sources of light can be classified according to their spectral characteristics, for example, *continuous sources* radiate over broad regions of the spectrum, and *line sources* radiate only at specific wavelengths. An incandescent object is an example of a continuous source. Electrical discharge lamps, fluorescent lamps, and the laser are examples of line sources. In this chapter each of these will be considered in turn. Some sources produce a combination of thermal radiation and a line spectrum, and these cases will be pointed out. But first we shall examine a number of important aspects of a common form of thermal radiation—called *black body radiation*—that tells us a great deal about incandescence.

■ 7-1 BLACK BODY RADIATION

A surface that absorbs all of the radiation incident on it appears black. There are few truly black surfaces in nature, but many surfaces come close to this extreme behavior. An object with such an absorbing surface is technically known as a *black body*.

What Makes a Surface Black

Two ways to form a nearly perfect black body are shown in Fig. 7-1. One simple way is to arrange razor blades in a stack. When seen edge-on the corrugated surface appears perfectly black. This is because virtually all of the incident light passes by the sharp edges and glances off the broad faces of the blades. The light is reflected inward, being partially absorbed on each reflection, until none remains to reemerge. Another way to make a black body is shown in Fig. 7-1b. It is the entrance of a hole leading into an enclosed cavity. If the hole is sufficiently small compared with the volume of the cavity, light that enters is totally absorbed before it can find its way out. The entrance therefore has the important characteristic of a black body.

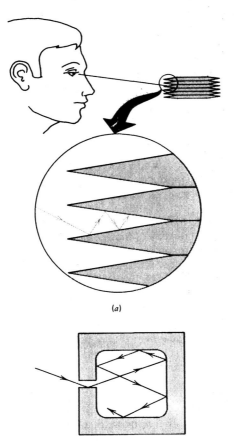

(a)

(b)

Fig. 7-1. Two examples of nearly perfect black bodies: (*a*) a stack of razor blades seen edge-on; (*b*) a small hole leading to a cavity. In both cases the incident light has little chance of being reflected outward before being totally absorbed.

It seems paradoxical but a material that strongly absorbs light in its interior is also strongly reflecting at its surface. Metals are an example. We know that silver and aluminum strongly reflect light, but it is also true that the correspondingly small amount of radiation that is not reflected but enters the material is rapidly absorbed. The semiconductor silicon also strongly absorbs the light that enters it, but its surface has a grayish appearance from the reflected light. By contrast a material such as glass reflects very little light that comes straight onto its surface, but the remainder that enters the glass is transmitted through with very little absorption. The reflecting property of absorbing materials can be understood from considerations of how an electromagnetic wave divides into refracted and reflected waves at the surface, but to discuss these details would take us away from our main point. Objects that are truly black such as a soot-covered surface do not rely on absorption alone but have an irregular structure near the surface that causes multiple reflection of the light before it can emerge. With some absorption taking place at every reflection, very little light is seen emerging. The stack of razor blades is an example, but so too is black velvet with its deep pile and flat black paint with its myriad particles of pigment. Thus the appearance of blackness results from the fact that essentially none of the light that falls on an object returns to the observer.

Thermal Radiation of a Black Body

It is significant that a black body has another characteristic property: it emits the maximum possible thermal radiation from its surface. This radiation is called *black body radiation*. It has an SPD that depends only on the temperature of the body and not on its specific composition.

A short logical exercise shows why a black body radiates the maximum amount of thermal radiation compared with all other objects. We start with the notion that the temperature of the walls of the cavity depicted in Fig. 7-1b is uniform. Moreover, we suppose that the rate at which the walls absorb radiant power matches the rate at which they emit it. Then the temperature of the walls remains constant. We say that the walls and radiation are in *equilibrium*. Imagine now that we place a black body of the same temperature in the cavity. In this condition of equilibrium it also emits from a given area of its surface as much radiant power as it absorbs. Because this given area absorbs *all* of the incident power, it must also emit more power than the same area of an object that is partially reflecting. Therefore the surface of a black body emits thermal radiation at the maximum possible rate. Since the radiant properties of a surface are not changed if the object is removed from the cavity, we conclude that a given area of a black body emits the maximum possible thermal radiation compared with the same area of any partially or completely reflecting object at the same temperature.

The hole in such a cavity constitutes a black body because all of the radiation falling on that hole from the outside is absorbed. Therefore we can study black body radiation by observing what emerges from the small hole of a cavity held at a uniform temperature. If the cavity is kept at room temperature, we know from experience that the hole seen from outside appears black. It is natural to accept the notion that it is a black body. But if the cavity is heated to a much higher temperature, say several thousand degrees, the hole glows and would be called "red hot." At still higher temperatures it would appear orange and then yellow, and at a very high temperature it would be called "white hot." The hole nevertheless would absorb all of the radiant energy that may be directed toward it from outside, so it is still a "black body" even though it does not appear black because of its incandescence. Clearly the SPD of black body radiation depends on the temperature of the body.

Temperature

How should the value of the temperature of an object be properly indicated? The modern method used by weather forecasters is to give temperature according to the Celsius scale, which uses the freezing point and the boiling point of water as reference points. The range between these points is divided into 100° (hence, this scale is sometimes called the "Centigrade" scale). The Celsius scale, however, is an arbitrary choice, since it is based on the properties of one particular liquid (albeit the most common one).

A less arbitrary scale was created by fixing the zero point at the absolute zero temperature, the lowest temperature that one could hope to approach by the most sophisticated modern techniques. This is at $-273°C$ (or $-459°F$). For convenience the same temperature interval is used for the degree on this scale as for the Celsius scale, so there are 100 divisions between the freezing and boiling points of water. We then arrive at what is called the *Kelvin scale*. This is the most appropriate physical scale for indicating temperature, and is alternatively known as the "Absolute Temperature Scale." In this scale the freezing point of water is 273 K and the boiling point 373 K. Because the degree symbol is redundant when the "K" for Kelvin is also present, the degree symbol is often omitted. It is conventional to read 373 K as "three hundred seventy three Kelvin."

Shape of the Black Body SPD

Developing a theory to explain the shape of the SPD of black body radiation was an important goal for physicists in the late 1800s. But its ultimate explanation had to await the notion of energy quanta introduced by Max Planck in 1901 and the concept of the photon proposed by Einstein in 1905. The key question was how to explain the observed property that the SPD of black body radiation exhibits a peak at a particular wavelength, and decreases dramatically at shorter wavelengths. The best classical theories predicted a steady increase with decreasing wavelength, and this discrepancy came to be known as the "ultraviolet catastrophe." The answer lies in the fact that a short wavelength photon (with a correspondingly high frequency) represents a large amount of energy, and thermal agitation cannot supply enough energy to one atom to produce many such photons. Therefore the SPD drops toward zero at short wavelengths.

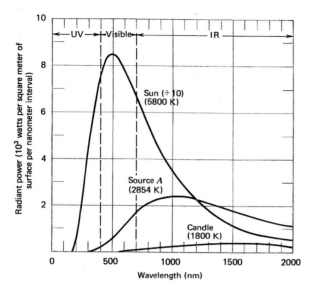

Fig. 7-2. Black body spectra for surfaces at three different temperatures. Values for the sun have been reduced by a factor of 10 in order to fit the curve into the figure. Source *A* is a black body surface maintained at 2854 K. A similar spectrum is produced by the filament of a 500 W incandescent lamp, but the output is only one-third of the values shown because its filament is not an ideal black body. The candle is such a weak source that its curve is barely visible on the figure. Its curve has a peak at 1600 nm. The vertical axis of the figure gives the radiant power emitted within a 1 nm portion of the spectrum centered at the wavelength given on the horizontal axis for each square meter of surface area of the black body.

Examples of the spectra of black body radiation are shown in Fig. 7-2 for bodies at three different temperatures. The peak of the SPD shifts toward shorter wavelengths as the temperature is raised. More photons are emitted each second at each wavelength, but proportionately more high-energy photons can be created by the greater thermal energy that is available. The wavelength λ_m denoting where this peak lies was first determined by measurements before Planck's theory appeared, and the formula describing its shift with temperature is known after its discoverer as the *Wien displacement law:*

$$\lambda_m = \frac{L}{T} \tag{7-1}$$

Wien found that the constant L has the value $L = 2.90 \times 10^{-3}$ m·deg. The temperature T inserted in this formula must be expressed in degrees Kelvin for λ_m to be given in meters. Wien's displacement law tells us that the wavelength for the peak in the SPD is inversely proportional to the temperature of the black body.

Example 7-1 Spectral Shift

To illustrate the use of Eq. 7-1, consider its application for predicting the location of the peak in the SPD of the sun. Of course the sun does not appear black and therefore one's first impression is that it could not be a black body. The point is, however, that we do not see the sun by virtue of light that is reflected off its surface; indeed, light that is incident on the solar surface from other stars is totally absorbed, and in this technical sense the sun is a black body. The light that we see from the sun is its thermal radiation, and its spectrum follows very closely the curve for a black body at a temperature of 5800 K, illustrated in Fig. 7-2. The peak in the spectrum is predicted by Wien's displacement law to lie at

$$\lambda_m = \frac{2.90 \times 10^{-3}}{5.80 \times 10^3} = 5.0 \times 10^{-7}\,m = 500\,nm$$

(we recall that 1 nanometer $= 10^{-9}$ meters). It is interesting that the human eye is most sensitive to light near this wavelength, a characteristic that no doubt evolved through the millenia as *homo sapiens* or his ancestors adapted to take maximum advantage of the environment.

Two other examples that we can deduce from Eq. 7-1 are the wavelengths for maximum thermal emission from the human skin and from the surface of the earth. Because neither is truly a black body, the indicated wavelengths must be considered rough approximations for the location of the actual peak. For humans ($T = 98°F = 310$ K), Wien's displacement law predicts $\lambda_m = 9400$ nm, and for the surface of the earth ($T = 288$ K), $\lambda_m = 10,000$ nm. Both are in the infrared region and, therefore, this radiation cannot be seen by the human eye.

Another remarkable aspect of black body radiation is the strong dependence of the *total* radiant flux ϕ_e on temperature. This is given by the area under each curve in Fig. 7-2. A quantitative relationship had been discovered empirically in 1879 by Joseph Stefan and was derived subsequently from general thermodynamic principles by Ludwig Boltzmann before the advent of the quantum theory. It is called the *Stefan-Boltzmann law* and is written:

$$\phi_e = \sigma T^4 \qquad (7\text{-}2)$$

This relationship says that the radiant flux ϕ_e from each square meter of a black body increases as the fourth power of the temperature. Consequently the power increases dramatically as the temperature increases: doubling the temperature increases ϕ_e by a factor of sixteen! The proportionality constant denoted by the Greek letter sigma (σ) has the value $\sigma = 5.67 \times 10^{-8}$ watt/meter^2·degree^4.

Example 7-2 Radiant Flux Varies with Temperature

Let us apply Eq. 7-2 to calculate the radiant flux given off by each square meter of the sun's surface. For a temperature of 5800 K we find that

$$\phi_e = 5.67 \times 10^{-8} \times (5.8 \times 10^3)^4 = 5.67 \times 10^{-8} \times 1.13 \times 10^{15}$$
$$= 6.4 \times 10^7 \text{ W/m}^2$$

This large amount would be enough to meet the power needs of a town of 4000 inhabitants! The total amount of power produced by the sun appears all the more impressive when we consider its immense size. Even with the loss of much of this energy as it spreads out from the sun in all directions, enough energy falls on the earth's surface to provide 1.3 kilowatts of radiant power for each square meter of area directly facing the sun. It is this incident thermal radiation that warms us on a sunny day, provides energy for plants to carry out photosynthesis, and drives the ocean currents and weather.

The human skin and the surface of the earth radiate considerably less power than the surface of the sun considered in Example 7-2, because their temperatures are much lower. For humans Eq. 7-2 predicts $\phi_e = 524$ W/m² and for the earth, $\phi_e = 390$ W/m². The actual amounts are somewhat lower because we are not dealing with true black bodies. Nevertheless with infrared detectors the pattern of radiant energy from surfaces of different temperatures can be measured, as illustrated in Color Plate 26, and the corresponding pattern of temperature can be deduced. In this manner it is possible to sense differences in temperature amounting to a fraction of a degree across the human skin, which is useful in several clinical applications, including circulation and tumor diagnostics. The amount of thermal radiation emitted by the human body is appreciable, and it is one reason why clothing helps to keep us warm on a cold day by absorbing this energy and radiating it back to us.

With Eqs. 7-1 and 7-2 it is possible to predict the amount of thermal radiation given off by a black body and where the peak of the SPD of the radiated power lies in the electromagnetic spectrum. This is one reason why the concept of a black body has proven so useful. Most common surfaces are not ideal black bodies. They do, however, absorb typically 90% of the incident energy and therefore approximate a black body. Therefore the radiant flux given off is somewhat less than would be predicted by Eq. 7-2. Nevertheless, the location of the peak predicted by Eq. 7-1 is not much in error. For example, the tungsten wire in a common incandescent lamp gives off only 32% of the radiant energy of a black body at the same temperature (typically 2800 K). Yet the shape of its spectrum in the visible follows the black body curve for a temperature that is only 50 K higher. Figure 7-3 shows a curve from which we can calculate the radiated power from each square meter of surface of a black body if the temperature of the body is known.

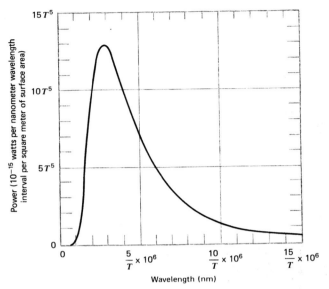

Fig. 7-3. Black body spectral power distribution as given by Max Planck's formula. The power radiated within a 1 nm wavelength interval from each square meter of surface is indicated for wavelengths near the peak in the spectrum. Numerical values for the power and corresponding wavelength can be deduced from this curve for any desired black body temperature T. For example, for $T = 500$ K the radiant power at a wavelength of $(5/T) \times 10^6 = 10{,}000$ nm is given by the curve as $7T^5 \times 10^{-15} = 0.22$ watts per nanometer wavelength interval from each square meter of surface. Thus, this amount of power would be found in the spectrum close to 10,000 nm, say between 10,000 and 10,001 nm, or between 9999 and 10,000 nm, etc.

■ 7-2 COLOR TEMPERATURE

Because the color of a black body depends systematically on its temperature, the notion of *color temperature* has been introduced to describe the colors of sources of light. This is the temperature of a black body whose light best matches the light from the source with respect to color. Because the peak of the SPD of a black body shifts toward shorter wavelengths if the temperature of the body is increased, the color of a black body depends on its temperature. As temperature increases to nearly 800 K the maximum emission is still in the infrared, but some visible light appears as a dull red glow. With further heating the color becomes bright orange at 2000 K and then goes through a very unsaturated yellow at 3000 K, becomes white at 6000 K, and blue at 10,000 K. In the last case, because the peak in the spectrum is found in the near ultraviolet, more power is emitted at short wavelengths than long. Except for very high or very low temperatures a black body has a very unsaturated color—so unsaturated in fact that when the peak of the SPD is in the green (for $T = 6000$ K), we do not see a green hue at all; rather we see white. This happens because the black body spectrum has such a broad peak that nearly the same amount of power is radiated in the blue and red as in the green. If a light source has a distinct hue other than red or blue, it cannot be matched in color by the broad spectrum from a black body, and therefore the notion of color temperature cannot be applied. The colors of black bodies at representative temperatures are shown on the chromaticity diagram in Color Plate 10. The unsaturated appearance of colors corresponding to temperatures of about 6000 K is evident.

The notion of color temperature is useful in photography, filmmaking, television, the printing and lighting industries, and astronomy because it allows a simple numerical specification of the color for many common light sources. Instead of the imprecise description such as orangish-yellow of which each of us has his own personal impression we can assign a simple number—the color temperature—to describe the color of a light source. Starlight ranges from somewhat reddish in the case of Antares, with a color temperature of 5000 K, to bluish-white for Sirius, with a color temperature of 11,000 K.

The color temperature need not describe the physical temperature of the source. The phosphers on the glass tube of common fluorescent lamps are rated at a color temperature of 6500 K; nevertheless the tube is cool enough to touch.

Sunlight can be described in terms of its color temperature. Because the atmosphere affects the light during its passage, the color temperature of the sun for an observer at ground level on earth does not exactly equal the black body temperature that describes the spectrum that the sun radiates. And the color temperature depends on whether the sky is clear or overcast. This is illustrated in Fig. 7-4, which shows the SPD of light falling on a horizontal surface in the two cases. The curve labeled C is typical of an overcast day and has more power in the blue end of the spectrum than the curve labeled B for direct sunlight at noon on a clear day. This is because on an overcast day much of the light has been scattered by fine water droplets or dust particles before reaching the observer, and blue light is scattered more effectively than red. Thus light on an overcast day is characterized by a higher color temperature than on a clear day, although of course, the physical temperature of the sun has not changed. Numerical data for the SPDs of curves B and C are listed in Table 3-3. These spectra can be produced by an incandescent lamp with different specified filters or with a fluorescent lamp having proper phosphors.

The graphic arts industry makes specific recommendations for the color temperature of overhead illumination when color matches of inks must be carried out. To represent north skylight on an overcast day, it recommends fluorescent lamps with a color temperature of 7500 K, and for noon sunlight 5000 K. These differ slightly from the actual color temperatures outdoors under these conditions. The illuminance on the printed surfaces during comparisons should be about 2000 lm/m^2 (or roughly 200 ft-cd).

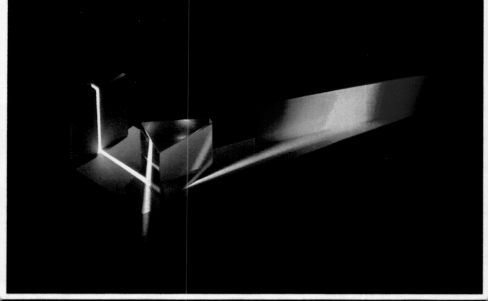

Plate 1

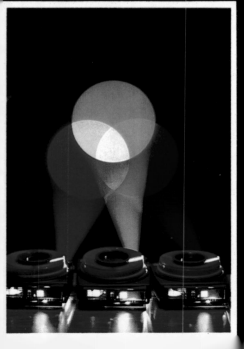

a

b

c

d

ate 4

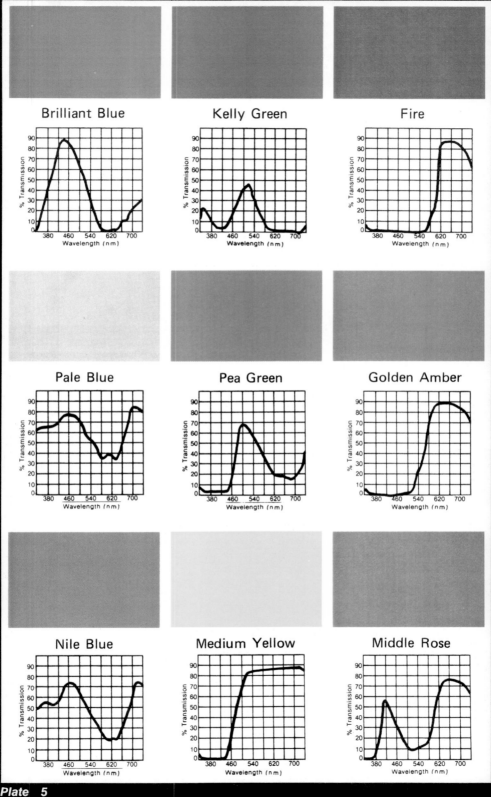

Brilliant Blue

Kelly Green

Fire

Pale Blue

Pea Green

Golden Amber

Nile Blue

Medium Yellow

Middle Rose

Plate 5

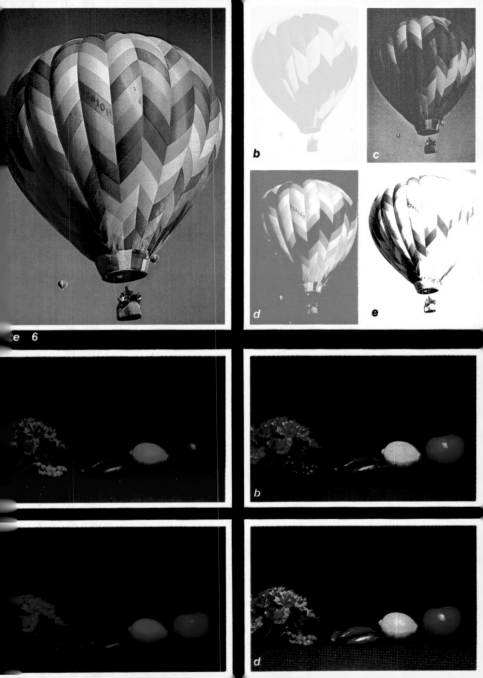

e 6

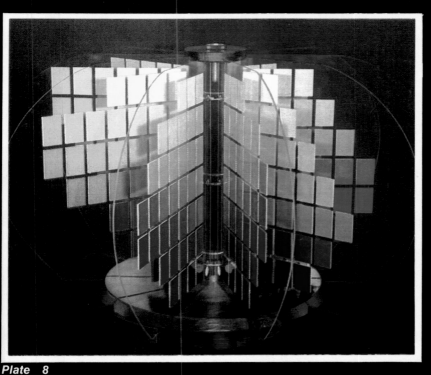

Plate 8

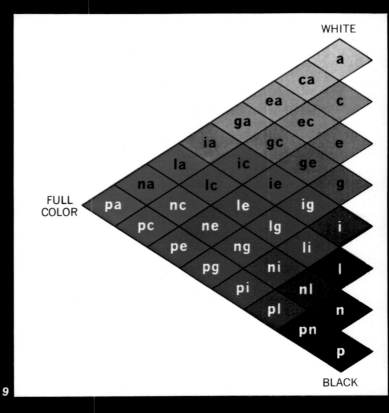

Plate 9

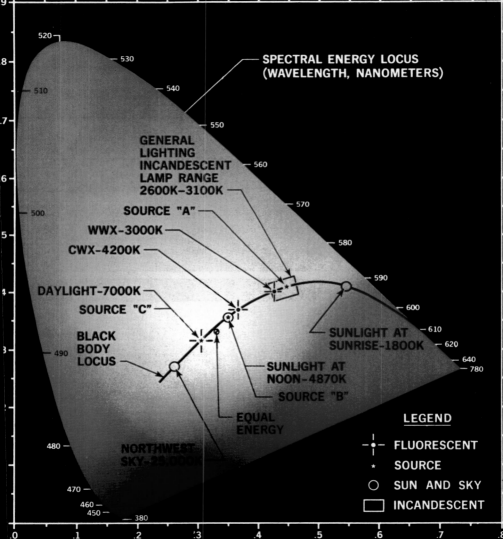

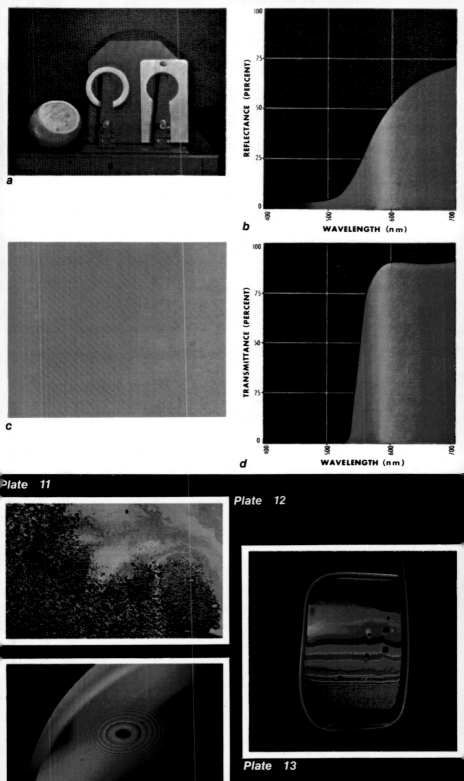

REFLECTANCE (PERCENT)

100

75

50

25

0

400 500 600 700

WAVELENGTH (nm)

a

b

TRANSMITTANCE (PERCENT)

100

75

50

25

0

400 500 600 700

WAVELENGTH (nm)

c

d

Plate 11

Plate 12

Plate 13

Plate 14

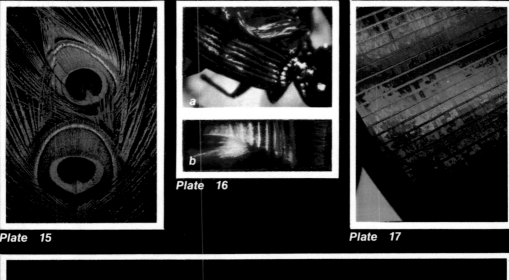

Plate 15

Plate 16

a

b

Plate 16

Plate 17

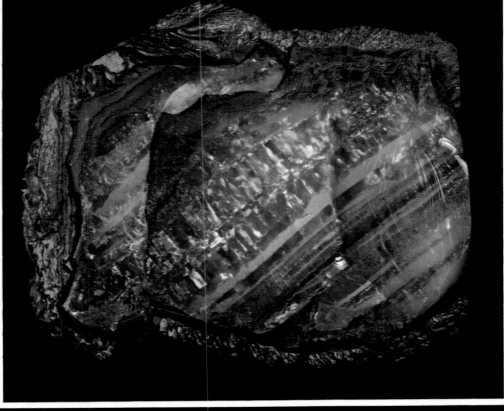

Plate 18

Legends for these color plates are opposite the first color page.

Plate 19

Plate 20 a

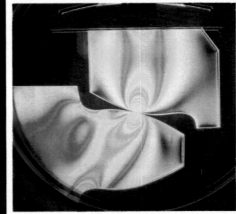

Plate 20 b

Legends for these color plates are opposite the last color page.

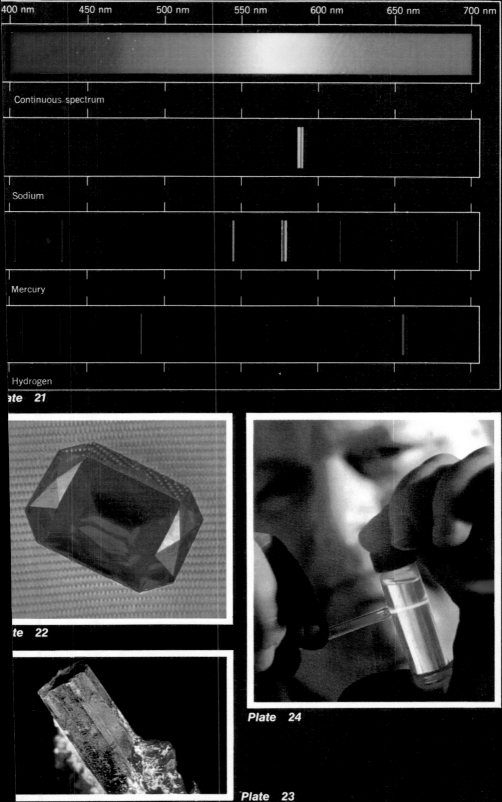

400 nm   450 nm   500 nm   550 nm   600 nm   650 nm   700 nm

Continuous spectrum

Sodium

Mercury

Hydrogen

Plate 21

Plate 22

Plate 23

Plate 24

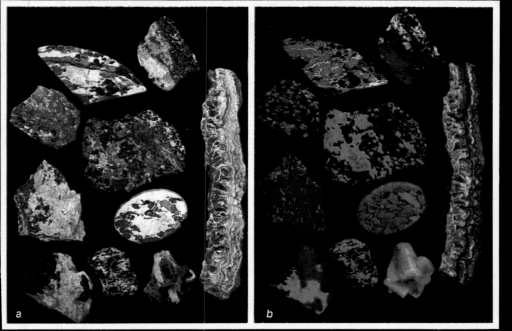

Plate 25

Plate 26

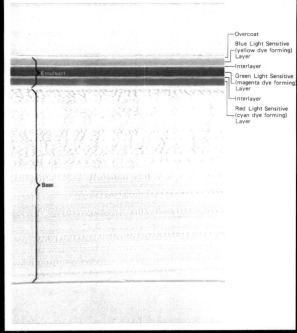

Plate 27

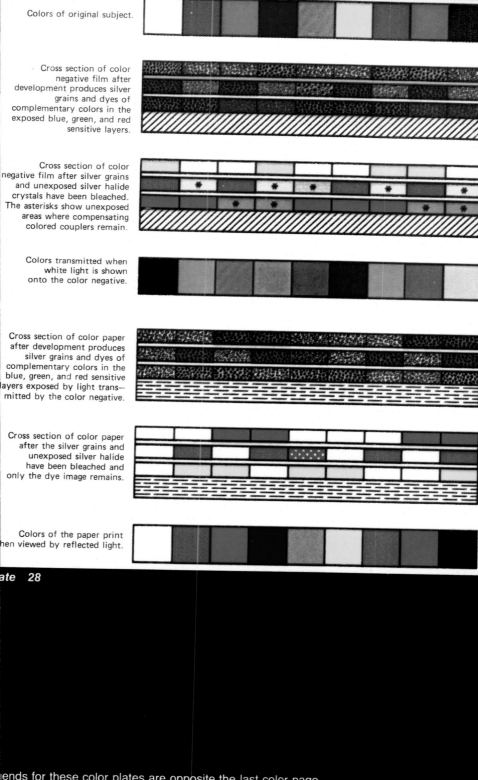

Colors of original subject.

Cross section of color negative film after development produces silver grains and dyes of complementary colors in the exposed blue, green, and red sensitive layers.

Cross section of color negative film after silver grains and unexposed silver halide crystals have been bleached. The asterisks show unexposed areas where compensating colored couplers remain.

Colors transmitted when white light is shown onto the color negative.

Cross section of color paper after development produces silver grains and dyes of complementary colors in the blue, green, and red sensitive layers exposed by light transmitted by the color negative.

Cross section of color paper after the silver grains and unexposed silver halide have been bleached and only the dye image remains.

Colors of the paper print when viewed by reflected light.

Legends for these color plates are opposite the last color page.

Plate 29

UER

NAIT

EST

NOB'

Et filius datus è no

bis cuius imperium

Plate   32

Plate   33

Plate 34

Plate 35

Plate 36

Legends for these color plates are on the facing page.

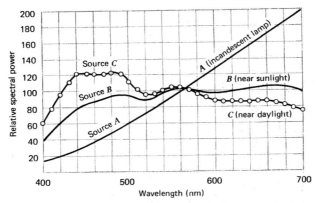

Fig. 7-4. Three SPDs adopted by the C.I.E. as standards to represent illumination from three different sources. The spectrum A represents power from an incandescent lamp with a color temperature of 2854 K; spectrum B represents direct sunlight at noon on a clear day with a color temperature of 4870 K; and spectrum C represents skylight on an overcast day with a color temperature of 6770 K. The differences between curve *B* and the smooth curve for the sun in Fig. 7-2 are produced by atmospheric absorption and scattering.

■ 7-3 THERMAL RADIATORS

An object heated to temperatures exceeding 800 K gives off thermal radiation that can be seen and is said to be incandescent. The flame of a candle, a glowing piece of charcoal, and the common incandescent electric light are examples. The electrodes of an arc lamp are still another. The spectral features of the light are determined almost entirely by the temperature, but in some cases are influenced by the nature of the hot element when it does not approximate a black body.

Incandescent Lamps

The hot element of the incandescent lamp is called the "filament," partly because it is made thin to keep its electrical resistance high to insure proper electrical heating. The first efforts to make an incandescent lamp were directed toward finding a material for the filament that could conduct electricity, withstand high temperature, and still radiate light with an efficiency as close as possible to a black body's. In the late 1800s Thomas Edison tried many materials, including fine platinum wire covered with carbon, but found them impractical. In 1879 he hit on the idea of a pure carbon filament, and found a way to make one by heating a fine strip of bamboo in a furnace to drive off the hydrogen and leave a carbon skeleton. Subsequent technological improvements in metallurgy have eliminated the disadvantages of short life and fragility of this filament by introducing one made from a coiled fine wire of tungsten. This has one of the highest melting temperatures of all metals.

The arrangement of elements in a modern incandescent lamp is shown in Fig. 7-5. Loss of heat from the filament to the environment can best be minimized by enclosing it in a glass bulb that maintains a vacuum. However, when tungsten is sufficiently warm to give appreciable light, the atoms at the surface of the metal vibrate so energetically that many of them overcome the forces that bind them together. These atoms are lost from the filament into the surrounding space, through which they travel until reaching the glass bulb. In about 1913 it was discovered that the rate of this process is appreciably decreased and the life of the lamp is prolonged if an inert gas, such as argon or nitrogen, surrounds the filament. Oxygen is avoided to keep the tungsten from oxidizing. The inert gas molecules collide with escaping tungsten atoms and cause a fraction of them to rebound back to the surface of the metal. Therefore a modern bulb contains an inert gas or a mixture of them at about 80% of atmospheric pressure. When heated during use the pressure increases to nearly atmospheric pressure. Because gases do not conduct heat efficiently, only a small amount of the filament's energy is conducted to the glass bulb. Nevertheless, despite these precautions some tungsten does collect on the inner surface of the bulb, causing it to darken with age and transmit less light. Eventually so much tungsten has been lost from the filament that a weak point develops and breaks, and the lamp "burns out." This usually occurs when turning the lamp on, because the rapid heating of the filament causes rapid expansion of the metal which puts it under mechanical stress.

Luminous Efficiency

A measure of the efficiency of a light source should tell how much light is emitted for each watt of power supplied to it. The *luminous efficiency* is correspondingly defined as the luminous flux emitted for each watt of *input* power. This is distinguished from the luminous *efficacy* defined in Section 6-4, which characterizes the luminous flux emitted for each watt of *emitted* radiant power. Luminous efficiency

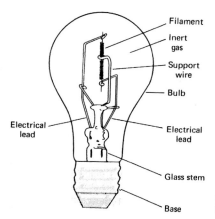

Fig. 7-5. Arrangement of elements in a tungsten filament lamp.

is inevitably lower than luminous efficacy, since some of the input power is lost in the form of heat and does not appear as emitted radiant power. With an incandescent source greater efficiency and efficacy are achieved by operating at higher temperatures. This is because the peak in the SPD then lies closer to the center of the visible spectrum where the eye's sensitivity is greater. The increase in efficiency is illustrated in Fig. 7-6 for a tungsten filament. However, in a tungsten filament, a higher temperature produces more rapid loss of metal from the filament, and the life of the lamp is shortened. A lamp can be made brighter by applying a higher electrical voltage across the filament, but the lamp life decreases dramatically (Fig. 7-7). On the other hand, by operating at a lower voltage than normal the temperature is reduced, the light output is also reduced, but the life is substantially prolonged. For example, a 150 W conventional lamp producing 2880 lumens of flux has an average life of about 750 hours. By comparison, a so-called "long-life" lamp may last twice as long but produces only 2350 lm because it operates at a lower temperature.

Tungsten when properly prepared can withstand temperatures up to 3643 K before melting, and at this point its color temperature is 3600 K. The slight difference between its physical temperature and color temperature results from the fact that tungsten is not a true black body. It emits about 47% of the amount of radiant flux expected of a black body at a wavelength of 400 nm, and this percentage smoothly decreases to 43% at 700 nm. Because the percentage is about the same across the visible spectrum tungsten could be called a "gray body."

At lower temperatures the color of tungsten changes from yellow through orange to red. For a standard filament at 2814 K, which is close to the temperature in a

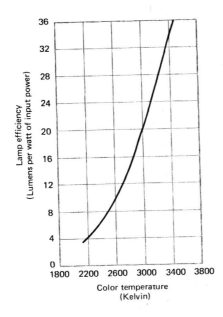

Fig. 7-6. The luminous efficiency of a tungsten filament increases dramatically when it is operated at higher temperatures, with correspondingly higher color temperatures.

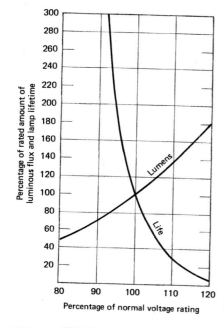

Fig. 7-7. An incandescent lamp provides more light when the electrical voltage is increased, but its life shortens precipitously. A 15% increase in operating voltage increases the luminous flux by 55%, but the lamp will last for only 15% of its rated life.

common 100 W lamp, the color temperature is 2854 K, and the emitted flux is only 33% of that of an ideal black body. The spectrum of this source is illustrated by the curve labeled A in Fig. 7-4. This is such a common source of illumination that it has been given the special name Standard Source A by the Commission Internationale de l'Eclairage. With such a low color temperature the light is distinctly yellowish compared with daylight. The best results for producing a color temperature matching daylight has been achieved with a higher wattage incandescent lamp filled with xenon gas.

The spectrum from common incandescent sources peaks in the infrared and progressively weakens toward the blue, the decrease being more pronounced the lower the color temperature of the source. The common *studio meter,* used by photographers to estimate the color temperature of their illumination, incorporates two spectral filters that pass light from two different portions of the spectrum to separate photodetectors. The ratio of the electrical signals from the two photodetectors when properly converted indicates the color temperature. We should not expect a meter of this type that does not sample the entire spectrum to give a correct reading if the illumination is provided by a source with a line spectrum.

Tungsten Halogen Lamps

Known during its early days as the *quartz-iodine lamp,* what is now termed the *tungsten-halogen lamp* offers substantially increased life compared with the standard incandescent lamp. These lamps are found in stage lights, slide projectors, television studios, and automobile head lights, wherever high intensity is desired. A small quantity of solid iodine, or now more likely bromine (both are members of the halogen group of elements that also includes chlorine), volatilizes when heated and reacts chemically with whatever tungsten is deposited on the bulb to form a gaseous halogen compound. This gas decomposes on making contact with the hot filament, and the tungsten is redeposited on the wire, providing longer life. It operates best at high temperature and thus must be made compact. The bulb is made from quartz to withstand both the heat and the chemical action of the vapor. When cold, the bulb should never be touched with the uncovered hand because skin oil when heated attacks the quartz and may cause it to crack. The bulb remains clear, as virtually all of the tungsten is returned to the filament. The life is thereby extended by a factor of 5 to 10 times over that of an incandescent lamp, although the SPDs are similar when the tungsten is operated at the same temperature. For the same operating life, the tungsten-halogen lamp can be operated at a much higher temperature and therefore provides substantially more light.

Combustion Sources

Some incandescent sources produce thermal radiation by virtue of combustion. The candle flame, gas lamp, and open fire are examples. A less obvious example is the photoflash bulb. These are generally not very efficient because the source of light (incandescent gases in the flame) does not even approximate a black body. Also, most of the energy goes into the infrared, as anyone who has felt the heat while watching a burning building can testify. Most of the yellow color of a flame actually comes from incandescent carbon particles, and there is a minor contribution from excited gas molecules. Clean-burning hydrogen gas, for example, produces a virtually invisible flame. A log fire, candle, or match has a color temperature of only 1800 K, because so much power is in the yellow and red end of the visible spectrum.

Combustion sources are therefore not very efficient from the standpoint of producing usable light from a given amount of energy. Most of the radiated power (about 95%) is found outside the visible spectrum in the infrared. Such a source can be called only 5% efficient. A candle is only 0.1% efficient!

A clever innovation introduced by Auer von Welsbach in 1866 substantially increased the utility of the open flame as a light source and enjoyed considerable success when applied to gas lamps. This invention is now known as the *Welsbach mantle* and is a white, open-mesh cover that encloses the flame. It is made by immersing a small woven cotton bag in a solution of a thorium salt. When dry, a match applied to the bag causes the cotton to burn away, leaving a fine replica of the mesh composed of thorium oxide. This mantle does not burn at high temperatures and usually has a little cerium added to increase its reflectance curve in the infrared. Thus when heated by a flame the mantle cannot cool as effectively by emitting thermal radiation in the infrared, and it warms to a temperature that is very close to the flame temperature of the burning gas. Much more radiation is given off in the visible by the Welsbach mantle at this higher temperature, and it provides an intense white light. A variation of the Welsbach mantle is found in modern gas lanterns for camping. The mantle, unfortunately, is fragile and deteriorates with use.

A similar idea was applied in the theater in the nineteenth century. To gain a bright source of light from burning gas, limestone (calcium carbonate) was placed in the gas flame and heated to incandescence. Its thermal radiation is somewhat stronger in the blue than today's incandescent lamps. From this practice comes the saying "in the limelight" for a star performer.

Flash Lamps

Photographers use flash lamps, which are designed to produce a short intense burst of light to properly illuminate a subject that otherwise could not be photographed. A wide variety of bulbs is available with differing flash intensity, duration, and spectral power distribution. All of these bulbs depend on the rapid combustion of a fine "wool" of a reactive metal such as magnesium or zirconium, which is surrounded by oxygen contained in a glass bulb. Ignition is produced by sudden electrical heating when a current is passed through the wire wool. The shortest flash duration is about $\frac{1}{50}$ sec. Most bulbs used with color films designed for indoor exposure have a color temperature of about 3500 K. For outdoor color film, designed for sunlight exposure with a higher color temperature, the excess energy in the yellow and red is reduced by a blue filter or blue coating over the outside of the bulb. In high intensity flash lamps for outdoor film, combustion occurs at higher temperature, and appreciable ultraviolet radiation may be emitted as well. One flash may produce as much as 10^6 lumens peak luminous flux.

Carbon Arc

It is not certain who discovered that an electrical arc between the ends of two carbon rods produces a bright light, but credit is usually given to the English chemist Sir Humphrey Davy (1801). This was the first light source to be powered by electricity and was put into limited use in the 1870s. Because of its extreme brilliance and small size, it is extensively used in commercial motion picture projectors, searchlights, and theater lights. Figure 7-8 shows an example of such a device. To start the arc an electrical current is passed from one carbon rod to the other when they initially touch; then by retracting one rod a short distance, the voltage between them pulls a few electrons out of the negative rod toward the positive rod. These free electrons, when moving toward the rod, strike air molecules with sufficient vigor to ionize them, releasing still more electrons with the net effect of producing a cascade of electrons flowing through the arc from one rod to the other.

As the electrons strike the end of the positive rod, they form a hot crater, and its surface produces most of the light. If the voltage applied from one rod to the other is steady, the direction of greatest luminous intensity is 40° away from the axis of the positive rod. On the other hand, if an alternating voltage is applied,

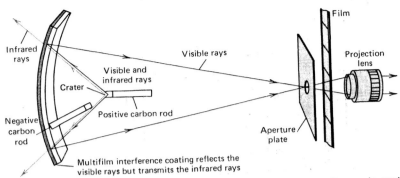

Fig. 7-8. Typical arrangement of carbon rods in a high-intensity movie projector. A mirror depends on interference to reflect light back through the film and to transmit infrared rays away to avoid heating the film. (Courtesy of Union Carbide Corporation.)

each rod is positive for half a cycle and negative for the next half. Then light is intensely emitted in both directions, with the greatest luminous intensity direction about 50° away from the axis of the rods.

The SPD of the radiation is close to a black body spectrum. The SPD can be altered by including different substances such as cerium in the carbon, commonly concentrated in the center or "core" of the rod. In this way, the blue end of the spectrum can be intensified to produce a white light that is essential for theater motion picture projectors.

The high temperature at the surface of the carbon rods facilitates chemical reactions with atmospheric gases. A narrow band at 389 nm in the emission spectrum has been attributed to light given off by cyanogen, a compound formed from nitrogen and carbon. With time, the carbon rods are oxidized and eroded and must be replaced rather frequently. The carbon arc is much less efficient than the modern incandescent bulb, but is still favored for its high brightness and color temperature. ■

7-4 ELECTRICAL DISCHARGE LAMPS

Some lamps depend on the passage of an electrical current through a suitable gas to provide the energy that the gas atoms then convert into light. Unlike the carbon arc lamp, in which the current was used to heat the top of the carbon electrodes, electrical discharge lamps share the common feature that their radiation is not determined by the temperature of the gas or electrodes. Rather, it is characteristic of the energy levels of the atoms comprising the gas. This type of lamp includes the *glow tube, sodium vapor lamp, mercury arc,* and *strobe lamp. Fluorescent lamps* also use an electrical discharge, but the light we see comes from a phosphor material on the glass envelope and not from the gas within; therefore we will consider these lamps separately.

Glow Tubes

Glow tubes are better known under their popular, but less precise name "neon lights." Neon gas was used in the first commercial tubes that appeared around 1913. But many other gases may be used instead of neon, depending on the desired color. Practically any gas that is stable will give off light if a current is passed through it. The electrical discharge is formed when a high voltage amounting to 1000 to 15,000 V (at the power line frequency of 60 Hz in North America and 50 Hz in Europe) produced by a transformer is applied between two electrodes at the ends of an enclosed tube containing neon or other appropriate gas at low pressure (Fig. 7-9). The accelerated electrons produce a cascade effect when they ionize gas atoms and release additional electrons, so that an electrical current is formed between the two electrodes. The SPD of the radiation is characteristic of the atoms in the gas, because the photon energy must match the difference in energy between two quantum states of the atom. Neon tubes have an orange-red color; krypton, pale blue; sodium, yellow-orange; and mercury, blue-green. A mixture of gases may be used for color addition, and sometimes a phosphor is applied to the inside surface of the tube for greater efficiency or special effects. Colored glass by itself provides subtractive mixing: yellow glass with mercury gas allows only green light to emerge.

Sodium Arc Lamp

The *sodium lamp* is a glow tube now widely used for illuminating public areas, because it is the most efficient lamp yet developed. A 500 W lamp produces about seven times more lumens per watt than an incandescent lamp. The name "discharge lamp" is commonly used to distinguish high intensity sodium and mercury lamps from their low intensity cousin, the neon lamp. The high intensity sodium lamp was made possible through the development of a translucent ceramic tube to contain the chemically reactive sodium

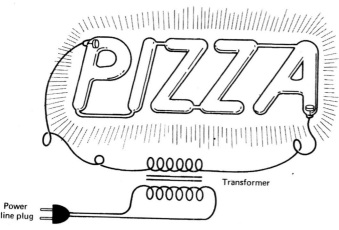

Fig. 7-9. Placement of electrodes in a neon glow tube. The transformer provides sufficiently high voltage to maintain the electrical current through the gas.

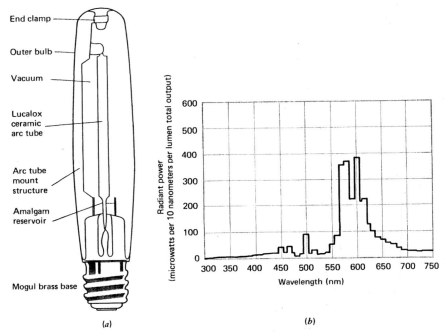

(a)   (b)

Fig. 7-10. Lucalox® 400 W high pressure sodium lamp. The light is produced by excited sodium atoms that sustain an electrical discharge inside the ceramic arc tube. The outer bulb protects the inner components and helps to maintain a constant temperature inside. (Adapted from General Electric Company publication TP-109R.)

vapor, and the discovery that this gas when excited at high pressure yields light of acceptable though rather yellow color. The color temperature is approximately 2100 K. The arrangement of the elements in the bulb is shown in Fig. 7-10.

The sealed ceramic tube in which an electrical discharge is established contains a small amount of a liquid amalgam of sodium and mercury, as well as a starting gas, xenon, at low pressure. An electric discharge is initiated in the xenon by a voltage pulse of about 2500 V, and the resulting arc warms the tube and evaporates the mercury and sodium. The light produced by the ionized xenon and mercury first appears bluish-white and then shifts to monochromatic yellow-orange as the sodium atoms become excited. The brightest lines by far are found at 589.1 and 589.6 nm, and these correspond to electrons dropping into the ground state of sodium from the next two higher excited states. These are called the sodium D lines and contain 90% of the radiant flux. As the sodium warms and its pressure increases to about atmospheric pressure, the spectrum broadens and develops a "hole" in the region of 589 nm (Fig. 7-10b). What happens is that the sodium effectively absorbs much of its own radiation at wavelengths corresponding to the D lines. But emission at nearby wavelengths—made possible by marked perturbations of the orbits when atoms collide frequently at high pressures—escapes. About 92% of the emitted light is transmitted through the ceramic tube of this lamp.

Mercury Arc Lamp

Mercury arc lamps offer exceptionally long life and fairly high efficiency. The mercury, together with an inert starter gas, argon, is enclosed in a quartz arc tube having an electrode mounted at each end, as shown in Fig. 7-11. The tube is enclosed within a protective outer bulb filled with nitrogen, which helps to maintain a nearly constant tube temperature of 1300 K. When started, the high voltage is first applied between one electrode and a "starting electrode" located nearby. The small gap between them sets up a strong electric field and initiates an arc in the argon. The path being so short, a "starting resistor" is included in the circuit to limit the current. The heat from the arc vaporizes a small quantity of mercury in the bottom of the tube, and when this additional gas reduces the resistance sufficiently, the main-arc forms. Current through the starting electrode is cut off automatically as the lower resistance main-arc diverts it. It takes five to seven minutes before temperatures reach their operating values and the color of the light is stabilized.

The history of high intensity discharge lamps traces back to the Cooper Hewitt low-pressure mercury vapor lamp developed in 1901 by Peter Cooper Hewitt. It was more than a meter long and produced a distinctly bluish-green light, characteristic of the wavelengths 405, 436, 546, and 578 nm, where virtually all of the light appears. A substantial amount of ultraviolet is also created, and subsequent models have been used for sunlamps, as "blacklight," and in research. Since glass absorbs ultraviolet, protective bulbs for such applications are made of quartz. When bulbs operate with mercury gas at higher

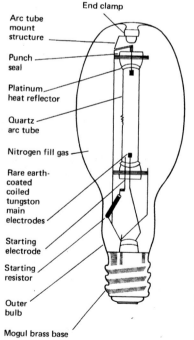

End clamp

Arc tube
mount
structure

Punch
seal

Platinum
heat reflector

Quartz
arc tube

Nitrogen fill gas

Rare earth-
coated
coiled
tungsten
main
electrodes

Starting
electrode

Starting
resistor

Outer
bulb

Mogul brass base

Fig. 7-11. Mercury lamp with inner quartz arc tube containing mercury and an inert gas. The main electrode is coated with a rare earth metal so that electrons will more easily boil off when heated. (Adapted from General Electric Company Publication TP-109R.)

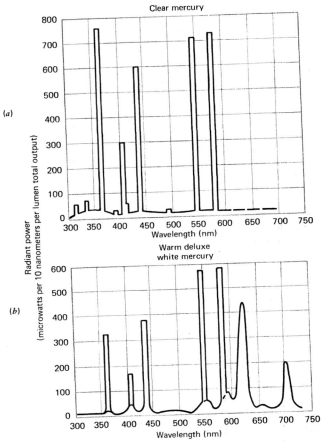

Fig. 7-12. (*a*) Spectral power distribution of a mercury arc lamp with a transparent bulb, showing the strong lines and much weaker continuous portion of the spectrum. (*b*) The SPD of a lamp whose bulb is coated with a phosphor that provides additional power at long wavelengths. (Adapted from General Electric Company Publication TP-109R.)

pressure (greater than 10^{-3} of atmospheric pressure) the arc gives off an intense bluish-white light and relatively less ultraviolet. Outer bulbs made of glass absorb the potentially harmful ultraviolet. Should the outer bulb be broken the lamp may still operate for some time and be a hazard. Contemporary high intensity bulbs operate near atmospheric pressure and produce continuous emission across the visible spectrum. Yet most of the power remains in the line spectrum (Fig. 7-12*a*). Lamps drawing 300 to 400 W may have their color temperatures altered by the addition of fluorescent light produced by phosphors coating the inner surface of the glass bulb. As in fluorescent lamps these phosphors absorb ultraviolet radiation, particularly the strong 365 nm line, and radiate lower energy photons. As illustrated in Fig. 7-12*b*, the color temperature of the light can be shifted downward to 3300 K by addition of fluorescent light in the red portion of the spectrum.

Strobe Lamp

One form of glow tube is operated as a flash lamp for photography or as warning lights on airplanes, high towers, and boats. Invented in 1931 by Harold Edgerton, the discharge is initiated by a high voltage when a charged capacitor (a device that stores electricity) is switched to make contact with the electrodes. The conducting path through the gas between the electrodes quickly discharges the capacitor (in as short a time as 10^{-6} seconds in fast systems) and creates a very intense light during that brief interval. Many times brighter than sunlight, the short flash of the light can substitute for the opening and closing of a camera's shutter. This rapid exposure "stops" motion of the subject, which may be a bullet in flight or the splash of a droplet (Fig. 7-13).

The first example of "instantaneous" photography was devised in 1851 by William Fox Talbot, who used light from an electrical discharge that formed in air when wires con-

Fig. 7-13. Milk drop splashing on a wet plate is "stopped" by the flash of a strobe lamp whose duration is less than a millionth of a second. (Reprinted from H. E. Edgerton and J. R. Killian, Jr., *Moments of Vision. The Stroboscopic Revolution in Photography,* by permission from M.I.T. Press, Cambridge, Massachusetts, copyrighted 1979.)

nected to the terminals of an array of Leyden jars (an early form of capacitor) were moved close enough together.

One advantage of the modern strobe is long life, because it can be recycled by recharging the capacitor. With a repetition of flashes, the images of a moving subject can be superimposed on a film to reveal the details of movement. If the motion of the subject is repetitive, as is the rotating propeller of an airplane, setting the strobe at exactly the same rate as the propeller makes the motion appear to stop.

The color of the flash depends on the mixture of gases used in the glow tube, and for white light the tube is commonly filled with xenon.

7-5 FLUORESCENT LAMP

The common fluorescent lamp is more efficient and economical than incandescent lamps because of its greater luminous efficacy and longer life. Light comes from the surface of a long glass tube sealed at both ends by fittings that hold electrodes. The tube, or bulb, contains a highly purified filling gas at low pressure—usually argon or an argon-neon mixture—and a small amount of liquid mercury. The arrangement is illustrated in Fig. 7-14. The filling gases ionize readily when sufficient voltage is applied across the electrodes (called "cathodes" in these tubes); and when a conducting path forms, the decrease in resistance allows an appreciable current to flow, thereby warming and vaporizing the mercury. The small amount of mercury causes a pressure increase of only 10^{-5} of atmospheric pressure, but the excitation of these atoms by the electrons in the current causes a strong radiation both in the visible and in the ultraviolet, especially at 254 nm. A coating of fluorescent powder on the inner wall of the tube absorbs the ultraviolet and emits light over a broad spectrum through fluorescence. This translucent powder allows the visible light from the excited mercury to pass with little absorption. The extended surface of the bulb produces fairly even illumination with few harsh shadows. Fluorescent bulbs should be treated with care, because they contain a small amount of toxic mercury.

The spectral power distribution of the lamp can be tailored to achieve a desired appearance through the selection of the type and amount of various phosphors in the coating (Fig. 7-15). Because relatively little of the excitation power is converted into heat, the outside of the tube remains at a relatively cool 40°C. Approximately 36% of the input power is radiated in the infrared, and most of this power comes from the bulb's wall.

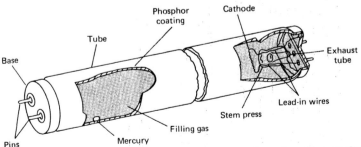

Fig. 7-14. Cutaway view of a fluorescent tube. When placed in a socket, two electrical contacts must be made to the two pins at each end of the tube, so that the cathode may be heated and a high voltage established between one cathode and the other.

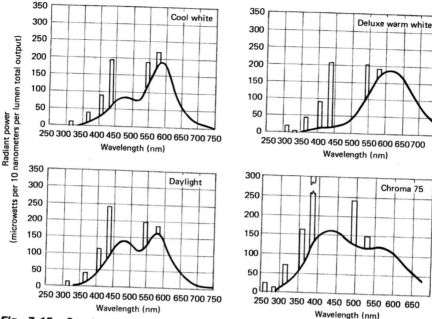

Fig. 7-15. Spectra emitted by four different fluorescent lamps. "Cool white" is designed for highest efficiency consistent with acceptable color rendition in many applications. Skin tones appear much more natural under "Deluxe warm white" because of the extra power in the red. A "cooler" appearance is rendered by "Daylight," which has a more even spectral power distribution. Light approximating north skylight must peak in the blue and is provided by "Chroma 75." The descriptive names are trademarks of General Electric. (Adapted from General Electric Company Publication TP-111.)

Fluorescent lamps emitting strongly in the blue, as is true for "cool white" versions, also yield a modest amount of ultraviolet, which can affect fabrics and certain paints after long exposure.

Less voltage is needed to strike the initial arc if the cathodes are heated to release some electrons; formerly, bulbs were started by first passing a current through the electrodes before applying the high voltage between electrodes. In such "preheat" circuits, there is a time delay between turning "on" the light and appearance of the light. Desk lamps still use this system, and a button must be held down for a few seconds until a sufficient number of electrons has boiled off to enable the arc to strike when the button is released. Because many new lamps have continuous heating of the electrodes, they will start immediately after the switch is turned on. A different instant start system merely applies a sufficiently high voltage between cold electrodes to form the discharge.

Fluorescent bulbs require a "ballast" in the lamp's electrical wiring to limit the current through the bulb. An arc discharge has the interesting property that the more current in the arc, the greater the ionization and the lower the resistance. This would lead to a runaway increase in current if a voltage source were connected between the electrodes. A ballast serves as an additional electrical load to prevent such an increase, and is used

TABLE 7-1 Efficiency of Various Light Sources[a]

Edison's first bulb	1.4 lumens/W
Carbon arc	5
Filament incandescent lamp	16–20
Mercury lamp	50
Fluorescent lamp	75–80
Sodium lamp (high pressure)	140

[a]Efficiency is expressed by the number of lumens of light output for each watt of power consumed.

with all types of electrical discharge lamps, including neon, sodium, and mercury lamps. A transformer provides sufficient voltage to start the discharge and, in the case of continuously heated electrodes, has a low voltage output for this purpose.

Since the resistance of the conducting path through the ionized gas in the bulb increases with the length of the path, the power rating of fluorescent bulbs is closely related to their length. A 40 cm bulb is rated at about 14 W, and a 120 cm bulb at 40 W.

Table 7-1 gives representative examples of light sources, which vary considerably in their luminous efficiency. A wide range of color temperatures can also be represented, as shown in Table 7-2.

7-6 LASERS

A source that produces light with some truly unique characteristics is the laser. The name is an acronym for *l*ight *a*mplification by *s*timulated *e*mission of *r*adiation. Historically, the basic principles on which this source operates were first exploited to produce a device that would amplify microwave radiation, the "maser," which is used in microwave communications and in radio astronomy.

The new feature that makes lasers unique is described by the words "stimulated emission" in the name. This is based on the notion that an excited atom more rapidly gives up its excess energy by radiating a photon when another photon of the same energy is present. This is a basic phenomenon of quantum theory. In effect, a photon "stimulates" an excited atom to radiate another photon, and the new photon has the same energy (and

TABLE 7-2 Color Temperatures for Representative Light Sources

Source	Temperature (K)
Candle flame	1900
Sodium arc lamp	2100
250 W photoflood	3200
Carbon arc crater (hard core)	4000
Colorarc 300	5000
Daylight photoflood	5000
High intensity carbon arc (sun arc)	5500
Sunlight	5500
Xenon short-arc	6400
Skylight	12,000–20,000

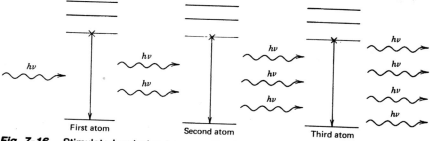

Fig. 7-16. Stimulated emission from a collection of identical excited atoms increases the intensity of radiant energy at one frequency as successive atoms are stimulated to radiate photons of the same frequency.

frequency) as the first one. Remarkably, the new photon produced by this stimulated emission is coherent with the first one. The result is depicted in Fig. 7-16 for a collection of identical atoms. The amplitude of the light beam grows as it moves past the atoms.

There are other processes that compete with this amplification. Some of the radiated photons are not given off in the same direction as the beam and thus do not help it to increase in intensity (spontaneous emission). But more important, some photons are absorbed by atoms that are not excited but are in their ground state. Thus there is a competition between this absorption and the stimulated emission. The only way to have a net gain in the number of photons in the beam is to always have more atoms in the excited state than ground state. This is said to be a condition of "inverted population," because it is a property of nature that normally we always have fewer atoms in the excited state than ground state. The laser must include a mechanism to excite the atoms and produce the inverted population. This is called a "pumping source" and may be a lamp that emits ultraviolet radiation to "pump" the atoms into the excited state (Fig. 7-17).

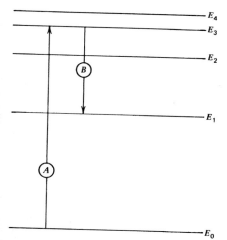

Fig. 7-17. One method to "pump" atoms into an excited state is by bathing them in ultraviolet light, so that a transition such as *A* occurs. One or more subsequent transitions such as *B* may quickly follow before the atom ends up in the excited state of interest (such as E_1). This state should have a lifetime that is sufficiently long to make it more likely that stimulated emission and not spontaneous emission will occur in a laser.

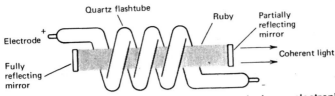

Fig. 7-18. An early design for a ruby laser employing an electronic flashtube for pumping the chromium atoms in the ruby.

The first successful laser used the light from chromium atoms in a crystal of ruby (Fig. 7-18). As was shown in Fig. 5-21, these impurity atoms that are responsible for this laser's red color absorb intermediate and short wavelength radiation. They quickly lose some of this energy to the crystal and eventually reach the excited state that lies just above the ground state. Since selection rules discourage a rapid transition from this excited state to the ground state, the atoms remain for a comparatively long time in this state before fluorescing. Consequently, if the ruby is strongly pumped with short wavelength radiation, there is a good chance that a photon emitted earlier by one of the atoms arrives at another atom before it has fluoresced, and stimulated emission takes place. A cascade effect proceeds as the two photons travel on to produce stimulated emission from other excited atoms.

To have the laser beam build up a high intensity within the ruby and enhance the chance for stimulated emission, mirrors are placed at each end of the crystal. The laser beam is reflected back and forth, as photons traveling parallel to the axis of the crystal increase in numbers while those not reflected directly back by the mirrors are lost. The mirror at one end is constructed so that it is not perfectly reflecting; and it transmits perhaps only 1% or 2% of the incident light. The amount that passes through is the "laser light." Because only photons traveling exactly parallel to the axis build up, the fraction that emerge form a well-collimated beam. An engineering decision fixes the percentage transmission of the mirror, taking into account the normal desire to have a high beam intensity emerge while still reflecting enough to sustain the stimulated emission properly.

The ruby must be pumped with such an intense source that it warms appreciably during the process. To avoid damage from overheating, the laser is operated only for short intervals of time. It is consequently an example of a "pulsed laser."

A second type of laser uses a gas as the lasing medium. In a gas laser the atoms to be pumped are contained within a glass tube. This gas is excited by an electrical discharge as in a glow tube. To avoid reflection, the ends of the tube are flat and are inclined at Brewster's angle. Consequently the light that passes through is polarized. This light is reflected back by a completely reflecting mirror at one end and a partially reflecting one at the other, as in the ruby laser.

The color of the laser light from a gas laser is characteristic of one of the constituents. If the gas is a mixture of helium and neon, a red light at 633 nm from the neon is produced. This is a particularly effective laser because it can be efficiently pumped and produces a radiant flux of several hundred milliwatts. The helium-neon laser can be operated continuously and, for that reason, is a favorite for many practical applications. Carbon dioxide provides another medium for laser action. The excited state that emits the laser "light" corresponds to a bending of this linear molecule. In this case the name "laser" is a misnomer, because the emitted radiation is at 10,600 nm, in the infrared. Carbon dioxide lasers (or "lrasers") are either continuous or pulsed, and the maximum radiant flux achieved to date is in excess of 10 kW for pulsed operation.

Argon is used in another gas laser that has found many applications. It can lase at several different wavelengths (Fig. 3-15). By inserting a prism between one end of the tube and the adjacent mirror, one of these wavelengths can be selected for laser action.

Several special characteristics of the light produced by a laser can be appreciated from the preceding explanations:

1. The light is *monochromatic.* One factor contributing to this feature is that the photon produced by stimulated emission has the same frequency as the stimulating photon; therefore all photons have the same frequency and wavelength.
2. The radiant flux at a select wavelength is great. This is because much of the pumping energy is converted through the laser action to photons at one frequency. Most lasers do not emit great overall power (say, 1 mW to 1 W) but, compared to other light sources, their flux is concentrated in a narrow portion of the spectrum.
3. The light is *coherent.* The reason for this unusual feature is that stimulated emission produces photons coherent with the original ones. Coherence is a useful property for investigating interference phenomena with large or widely spaced slits, or for producing holograms.
4. The beam is well *collimated.* The origin of this adjective is the same as for the word "column," and it means that the emerging photons all travel in virtually the same direction, that is, the beam has very little divergence. The end mirrors of the laser are responsible for this collimation, because photons not traveling perpendicular to the mirrors' surfaces leave through the side of the laser after several reflections and therefore do not contribute to subsequent stimulated emission. Good collimation is useful in surveying. A laser recently directed toward the moon (3.8×10^5 km away) left the laser with a beam diameter of 0.5 cm and illuminated a circular region on the moon with a diameter of only 1.5 km. The divergence of the beam amounted to 5 sec of arc, or 4×10^{-6} radians. Another advantage of collimation is that a lens can focus the light to a very small point. This produces an exceptionally intense region of light, and the radiant energy is sufficient to melt steel or re-attach a detached retina to the eye.

Anyone who has seen a laser beam illuminating a wall or other surface notices a curious phenomenon: the wall has a granular or speckled appearance. If the observer moves his head while continuing to look at the illuminated spot, the speckles move about. This is a clue to the fact that the phenomenon comes from an interference effect. Because the wall is not really smooth but has fine irregularities, the reflected light waves must travel different path lengths before reaching the eye where they form an interference pattern on the retina.

7-7 LIGHT-EMITTING DIODE

In Chapter 5 we briefly mentioned the fact that a semiconductor, unlike a metallic conductor, is a material in which the lowest excited state of the electrons lies at a measurable energy above the ground state. This difference in energies is called the *energy gap* of the particular semiconductor. When a device called a *diode* is fabricated from an appropriate semiconducting material (such as gallium arsenide), the presence of this energy gap can be exploited to create light. On passing an electrical current through such a diode, electrons enter one side in states above the gap and are removed at the other side from states below the gap. At one point in the diode electrons drop from the excited state into the ground state, and many give up their energy by emitting photons. The energy of the photon corresponds approximately to the energy of the gap. Light from gallium arsenide is a saturated red, associated with emission at the extreme long wavelength end of the spectrum. The energy gap itself corresponds to a wavelength of 870 nm, just outside the visible spectrum. The light-emitting diode (called an LED for short) is an efficient device for producing light and is popularly used in digital watches and in scientific equipment such as hand calculators where a fairly bright but small display is wanted. Light-emitting diodes can also be used as the basic element of a laser. Diode lasers are currently being used in optical communications systems in which many telephone messages are carried by a single optical fiber. (See Section 8-1.)

SUMMARY

All objects emit electromagnetic radiation as a consequence of the thermal agitation of their atoms and electrons. A black body—which absorbs all radiation falling on it—emits radiation with a characteristic SPD that is determined entirely by its temperature. The SPD of the sun, which is approximately equal to that of a black body at 5800 K, peaks near a wavelength of 500 nm. Bodies of higher temperature emit radiation peaked at shorter wavelengths, whereas the radiation of bodies at lower temperature is peaked at longer wavelengths. Because the total radiant flux emitted from a given area of a black body increases with the fourth power of the temperature, a slight increase in temperature causes a substantial increase in radiant flux. Human skin radiates very weakly, with its SPD peaked near 10,000 nm in the infrared. The SPD of a light source that is not too different from a black body is conveniently characterized by the color temperature, which is the temperature of a black body whose color best matches the color of the source. The color temperatures of stars range from 5000 to 11,000 K, fluorescent lamps typically have color temperatures near 6000 K, and incandescent lamps near 3000 K. The incandescent lamp is an example of a thermal radiator that has an SPD close to that of a black body. Other light sources such as electrical discharge lamps, light-emitting diodes, and especially lasers have spectra containing narrow lines, and their SPDs do not resemble those of black bodies.

QUESTIONS

7-1. Describe the process by which a warm object emits light.

7-2. Why does a rough surface better approximate an ideal black body than a smooth surface?

7-3. When an object is heated, how does the wavelength of the peak thermal radiation change? How does the power emitted by its surface change at wavelengths that are greater and less than where the peak is found?

7-4. At what wavelength is the maximum thermal radiation from a tungsten filament of an incandescent lamp if it has a temperature of 3200 K and acts as a black body?

7-5. Describe the sequence of colors of an incandescent object as it is heated from 2500 to 10,000 K.

7-6. How much thermal energy is emitted from 1 square centimeter of a plant's leaf at room temperature of 22°C if it radiates like a black body? In what portion of the electromagnetic spectrum is this found?

7-7. What sources are represented by the C.I.E. standard illuminants A, B, and C? Why are these sources chosen for special recognition?

7-8. Why does an incandescent lamp produce more luminous flux if supplied a higher voltage? Why does the lamp then have a shorter life?

7-9. A "long life" incandescent lamp has a filament that operates at a somewhat lower temperature than usual. If it operates at a color temperature of 3000 K instead of 3200 K by what percentage is its luminous flux reduced?

7-10. Explain why a tungsten-halogen lamp has a longer life than a standard incandescent lamp operating at the same filament temperature.

7-11. What is the principle of the "neon light" that is popularly used in displays?

7-12. What gas is confined in a fluorescent lamp for the purpose of exciting the phosphor coating on the tube?

7-13. Why is the color of a sodium arc lamp less saturated when operated at high pressure than low pressure?

7-14. Explain why light from a laser is more coherent than light coming from an ordinary incandescent lamp.

7-15. How is the light from a laser made so well collimated.

7-16. Does the light-emitting diode emit coherent or incoherent light? Explain.

FURTHER READING

A. Stimson, *Photometry and Radiometry for Engineers* (Wiley, New York, 1974). Readable description of light sources and light-measuring instruments (even for the nonengineer!) Many useful tables.

I.E.S. Lighting Handbook, 1981 Applications Volume; IES Lighting Handbook 1981 Reference Volume (Illuminating Engineering Society, New York, 1981). How to design lighting installations for a wide variety of spaces, including residences and public rooms.

Light and Its Uses, Making and Using Lasers, Holograms, Interferometers and Instruments of Dispersion (Freeman Press, San Francisco, 1980). Readings from the "Amateur Scientist" column of *Scientific American*. It contains step-by-step instructions for building and using lasers, holograms, and devices for demonstrating the wave aspects of light.

H. E. Edgerton, *Electronic Flash, Strobe* (The M.I.T. Press, Cambridge, Massachusetts, 1979). The technical aspects of strobe lamps, circuits, and applications.

G. Wyszecki and W. S. Stiles, *Color Science, Concepts, and Methods—Quantitative Data and Formulas* (Wiley, New York, 1967). Detailed, professional level coverage of light sources.

J. C. Dainty, Ed., *Laser Speckle and Related Phenomena* (Springer-Verlag, Berlin, 1975). Technical discussion of laser beams, interference, and coherence.

S. T. Henderson, *Daylight and Its Spectrum* (Halsted Press, New York, 1977). Detailed description of sunlight and the light from comparable artificial sources.

A. G. Gaydon and H. G. Wolfhard, *Flames: Their Structure, Radiation and Temperature* (Methuen, 1979). Introduction for the junior or senior level in college science or engineering.

Encyclopedia Brittanica, 11th Edition (Cambridge University Press, 1911). For the historically minded a delightful account of early sources of light, including oil and incandescent lamps, as well as lighthouses and their lenses.

D. C. O'Shea and D. C. Peckham, "Resource Letter L-1: Lasers," *American Journal of Physics,* Vol. 49 (1981), p. 915. An annotated list of books, review articles, pedagogical articles and films on lasers, demonstrations with lasers, and background topics.

From the *Scientific American*

W. C. Gardiner, Jr., "The Chemistry of Flames" (February, 1982), p. 110.

M. Josephson, "The Invention of the Electric Light" (November, 1959), p. 98.

A. L. Schawlow, "Laser Light" (September, 1968), p. 120.

C. K. N. Patel, "High-Power Carbon Dioxide Lasers" (August, 1968), p. 22.

J. Walker, "The Amateur Scientist: The "Speckle" on a Surface Lit by Laser Light Can Be Seen With Other Kinds of Illumination" (February, 1982), p. 162.

M.B. Panish and I. Hayashi, "A New Class of Diode Lasers" (July, 1971), p. 32.

■ CHAPTER 8 ■

RAY OPTICS

The phenomena of reflection and refraction are the basis for a great variety of important optical devices. Eyeglasses, magnifying lenses, microscopes, and cameras are but a few examples. The way that lenses and mirrors alter the direction of light and form images is the subject of this chapter. The name *ray optics* describes our approach, since we shall trace the path of a narrow beam of light as it is successively refracted or reflected on passing through the lenses of an instrument. Hero's law of reflection and Snell's law of refraction introduced in Section 5-1 are fundamental. Although the wave phenomenon of diffraction certainly occurs, we assume that the lenses are sufficiently large to make it a negligible effect.

■ 8-1 REFRACTION PHENOMENA

Because refraction bends light's path, the apparent location and size of a subject may be altered as light travels toward the observer. We shall first consider a simple situation where refraction occurs at a flat surface.

Refraction in Water

In Fig. 8-1 we consider a person standing on the bank of a river or lake, looking down into the water at an angle. A light ray leaving a fish at position F will be refracted at the surface of the water and reach the person along the path indicated by the solid line. The amount of refraction is governed by Snell's law (Eq. 5-2), which relates the incident angle θ_1 and refraction angle θ_2 by :

$$n_1 \sin \theta_1 = n_2 \sin \theta_2 \tag{8-1}$$

Recall that n_1 is the refractive index of the medium from which light is incident (water), and n_2 is the index of the medium into which it is refracted (air).

But the psychological processes in visual perception are conditioned to "see" an object in a direction straight back along the arriving ray, and the fish is erroneously perceived as being at the point F'. Only if viewed from directly overhead

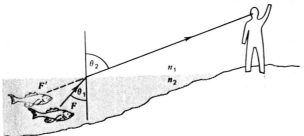

Fig. 8-1. The fish F appears to be at position F' as a result of refraction at the surface of the water.

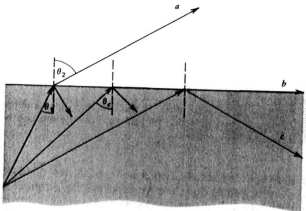

Fig. 8-2. When light passes into a medium of lower refractive index, it travels at a greater angle from the perpendicular direction (*a*). A small portion of the light is reflected at the surface, except when the incident angle exceeds the critical value θ_c, in which case *all* of the light is reflected (*c*).

where $\theta_1 = \theta_2 = 0°$ does the fish appear to be in the correct direction. A simple extension of this argument also explains the perception of an apparent bend at the waterline in a straight stick protruding from a pond.

Snell's law predicts that light entering a medium of lower refractive index (from water into air in this case) is refracted into a direction further away from the perpendicular. As shown in Fig. 8-2, this fact leads to a remarkable phenomenon if we consider rays that are incident at successively greater angles. At a particular angle the refracted ray (*b*) goes off parallel to the surface! This special condition is described by a refraction angle of $\theta_2 = 90°$. The incident angle θ_1 corresponding to this condition is called the *critical angle* and is denoted by the symbol θ_c. By using the fact that sin 90° = 1 in Eq. 8-1, Snell's law gives the value of θ_c by the formula

$$\sin \theta_c = \frac{n_2}{n_1} \qquad (8\text{-}2)$$

For water the refractive index is $n_1 = 1.33$, and for air $n_2 = 1$. The sine of the critical angle between water and air is thus sin $\theta_c = 1/1.33 = 0.75$. Figure 5-4 shows us that the angle θ_c with this value for its sine is $\theta_c = 49°$.

Total Internal Reflection

What happens to the light if it is incident at an angle that exceeds the critical angle? The light is completely reflected so that it remains in the first medium (ray *c* in Fig. 8-2). This effect is called *total internal reflection*. It is predicted by Snell's law when-

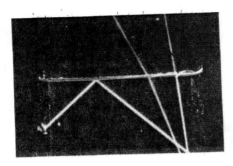

Fig. 8-3. Refraction and total internal reflection of light rays at the surface of water.

ever the calculated value of sin θ_2 exceeds 1. This prediction is nonsense, because the sine of an angle can never exceed 1. Therefore we conclude that a refracted ray does not exist. Instead, the light is totally reflected at the surface, as shown in Fig. 8-3.

Let us turn around the situation in Fig. 8-2 and imagine a submerged swimmer looking up at the water's surface as in Fig. 8-4. Because of refraction, the swimmer sees a displacement in the apparent position of subjects above the water's surface. The light seen by the swimmer now goes from the air into the water. The relationship between the angles of a ray in the air and in the water remains as before, because Snell's law fixes it by the ratio of the refractive indices, and not by which direction the light travels. A bird at position B appears to be at B', while the duck at D on the surface of the water appears to be at D'. Since the ray from the duck strikes the surface of the water nearly horizontally, that is, at nearly an angle of 90° from the vertical, on entering the water the angle that this ray makes with the vertical is very close to the critical angle θ_c.

Suppose that the swimmer in Fig. 8-4 looks further to the right, beyond point M where the ray refracted at the critical angle entered. All rays reaching the swimmer from these more distant points leave the water's surface at angles greater than θ_c, and cannot correspond to rays that had been refracted on entering from the air. Instead, these rays coming from the surface have been *reflected at the surface* by the phenomenon of total internal reflection. To the swimmer looking to the right of point M, the water's surface will appear to be a mirror in which will be seen the image of subjects such as the fish located in the water at F. In other words, the world above the water level is compressed into a cone of visual directions extending as far as $\theta_c = 49°$ (corresponding to point M). At larger angles to the vertical the light reveals subjects within the water itself. Look for yourself when you are at the bottom of a swimming pool!

Lightguides

An important application of total internal reflection is the *light pipe* or *lightguide*. The best-known example is found in fiber optics. A lightguide in that case is a small diameter fiber of a transparent material having a refractive index greater than air's.

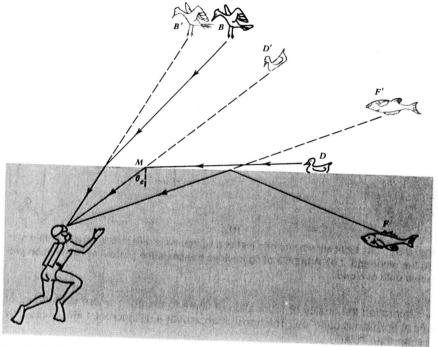

Fig. 8-4. The mirror illusion. To a swimmer looking up from below the water's surface, the region of the water's surface at an angle greater than the critical angle θ_c (at point M) looks like a mirror.

When light is directed into one end, it travels in a straight line until meeting the surface (Fig. 8-5a). If the fiber is not bent too sharply, this beam meets the surface with an incident angle that exceeds the critical angle, and it is totally reflected. This may occur several times before the beam reaches the other end of the fiber, where it emerges from the flat surface. Physicians use optical fibers as a catheter that can be inserted into the body for providing illumination and visual inspection of areas that cannot be seen directly (Fig. 8-5b). A bundle of these fibers, properly arranged, transmits light patterns with a spatial resolution determined by the size of the individual fibers.

Lightguides promise advantages also in communications, because the amount of information or number of telephone calls that can be encoded on an electromagnetic wave increases with the frequency of the wave. Microwaves with wavelengths of about 3 cm and frequencies of approximately 10^{10} Hz commonly carry signals over long distances now. Imagine the advantage of a lightguide where the wavelength can lie in the visible spectrum and the frequency is 10^{14} Hz! The advance in materials technology during the past decade has increased the purity

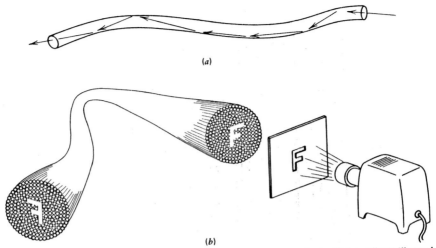

(a)

(b)

Fig. 8-5. (*a*) Light shown into one end of a lightguide is reflected internally until reaching the other end. (*b*) A bundle of lightguides transmits the pattern of illumination projected onto one end.

and improved the quality of fabrication of fibers to the point where the amount of light absorption is quite low. The transmittance over a distance of 1 km is presently greater than 70%.

Sparkle of a Gem

A cut diamond is a splendid example of total internal reflection. A diamond cutter's art is in choosing how to cut the crystal along the best directions so that light is internally reflected two or more times until it is finally refracted out to the viewer. This accounts for its brilliance. At the same time the facets of the cut stone are chosen with angles to enhance the dispersion of the emerging light and better separate its spectral components. A diamond's sparkle comes from these spectra being rapidly swept past the eye as the diamond is moved.

■ 8-2 SIMPLE LENSES

The phenomenon of refraction has found its most useful application in the optics of lenses. It was already known to the Franciscan friar Roger Bacon in the mid-thirteenth century that people with poor eyesight could often improve their vision by looking through an appropriately shaped piece of glass. Not long after, the Italian Salvino D'Armate exploited this notion by making the first pair of eyeglasses.

We shall show that the key to this possibility is the fact that light suffers a net change in direction on passing through a piece of glass whose surfaces are inclined to one another as in a prism. We note that the main effect of a prism is to refract or bend light passing through it. For now we ignore the much smaller effect of dispersion, which we shall consider in Section 8-5.

With two prisms we can illustrate the effect of a lens when light comes from a small distant source, such as a star. These rays are nearly parallel on reaching the prisms, as shown in Fig. 8-6a, because they come from such a great distance. With the prisms arranged as shown, two rays passing through corresponding points on the prisms are brought together at a single location. This is called the *focal point*. The distance between the focal point and the center of this crude lens is called the *focal length*. However, this is an unsatisfactory lens, in the sense that it cannot bring all the intercepted rays from a star to a focus at a common point. The

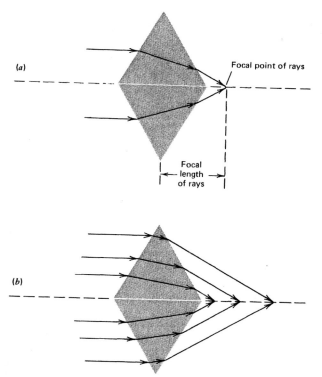

Fig. 8-6. (*a*) Two prisms with bases joined form a crude "lens" because two parallel rays from a distant source are brought together as a result of refraction. (*b*) This arrangement suffers from the problem that not all rays arriving from the source are brought to the same focal point.

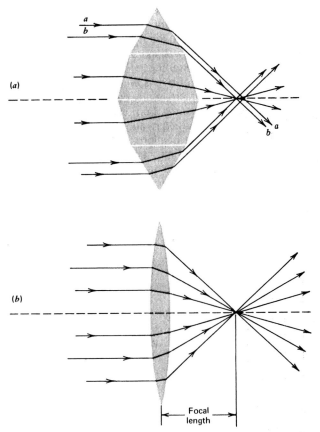

Fig. 8-7. (*a*) An improved "lens" constructed from sections of four prisms. The more sections that are included, the better is the quality of focus. (*b*) The limiting situation of a curved surface focuses all the light to a common point.

focal length for each set of rays depends on where they strike the prisms (Fig. 8-6*b*).

To improve the focusing quality, each half of the "lens" could be made from two prisms of differing angles as shown in Fig. 8-7*a*. Then rays arriving near the center are refracted less than those farther out. Still, rays coming through different regions of the same prism will not be focused at the same point (rays *a* and *b* of Fig. 8-7*a*). Continuing the development of the "lens" we could make each half from even more sections of prisms with differing angles. The limiting case is reached when the surfaces of the lens have smooth curvature (Fig. 8-7*b*). In practice it is easy to grind smooth surfaces with a spherical curvature and with this

Fig. 8-8. Adaptation of a sixteenth-century illustration of a "burning glass" that focuses the sun's rays one focal length *f* behind the lens to ignite a fire.

shape a focus of adequate quality can be achieved. The distance between the center of the lens and the focal point is called the *focal length* of the lens.

Figure 8-8 is a sixteenth century woodcut showing how such a lens concentrates the sun's rays at the focal point to ignite a fire. When the rays of the sun, or any other distant small source, are focused to produce the smallest image, the focal length (f) of the lens can be determined. The image is formed at a distance of one focal length behind the lens.

Lenses are made in various shapes. A lens of the type just discussed is an example of a *converging lens,* because it brings light rays from a distant source together at the focal point. A converging lens with both surfaces bowed outward in the middle is said technically to be *double convex* (Fig. 8-9*a*). A variation of this shape has one surface flat and is called *plano-convex* (Fig. 8-9*b*).

By contrast, if both surfaces of a lens are bowed inward, it is an example of a *diverging lens.* As shown in Fig. 8-10, parallel incoming rays are not brought to a focus but are refracted away from the center to emerge in a spreading pattern. The spreading rays appear to originate from a common point on the lens' axis. The distance from this point to the lens is called the *focal length* of the diverging lens. The shape of this particular lens is *double concave* (also shown in Fig. 8-9*c*). A *plano-concave* lens (Fig. 8-9*d*) is also diverging. The simplest practical photographic lens has one convex and one concave surface and is called a *meniscus* lens (Fig. 8-9*e*). Common eyeglasses for reading also have this shape. If the convex side is more sharply curved (as shown), it is a converging lens; if less sharply curved, it is a diverging lens. Many of the preceding types of lenses are used in combination in optical systems of high quality, such as cameras.

All of the preceding lenses are shaped symmetrically about an imaginary line through the center called the *optical axis.* When a converging lens focuses light from a star located on the optical axis, the rays—all of which travel parallel to the axis—are brought to a focus at the focal point, which lies on the optical axis behind the lens. Its distance from the lens, or focal length, depends on the curvature and refractive index of the lens. The greater the curvature or refractive index,

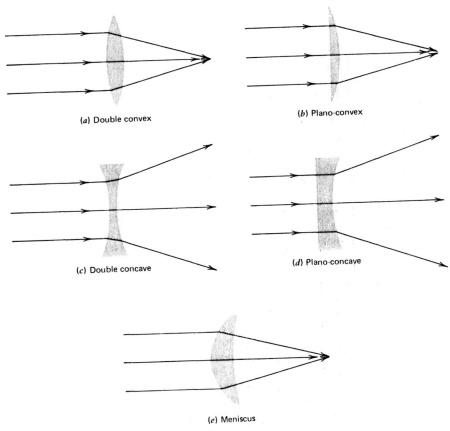

(a) Double convex

(b) Plano-convex

(c) Double concave

(d) Plano-concave

(e) Meniscus

Fig. 8-9. Shapes of common simple lenses.

the shorter the focal length and the greater the "strength" the lens is said to have. The diameter of a lens has no effect on focal length, or strength. The square of the diameter determines the area of the lens and the corresponding amount of light that is intercepted. The square of the diameter is thus a measure of the lens' light-gathering ability.

Real Images

In most applications the purpose of a lens is to form an image. The image will be either *real* or *virtual*, depending upon the type of lens and how close it is to the source of light. We use the word "subject" to denote the source. The subject may

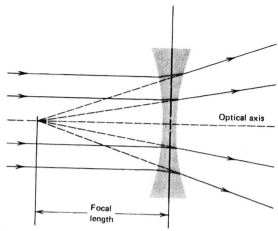

Fig. 8-10. A diverging lens refracts rays from a distant source so that they appear to come from a common point in front of the lens. The distance from the lens is said to be the focal length (*f*) of the lens.

be self-luminous as with the sun or the filament of an incandescent lamp, or it may be a surface that reflects light from another source toward the lens.

A real image is formed if light from each point of the subject is brought to a focus at a corresponding point behind the lens. Figure 8-11 illustrates this for a point on the subject that lies above the optical axis. By placing a piece of white paper perpendicular to the axis behind the lens where the light is focused, we would see a spot of light on the paper whose brightness is proportional to the brightness of the corresponding spot on the subject. The continuous pattern of light

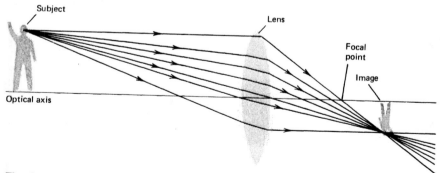

Fig. 8-11. A converging lens refracts all the rays leaving a given point on the subject so that they are brought together again at another point behind the lens.

across the paper provides an image that duplicates the appearance of the subject, although it may be of different size and inverted (top to bottom and left to right). If we were to hold the paper closer to the lens or further away, the rays from any point on the subject would no longer be focused on the paper. They would strike the paper at slightly different points, so that the spot they form would be enlarged. Thus one spot would overlap its neighbors, and the clarity of the image would be lost. The image would be "out of focus."

Whether exactly focused or not, the image produced by the setup depicted in Fig. 8-11 produces a real image. That is, we can place a paper or screen at the position of the image and see the pattern of light that is formed. Virtual images are quite different, as we shall explain shortly. Figure 8-12 illustrates how a real image can be produced simply by placing a magnifying lens far from the subject, between the subject and a viewing screen, in this example a white paper taped to a wall. Ordinarily the image is quite dim, but it is easily seen if the subject is illuminated by a bright light.

It is fascinating how the geometry of a lens sorts out rays arriving from different directions at various locations on its surface and refracts them in appropriate directions to establish an image. No image is found at the lens itself, of course, because each location on the lens receives light from every point on the subject. If the lens is relatively thin, it is easy to predict the size and position of the image. A ray diagram helps us to make such predictions quickly. We need only follow the path of two carefully chosen rays emitted from the same point on the subject and see where the lens brings them back together again.

Start with a point on the subject well off the optical axis as in Fig. 8-13. One of the rays we choose to follow is headed straight through the center of the lens. This

Fig. 8-12. (a) A simple demonstration of how to produce a bright real image with a common magnifying lens. A floodlamp is placed to strongly illuminate the subject. (b) An inverted, real image is formed on the wall if the lens is placed at the correct position between the subject and wall.

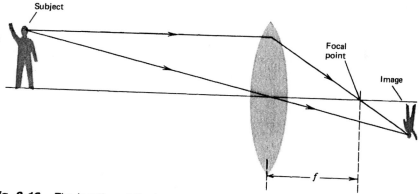

Fig. 8-13. The location of the image can be predicted by following two rays from a point on the subject and determining where they are brought together again.

ray passes through undeflected, because the lens' surface at the point of entry is parallel to the surface at the point of exit. The emerging ray is slightly displaced (see Fig. 5-5b), but this is a negligibly small effect for a sufficiently thin lens. The second ray we choose to follow heads toward the lens running parallel to the optical axis. This ray, is refracted so that it passes through the focal point behind the lens. The image of the light-emitting point of the subject is found where this ray intersects the first one.

Other rays could be traced by applying Snell's law to predict the refraction angles at the two surfaces of the lens. All the rays leaving the same point on the subject are refracted so that they pass through a corresponding point behind the lens. This is where the image of the point on the subject is formed. Because a point on the subject lying above the optical axis is focused at a corresponding point below the axis, and a point lying to the right of the axis is focused at a corresponding point to the left, the image is inverted. However, to locate the image it suffices to draw only two rays and, by using the procedure just explained in the preceding paragraph, we can avoid calculations with Snell's law.

Figure 8-14 shows for various positions of the subject how the position of an image is determined by drawing just two rays, once the focal length f of the lens is known. If the subject is moved a great distance away from the lens, say more than 10 focal lengths away, the image is focused at a distance that is very close to one focal length behind the lens. In this case, regardless of the precise distance of the subject, for all practical purposes we can say that the image lies one focal length behind the lens. The image is very small, having a size that is inversely proportional to the subject's distance (this will be proven in Section 8-3). The further away the subject, the smaller the image. If we draw a series of ray diagrams we can also convince ourselves that the size of the image is proportional to the

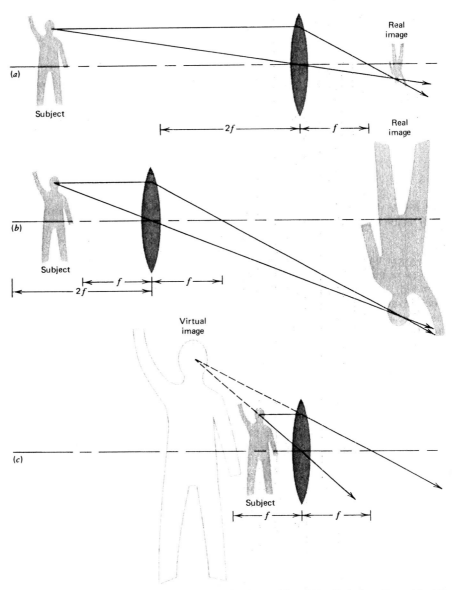

Fig. 8-14. Images produced by a converging lens of focal length *f* when the subject is at various distances from the lens: (*a*) subject further than 2*f*, the image is real, inverted, and smaller than the subject; (*b*) subject between *f* and 2*f*, the image is real, inverted, and larger than the subject; (*c*) subject closer than *f*, the image is virtual, upright, and larger than the subject.

lens' focal length. A lens with a longer focal length produces a larger image, when the subject is far away. This is the principle of the telephoto lens used on cameras to magnify a distant subject.

By following the procedure of drawing two rays, we find that as the subject is brought closer to the lens, its image moves further away and increases in size. When the subject is two focal lengths in front of the lens, the image is also two focal lengths behind the lens. The image is the same size as the subject. Once the subject is less than two focal lengths from the lens, the image is more than two behind, and the image is larger than the subject (Fig. 8-14b). When the subject reaches one focal length away from the lens, the image has been projected back an infinite distance; that is, the rays refracted by the lens are all running parallel to the optical axis and so will never come to a focus (this is just the reverse to the situation in Fig. 8-7b). Now when the subject is moved closer than one focal length from the lens a remarkable change occurs: the image becomes a virtual one. The concept of a virtual image will be explained shortly.

Strength of a Lens

A converging lens with sharply curved surfaces forms the image of a distant subject closer to the lens than one with less curvature. Its focal length therefore is shorter. Thus the *strength d* of a lens is defined logically as the *reciprocal* of its focal length: $d = 1/f$. A small value of f implies great strength. When f is expressed in meters, the strength d is expressed in a unit called the *diopter*. For example a lens with a focal length of $f = 50$ mm, or 0.05 m, has a strength of $d = 1/0.05 = 20$ diopters.

As will be explained, a diverging lens is indicated by assigning a negative value for its focal length and corresponding negative value for its strength. Eyeglass prescriptions are given by the required strength in diopters, with $+$ or $-$ indicating a converging or diverging lens. When two or more lenses are placed close together the total strength is the sum of the individual strengths.

f-number

The earlier discussion of ray optics showed that when a subject is more than about 10 focal lengths from a lens, the image lies one focal length behind the lens. Also, the height of the image is proportional to the focal length f. The increase in size with increasing f is accompanied by a decrease in brightness of the image, because the same amount of luminous flux intercepted by the lens is spread over a larger area. The concentration of light (the illuminance) forming the image is reduced, and correspondingly the brightness of the image is diminished. On the other hand, the illuminance and brightness can be increased if the diameter D of the lens is increased, since the luminous flux intercepted by the lens is proportional to the area of the lens. The ratio f/D is called the *f-number* of the lens:

$$f\text{-number} = \frac{f}{D} \tag{8-3}$$

From the foregoing we surmise that a lens with small f-number (short f and large D) yields a brighter image than one with a large f-number. In a camera, an adjustable aperture masks the outer portion of the lens and controls the luminous flux admitted through it. A high quality lens may permit a range in f-numbers from 1.2 (denoted "f/1.2") for the brightest image to f/22 for the dimmest.

Virtual Images

A virtual image is produced when rays from any point on a subject are refracted or reflected in such a way that they appear to originate from somewhere the light has never been. The image formed by a mirror is a familiar example (Fig. 8-15). The reflected rays appear to originate at a position *behind* the mirror, where of course they never were! From this effect arises the distinctive term "virtual" image.

A virtual image is also produced by a converging lens if the subject is sufficiently close, as shown in Fig. 8-14c. We draw two rays as before, with one going straight through the center of the lens and the other parallel to the optical axis and then through the focal point. With the subject less than a focal length in front of the lens, the two emerging rays diverge! Tracing back these rays we identify the point where they appear to originate. When these rays enter the eye our visual processes give

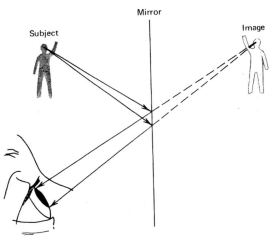

Fig. 8-15. An observer judges the location of a subject by the direction of light that is incident on the eyes. With a mirror, the light appears to come from a location behind the mirror, the same distance behind that the subject is in front.

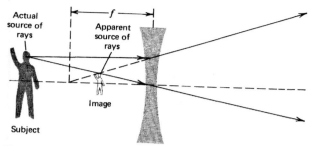

Fig. 8-16. A virtual image produced by a diverging lens. The image is located where rays from one spot on the subject appear to originate.

us the impression that the source of light lies at that point. Of course the rays were never physically at that position, so the image we see must be a virtual one. The image lies on the same side of the lens as the subject. The image lies further from the lens and is correspondingly larger. It is also upright. The preceding arrangement is exactly how a magnifying glass is commonly used, as will be explained in Section 8-6.

A diverging lens always produces a virtual image of a subject. When the subject is viewed through such a lens (Fig. 8-16), the observed image is on the same side of the lens as the subject; it is upright; and it appears to be much closer than the subject actually is. To determine where the image lies we follow two rays whose paths can be traced easily. One travels toward the center of the lens and continues straight through. The second travels parallel to the optical axis and is refracted away from the axis by the lens. It appears to have passed through a point on the optical axis one focal length in front of the lens. Tracing back along the directions of the emerging first and second rays we find a common point from which both appear to come. This is where the image is located. Because in fact one of the rays was never physically at this point, the lens has formed a *virtual* image. ■

8-3 THIN LENS FORMULAS

The relationships between image and subject positions and their sizes have simple mathematical expressions when the thickness of the lens is small compared with its focal length. This is equivalent to saying that the lens is so thin that a ray passing through its center is not appreciably shifted in the way illustrated in Fig. 5-5b. In this section we derive the formulas describing the position and size of an image and apply them to several examples.

Magnification

The size of an image (say, its height h_i) compared with that of the subject (of height h_s) gives the magnification M provided by the lens:

$$M = \frac{h_i}{h_s} \tag{8-4}$$

Figure 8-17 shows how geometrical relationships for similar triangles allow us to reexpress this ratio in terms of the distance i of the image from the lens and the distance s of the subject from the lens:

$$M = \frac{i}{s} \tag{8-5}$$

Thus, if the image is further from the lens than the subject, M is greater than 1, and the image is larger than the subject. But if the image is closer to the lens than the subject, M is less than 1, and the image is smaller than the subject.

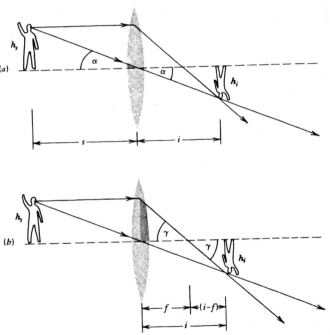

Fig. 8-17. Geometrical relationship between subject and image positions and sizes: (a) because the angles labeled α are equal, the shaded right triangles are geometrically similar and, therefore, the ratio of their heights is equal to the ratio of their bases: $h_i / h_s = i / s$; (b) because the angles labeled by the Greek letter gamma (γ) are equal, the shaded right triangles are geometrically similar and the ratio of their heights is equal to the ratio of their bases: $h_i / h_s = (i - f) / f$.

Image Position

The geometry of the rays in Fig. 8-17 also shows how the ratio of image height to subject height is related to the image distance i and focal length f of the lens:

$$\frac{h_i}{h_s} = \frac{i - f}{f}$$

(8-6)

To recast this expression in a form that involves only the subject and image distances, we use Eqs. 8-4 and 8-5 to replace the left-hand side and obtain:

$$\frac{i}{s} = \frac{i - f}{f} = \frac{i}{f} - 1$$

(8-7)

Then, by dividing the left side and right side by i and rearranging terms, we arrive at the *thin lens formula:*

$$\frac{1}{s} + \frac{1}{i} = \frac{1}{f}$$

(8-8)

If we know the values of any two quantities in the thin lens formula (8-8), we can predict the value of the third.

Example 8-1 Projection of a Slide

A lens of 10 cm focal length is positioned 10.5 cm from a photographic slide in a projector. Where is the image formed and how much larger is it than the scene in the slide? To answer this we first note that the image is obtained by transmitting light through the slide whereupon it is then gathered by a lens and brought to a focus. The surface of the slide serves as the subject. We solve Eq. 8-8 for the image distance i. We can express all lengths in meters, in which case the answer is also expressed in meters; or we can express all lengths in centimeters and obtain the answer in centimeters. Choosing the latter alternative we find that

$$\frac{1}{i} = \frac{1}{f} - \frac{1}{s} = \frac{1}{10} - \frac{1}{10.5} = 0.100 - 0.095 = 0.005$$

$$i = 200 \text{ cm}$$

Consequently the image is 200 cm from the lens. The magnification can now be found by using Eq. 8-5 with the information we just obtained:

$$M = \frac{i}{s} = \frac{200}{10.5} = 19$$

For a typical application, projecting a 35 mm slide (whose dimensions are actually 24 mm × 36 mm) would produce on a screen 200 cm away an image that is 46 × 68 cm.

If the screen when first set up is not at the proper distance for focus, either the distance between screen and projector (i) or the distance between lens and slide (s) must be adjusted. The focusing adjustment on most slide projectors changes s. A more elaborate projection system is discussed in Section 8-11.

Example 8-2 Image of a Distant Subject

When light arrives at a lens from a distant subject, the image is focused nearly one focal length from the lens. This is easily demonstrated when the distance of the subject is infinite, a value indicated by the symbol $s = \infty$. The thin lens formula predicts for this case:

$$\frac{1}{i} = \frac{1}{f} - \frac{1}{\infty} = \frac{1}{f} - 0 = \frac{1}{f}$$

Notice that the fraction $1/\infty$ is zero, because the denominator is infinitely large. From the above expression we find that $i = f$. The image remains one focal length from the lens so long as the subject is sufficiently far from the lens. Even when the subject is as close as 10 times the focal length, the image distance behind the lens is only 10% greater than the focal length.

The conclusion reached in Example 8-2 leads to a simple prediction for the magnification. We deduce that the magnification $M = i/s$ for a distant subject can be rewritten to a good approximation as

$$M = \frac{f}{s} \qquad (8\text{-}9)$$

Therefore for a fixed subject distance s, the image size is proportional to the focal length of the lens. Compared with a 50 mm lens, one with 100 mm focal length provides twice the image size; and a 200 mm lens, four times the size. This equation gives the magnification for a telephoto lens when photographing distant subjects.

The thin lens formula also predicts when a virtual image is formed. This is indicated when the value calculated for i turns out to be negative. The negative sign tells us that the image is on the same side of the lens as the subject, instead of the opposite side, and therefore is virtual.

The thin lens formula also applies to a diverging lens, but the focal length describing the lens is assigned a negative number. Then the formula always predicts an image distance i that is negative. Again, this says that the image is on the same side of the lens as the subject, and the image is therefore always virtual.

8-4 COMPOUND LENSES

When two or more lenses are arranged in series, the assembly is called a *compound lens;* and the individual lenses, the *elements.* To predict the image position and size, the thin lens formula (Eq. 8-8) can be applied to calculate the position and size of the image produced by the first lens. This image serves as the subject for the second lens. Its distance from the second lens is the new subject distance for this lens. Then the thin lens formula is applied again to deduce the features of the image of this second lens. The procedure is repeated to predict in succession the effect of each lens. Alternatively, a pair of rays can be "traced" through the system to locate the image.

Virtual Subjects

In our previous use of the thin lens formula the light from the subject was diverging as it approached the lens. The distance s was given a positive sign to indicate that the subject was positioned in front of the lens. However, in a compound lens the light to be focused by a given lens may already have passed through one or more elements and be converging as it approaches the lens of interest. As shown in Fig. 8-18, the convergence in the absence of the lens would have produced a real image behind our lens' position. To distinguish this converging aspect of the incoming rays, we say that the lens has a *virtual subject*. The virtual subject is located where the rays would have focused. Its distance from the lens is assigned a negative value when used in the thin lens formula to predict where the image lies when the lens is inserted.

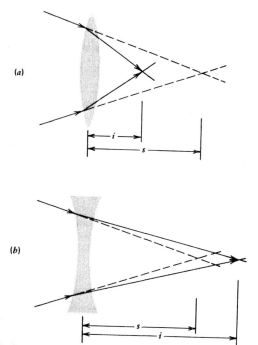

(a)

(b)

Fig. 8-18. If light rays coming toward a lens are converging to a focus at a distance s behind the lens, the subject of the lens is said to be a virtual subject. For a converging lens (*a*) the image distance i is shorter than s; but for a diverging lens (*b*) the image distance is longer.

Addition of Lens Strengths

If the elements of a compound lens are placed close together, so that the distance between the first and last is greatly exceeded by both the subject distance and image distance, the thin lens formula has a simple generalization to predict features of the image. This is easily deduced in the case of two converging lenses by noting that the image of the first is the subject of the second and by applying Eq. 8-8 to each lens in succession. For the first lens of focal length f_1, subject distance s_1, and image distance i_1, we have

$$\frac{1}{s_1} + \frac{1}{i_1} = \frac{1}{f_1} \tag{8-10}$$

The image of this first lens lies nearly the distance i_1 *behind* the second lens. The distance of the subject for the second lens, of focal length f_2, is therefore $s_2 = -i_1$, the minus sign appearing because this subject is virtual. The image distance i_2 of this second lens, of focal length f_2, is then predicted by the thin lens formula:

$$-\frac{1}{i_1} + \frac{1}{i_2} = \frac{1}{f_2} \tag{8-11}$$

Adding the left sides of Eqs. 8-10 and 8-11 and setting that equal to the sum of the right sides, we arrive at the conclusion:

$$\frac{1}{s_1} + \frac{1}{i_2} = \frac{1}{f_1} + \frac{1}{f_2} \tag{8-12}$$

The subscripts for distances can now be dropped, since with little error all distances can be measured from the center of the compound lens instead of from the individual elements:

$$\frac{1}{s} + \frac{1}{i} = \frac{1}{f_1} + \frac{1}{f_2} \tag{8-13}$$

The form of this equation implies that the right-hand side is an expression for the inverse of the focal length f of the compound lens:

$$\frac{1}{f} = \frac{1}{f_1} + \frac{1}{f_2} \tag{8-14}$$

Since the strength of a lens is just the inverse of its focal length (expressed in meters), the net strength d according to Eq. 8-14 is the sum of the strengths of the individual elements:

$$d = d_1 + d_2 \tag{8-15}$$

Following this logic, the net strength of three lenses placed close together would be $d = d_1 + d_2 + d_3$, and so on. If, for example, three lenses are used, one with $+2$ diopter

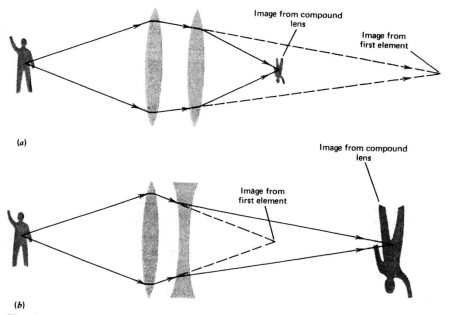

Fig. 8-19. (*a*) Two converging elements provide greater strength than either used separately. (*b*) A diverging element reduces overall strength.

strength, one with $+3$ diopter, and one with -1 diopter, the net strength is $d = +2 + 3 - 1 = +4$ diopter. Figure 8-19 gives two examples.

8-5 LENS ABERRATIONS

A simple lens no matter how well crafted suffers from several defects. Only the simplest optical instruments employ a single lens. Many defects inherent in a single lens can be overcome by a compound lens for which the designer has the freedom to choose the shape of the individual elements.

Chromatic aberration stems from inherent dispersion in the refractive index of the lens. Because dispersion causes blue light to be refracted by a greater angle than red, the focal length varies with wavelength (Section 5-2). An image formed by a converging lens will appear blurred, the most noticeable feature being colored fringes at the edges of the image. This defect is commonly overcome by replacing the simple lens with a compound one of two or more elements formed from glasses of differing refractive index and dispersion. Such a lens is described as "color corrected" or "achromatic." The two element lens of Fig. 8-20 has a converging element of crown glass and a weaker diverging one of flint glass. Lead oxide contained in the latter enhances its dispersion by a factor of two and substantially increases the refractive index at short wavelengths. The enhanced convergence of blue light provided by the first lens can be practically cancelled out by greater divergence by the second, thus endowing the compound lens with minimal chromatic aberration. To maintain the same focal length as a simple converging lens, the first ele-

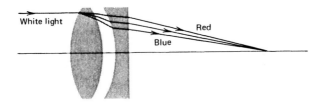

Fig. 8-20. Chromatic aberration can be corrected with a two-element lens. The diverging element is weaker but has greater dispersion to compensate for the dispersion of the first element.

ment of the compound lens must of course have stronger convergence to overcome the divergence of the second element. This lens with the two elements glued together is called an *achromatic doublet,* or simply an *achromat.*

Other aberrations affect lens performance even when focusing monochromatic light. These are spherical aberration, coma, astigmatism, curvature of field, and distortion. They occur with any simple lens having spherical surfaces, no matter how precisely shaped. The principal contribution from each of these as well as chromatic aberration can be avoided with a properly designed compound lens of three or more elements. We shall consider each aberration in turn.

Spherical aberration is the first example of image degradation owing to rays passing through different areas of the lens being brought to different focal points. The thin lens formula (Eq. 8-8) is appropriate only for rays that are incident near the center of the lens. It does not predict the fact that rays from a subject on the optical axis that pass through the lens far from the center come to a focus at a slightly different position behind the lens (Fig. 8-21a). Specially shaped lens surfaces can overcome spherical aberration, but their fabrication is too expensive for popular use. Spherical aberration can be minimized even with surfaces of spherical curvature by making the second surface of the lens concave, with about six times less curvature than the convex first surface. This is one reason why eyeglasses frequently have this shape of a meniscus lens (Fig. 8-9e). Spherical aberration also is minimized by blocking off the outermost portions of the lens by an aperture, and allowing light to pass only through the central region. Photographers know that this is one reason a sharper image is obtained with larger f-numbers.

Coma affects portions of an image that lie off the optical axis. The distortion arises from rays passing through the outermost region of the lens being focused at a different distance from the axis than those passing through the center (Fig. 8-21b). The word "coma" derives from the same Greek root as "comet," which describes the shape of the image of a point source of light off axis. Rarely does a lens exhibit pure coma, because other aberrations usually mask this characteristic pattern.

Astigmatism also arises with off-axis images, but the term is restricted to locations lying at a considerable distance from the optical axis. It is caused when the curvature along one diameter of the lens differs from the curvature along other diameters. Figure 8-21c shows that rays incident on the lens near the outer ends of the imaginary vertical line *BD* are brought to a focus at a distance i_1 behind the lens. But rays that are incident along the outer edges of the line *AC* converge to a focus at a further distance i_2. Because the latter have not yet converged to a point at the distance i_1, these rays form a horizontal band of light if a screen is placed at the distance i_1 to view the image. Similarly, if the screen is placed instead at i_2, the diverging rays from the focus at i_1 form a vertical band of light.

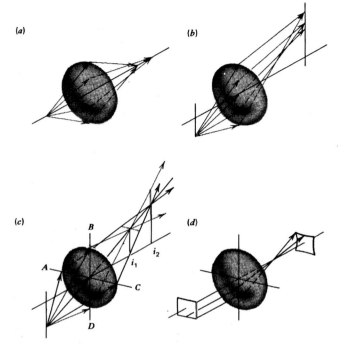

Fig. 8-21. Common aberrations produced by a simple double convex lens with spherically curved surfaces: (*a*) spherical aberration, (*b*) coma, (*c*) astigmatism, and (*d*) distortion.

Compound lenses especially designed to minimize astigmatism are said to be *anastigmatic*. This is a particularly important quality for projection lenses such as those in slide projectors and photographic enlargers, as the closeness of the subject to the lens places most of its area far from the optical axis. The human eye often suffers from astigmatism because of the nonspherical shape of its optical elements. This is corrected by using a cylindrical curvature in the surface of an eyeglass lens. That is, the curvature is made greater in the direction of one diameter than in the direction at right angles to it.

Distortion occurs when straight lines on the subject become curved lines on the image. It results when the magnification of the image depends on the distance of the subject from the optical axis. If it increases with distance, the image of a square subject has sides bowed inward (Fig. 8-21*d*); if it decreases they are bowed outward. These are called "pincushion" and "barrel" distortions.

Curvature of field accounts for photographs that are sharp in the center but blurred at the edges. Positions where the image is focused do not lie on a plane perpendicular to the optic axis but rather on a curved surface. It is particularly emphasized for rays coming from points far off the optical axis.

A simple lens suffers from most of the above defects to a certain degree. Modern lens design leans heavily on computers to design compound lenses with surfaces of the elements chosen to minimize overall degradation of the image quality.

8-6 MAGNIFIER

A subject appears larger when brought closer to the eye. A practical limit size is imposed by the closest distance at which the eye can focus on the subject. This is about 25 cm, although considerable variation is found from one individual to another. A converging lens of short focal length can serve as a *magnifier* if placed close to the eye, because it permits the subject to be brought much closer. The arrangement for normal viewing with a magnifying lens is illustrated in Fig. 8-22a. The subject is brought to a distance that is slightly less than one focal length from the lens, close enough to the focal point that the virtual

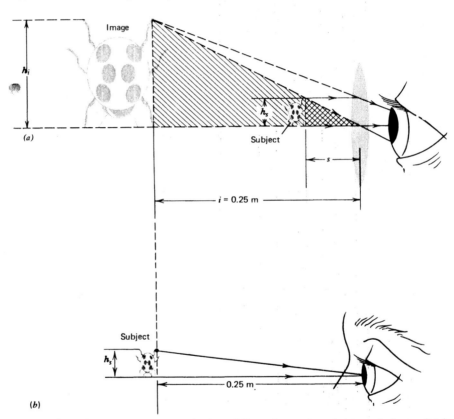

Fig. 8-22. The image of a subject viewed through a converging lens (a) remains in focus when the subject is brought closer to the eye than is possible without the lens (b). Rays coming from the top and bottom of the subject are shown by solid lines, while those appearing to come from the image are shown by dashed lines. The geometry in (a) shows that the right triangle of height h_s and base s is similar to the right triangle of height h_i and base 0.25 m, because they share a common angle. Consequently their corresponding sides are proportional: $h_i/h_s = 0.25/s$, when lengths are expressed in meters.

Fig. 8-23. Images from a lens: (*a*) the subject with a simple converging lens off to the side; (*b*) the magnified virtual image when the lens is placed in front of the subject; (*c*) the inverted real image seen on the observer's side of the lens when the subject is moved further than the focal length away from the lens.

image is about 25 cm in front of the lens. With the eye held close to the lens, this image is far enough away to be seen in focus. The image seen at this distance is considerably larger than the subject would be at that distance without the magnifying lens (Fig. 8-22*b*).

The effect of the magnifying lens is shown in Fig. 8-23. When the subject is closer to the lens than the focal length (Fig. 8-23*b*), a magnified virtual image appears on the subject's side of the lens. But when the subject is farther than the focal length (Fig. 8-23*c*), we notice on the observer's side of the lens an inverted real image, which can be seen if the observer is further away from the lens than the image is.

When a lens is used as a magnifier, its magnification can be expressed as the ratio of the height h_i of the image to the height h_s of the subject at the same distance from the eye (see Fig. 8-22):

$$M = \frac{h_i}{h_s} \tag{8-16}$$

In order to compare the magnifying ability of different lenses it is more useful to express the magnification as a characteristic of the lens, not of the subject and image size. The geometry of Fig. 8-22*a* shows how to replace h_s and h_i in Eq. 8-16 by more useful quantities, so that the ratio on the right side can be expressed as $0.25/s$ if lengths are expressed in meters. Then Eq. 8-16 becomes

$$M = \frac{0.25}{s} \tag{8-17}$$

We now recall that to have the image formed far in front of the lens the subject had to be placed nearly one focal length from the lens. We can then replace s in the above expression by f, to an accuracy of 20% or so. The magnification of the lens is then given by

$$M = \frac{0.25}{f} \tag{8-18}$$

Thus the magnification of a simple lens is determined solely by its focal length (in meters). A magnification of 5, for example, is normally indicated by the symbol "5X," where the "X" represents the word "times." A 5X lens according to Eq. 8-18 has a focal length of 0.05 meter (5 centimeters).

The practical limits on magnification that can be achieved by a lens are imposed by the aberrations it produces, which become more pronounced as the focal length is shortened. To minimize this problem the lens diameter is made smaller for lenses with more curvature (and shorter focal lengths). A lens with 5X magnification can typically be used to examine a subject 4 cm long, but a 10X lens provides only a 1.5 cm field of view. Even so, aberrations become so important that a simple lens is adequate only for magnification of 10 or less.

8-7 MICROSCOPE

The first practical microscope, devised by the Dutchman Zacharias Jenssen in 1590, had a compound lens. It was significantly improved in 1610 by Galileo Galilei but did not become a practical instrument for another century owing to the inadequacy of the theory of optics in the seventeenth century. However, the technique of grinding simple lenses of high magnification improved rapidly, and by 1650 Dutch lensmakers were producing single lenses of great strength. It was with these simple lenses (actually strong magnifying glasses) rather than the still unsatisfactory compound microscopes that the first great successes of microscopy were achieved. In 1658, red blood cells were first observed by the Dutch biologist Jan Swammerdam, and soon after Antony Van Leeuwenhoek became the first person to observe bacteria, protozoa, and spermatazoa.

Innovations by Johannes Kepler and Christiaan Huygens finally endowed the compound microscope with unsurpassed advantages when used for high magnification. The modern microscope consists of two lenses (Fig. 8-24a). The first lens, called the *objective,* has a very short focal length, so that with a subject placed close to the lens the image is found relatively far behind the lens, with substantial magnification (Eq. 8-5). The second lens known as the *eyepiece* (or *ocular*) serves as a simple magnifying lens to enhance the perceived size of the real image that is provided by the objective. The long barrel of the microscope permits a large image distance i_o for the objective compared with the subject distance s_o; consequently, most of the overall magnification comes from the objective's magnification: $M_o = i_o/s_o$. Values as large as 50X are attainable. The additional magnification from the eyepiece of $M_e = 0.25/f_e$, where f_e is its focal length in meters, provides another factor that typically ranges from 5X to 15X. The overall magnification is the product of the individual contributions:

$$M = M_o M_e = \left(\frac{i_o}{s_o} \right) \left(\frac{0.25}{f_e} \right) \tag{8-19}$$

Magnifications as great as 1000 are possible. For example, suppose that the focal length of the objective is 5 mm, so that to a good approximation we can also say that the subject distance is $s_o = 5 \times 10^{-3}$ meters. A barrel length that limits the image distance to $i_o = 30$ cm, together with an eyepiece of 5X magnification, then provides an overall magnification of

$$M = \left(\frac{30 \times 10^{-2}}{5 \times 10^{-3}} \right) (5) = 300$$

Microscopes are commonly provided with an assortment of interchangeable objectives and eyepieces so that the magnification can be selected (Figs. 8-24b and c). For example, one with a 10X eyepiece may have 5X, 15X, and 30X objectives. Notice that since the objective's real image is inverted with respect to the subject, the virtual image seen through the eyepiece is also inverted.

Normally the optical quality of a microscope is limited by aberrations produced by the objective. Another problem is image brightness, because the magnified image has less light coming from each square centimeter of its surface than the subject. Having the objective close to the subject helps by intercepting as much light as possible, but a bright lamp is normally required as well to increase the illuminance on the subject.

Anyone who has looked into a microscope quickly learns that it has a remarkably shallow *depth of field*. An object moved just slightly from its position for optimal focus produces a blurred image. This is because the image distance i_o changes dramatically for only small changes in the subject distance s_o, when s_o is nearly equal to the focal length of the objective. Large changes in i_o make it impossible for your eye to focus on the image produced by the eyepiece. We shall discuss depth of field further in Section 9-3.

Reticle

When using a microscope to gauge the size of a subject, we can take advantage of the fact that the objective produces a real image (called the *principal image*) at a position that is inside the structure of the instrument. This real image lies on a plane that is perpendicular to the optical axis (if the aberration known as curvature of field is not significant), and this is called the *focal plane*. The eyepiece when adjusted for proper focus of this image shows us the pattern of light coming from this focal plane. Thus, if a glass plate with a scale etched on it is mounted at the focal plane, the scale will appear superimposed on the image. This scale is called a *reticle (Fig. 8-24c)*. When its size is properly selected for the magnification M of the objective, the reticle indicates distances across the field of view. In other applications an observer may wish to restrict the field of view with an adjustable aperture. If placed at the focal plane, the aperture provides a sharp edge for the field of view without affecting the brightness of the image inside. On the other hand, if placed before the objective the aperture would not be seen in focus but would

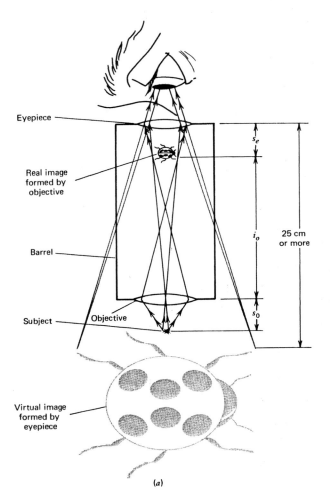

Eyepiece

Real image
formed by
objective

Barrel

Subject

Objective

Virtual image
formed by
eyepiece

s_e

i_o

25 cm
or more

s_0

(a)

(b)

Fig. 8-24. (a) Essential features of a compound microscope, showing the arrangement of objective and eyepiece. Only rays coming from the extremities of the subject and passing through the center of either the objective or eyepiece are shown. The barrel of a microscope is usually so short that the virtual image seen by the observer lies in front of the objective. (b) Modern binocular microscope (Courtesy American Optical Corporation.) (c) Optical system of the microscope shown in (b) (Courtesy American Optical Corporation.)

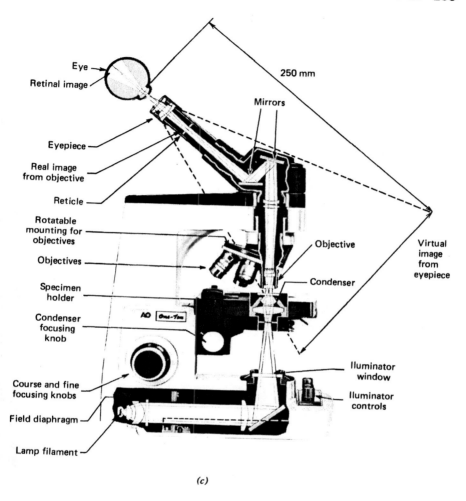

(c)

just restrict the amount of intercepted light, decreasing the brightness of the image without having any effect on the field of view.

8-8 REFRACTING TELESCOPES

The telescope was invented in Holland about 1608, probably by the spectacle maker Hans Lippershey. The earliest devices had a compound lens made by fitting a convex lens in one end of a metal tube and a concave lens in the other (Fig. 8-25a). Galileo Galilei made one of this type and captured the popular imagination with his observations of the mountains on the moon, satellites of Jupiter, and spots on the sun. This type of optical

(a)

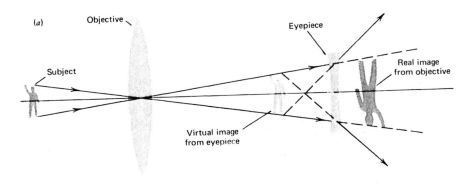

(b)

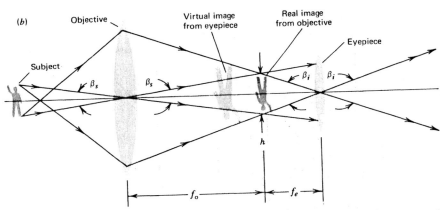

Fig. 8-25. Refracting telescopes. (a) The Galilean telescope provides an erect image, because the focal length of the objective is greater than the distance between the objective and eyepiece. Rays from the extremities of the subject that pass through the center of the objective are shown. (b) The Keplerian telescope has an objective with a shorter focal length and presents an inverted image. Rays from the extremities of the subject that pass through the center of either the objective or eyepiece are shown.

device with a convex objective and a concave eyepiece has since been christened the *Galilean telescope*. With the subject essentially an infinite distance away, the real image formed by the objective lies one focal length behind it. To reduce aberrations and enhance magnification, the objective has a long focal length. The concave eyepiece serves to produce an upright image.

A modification of the Galilean telescope in which the eyepiece is convex was suggested by Johannes Kepler (Fig. 8-25b). One advantage of this telescope is the additional magnification obtainable when the eyepiece serves as a magnifying lens. Of course, the inverted image from the objective remains inverted in the eyepiece's image. This is accepted for astronomical observations but may be undesirable for terrestrial use.

The performance of a telescope is characterized by its *angular magnification, field of view, image brightness,* and *resolution.* The last-mentioned characteristic of an instrument of high quality is imposed by diffraction (Section 4-6); the others will be discussed in turn below.

Angular Magnification

To predict the magnification of the Keplerian telescope shown in Fig. 8-25*b*, we cannot simply apply the formula in Eq. 8-9 for the objective, since the exact subject distance is unknown. A more convenient criterion is the *angular magnification* M_{ang} relating the angular size β_i of the perceived image to the angular size β_s of the subject:

$$M_{ang} = \frac{\beta_i}{\beta_s} \qquad (8\text{-}20)$$

We can recast this formula in a more useful format involving only the focal lengths of the objective (f_o) and eyepiece (f_e) by taking advantage of the fact that for most applications the height h of the real image is short compared with f_o. Therefore the angle β_s is small and, to a good approximation, its value is proportional to the image height divided by its distance from the objective: h/f_o. Similarly, because the eyepiece is placed so that this image is essentially a distance f_e away, the small angle β_i is proportional to h/f_e. The angular magnification in Eq. 8-20 then can be rewritten as:

$$M_{ang} = \frac{h/f_e}{h/f_o} = \frac{f_o}{f_e} \qquad (8\text{-}21)$$

The magnification depends only on the ratio of the two focal lengths. For example, a telescope with a barrel sufficiently long to accommodate an objective with $f_o = 60$ cm and with an eyepiece having $f_e = 2$ cm has an angular magnification of $M_{ang} = 60/2 = 30$. Galileo's most advanced telescope had a magnification close to this, although his first telescope had a magnification of only 3.

It was generally realized in the seventeenth century that the quality of the image is limited by chromatic aberration. A solution was found by the Englishman Chester Hall who in 1733 used lenses of different refractive indices. The common Galilean telescope still used today has a convex objective lens of crown glass and a concave eyepiece of flint glass.

Field of View

The angular diameter β_i of the image (Fig. 8-26) that just fills the field of view of an observer looking through a telescope is called the *apparent field of view.* It is determined by the size of the objective lens. The corresponding angular diameter β_s of the subject itself is called the *field of view.* By Eq. 8-20 they are related as $\beta_i = \beta_s/M_{ang}$. This indicates how much of the scene the aided eye takes in at one time. Galilean telescopes suffer from having a small field of view, ranging from 30° for low magnification opera glasses to 5° for high magnification astronomical instruments. The Keplerian telescope with convex eyepiece provides a larger field, but the image is somewhat degraded near the edge. An eyepiece consisting of a compound lens can avoid these aberrations.

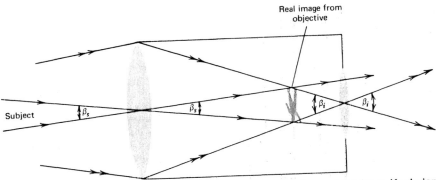

Fig. 8-26. Field of view β_s and apparent field of view β_i provided by a Keplerian telescope.

Entrance and Exit Pupils

Telescopes, particularly for astronomical studies, are designed with objectives of large diameter to maximize their light-gathering ability, in addition to minimizing diffraction blurring of the image. The large diameter increases the illuminance on a photographic film if the film is placed at the focal point to record the image. But it may not increase the brightness of the image seen through an eyepiece. This is a subtle notion that is worth exploring further.

The luminous flux allowed to emerge from an optical instrument is determined by the size of an aperture somewhere within the system; usually the designer makes the rim of the objective be this aperture. Then the area within the rim, or area of the objective, is called the *entrance pupil* of the instrument. The image of this entrance pupil as formed by the rest of the optical elements is called the *exit pupil* (Fig. 8-27). For most instruments

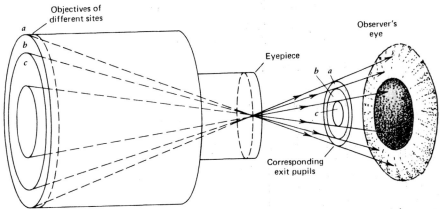

Fig. 8-27. The exit pupil is the circular region through which light leaves a telescope. It is identical to the image that the eyepiece forms of the objective.

such as telescopes the diameter of the eyepiece is made slightly larger than the exit pupil. The exit pupil can be measured by pointing the telescope at the daytime sky and observing the size of the bright focused disc projected onto a sheet of white paper held just under the eyepiece. The exit pupil determines the range of angles of the emerging light.

The exit pupil of a telescope should be larger than the entrance pupil of the observer's eye; otherwise, the image of the subject will be dimmer than when the subject is viewed directly. To understand why, notice first that the objective labeled *b* in Fig. 8-27 provides an exit pupil that just matches the size of the observer's pupil. Increasing the objective size (*a*) leaves the brightness unaffected, because no additional light enters the eye. Decreasing the size (*c*) reduces brightness, because less luminous flux enters the eye. The details are discussed in Advanced Topic 8A.

8-9 REFLECTING TELESCOPES

Isaac Newton recognized that chromatic aberration limited the quality of early Galilean telescopes but believed it to be an insolvable problem. For this reason he invented and built a reflecting telescope to avoid it. He appreciated that the law of reflection whereby the incident angle matches the reflection angle holds regardless of wavelength. As shown in Fig. 8-28, a concave mirror in his telescope served to form the image. From geometrical

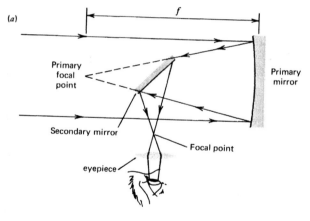

(*a*)

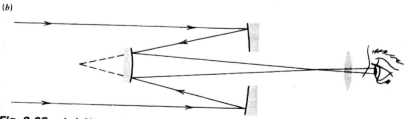

(*b*)

Fig. 8-28. (*a*) Newtonian reflecting telescope. (*b*) Cassegrain reflecting telescope.

considerations we could deduce that a mirror whose surface has the form of a parabola will bring to its focal point all the light from a point source a very great distance away on the optical axis. Newton realized that spherical curvature, which is more easily produced, does nearly as well, provided that the mirror has a long focal length and the subject is much further away than the focal length. Reflecting telescopes have other significant advantages over refracting ones. The quality of the glass is not important, only one surface needs to be shaped and polished, and a large mirror is easier to support rigidly than a large lens.

The mirror for modern reflecting telescopes is a thin coating of aluminum deposited on a shaped piece of glass and overlaid by a protective transparent film. The glass serves as the structural base and is favored for its good dimensional stability. In the Newtonian telescope a small plane mirror mounted on the optical axis before the light reaches the focal point is tilted at 45° to reflect the beam to the side for convenient observation (Fig. 8-28a). The shadow of this mirror of course reduces image brightness, but by a negligible amount. An eyepiece may provide additional magnification for visual observations. The Hale telescope on Mount Palomar has a mirror diameter of 5 meters and focal length of 16.8 meters, giving it an f-number of f/3.4.

The focal length may be extended to provide greater magnification by replacing the plane mirror with a slightly convex one mounted just short of the focal point on the optical axis (Fig. 8-28b). The light reflects back through a hole in the center of the principal mirror to reach its focal point behind the mirror. Named the *Cassegrain telescope* for its inventor, it can be designed with a focal length that is four or five times greater than for the Newtonian configuration.

Smaller Newtonian telescopes for personal use have a diameter ranging up to 20 cm, with the mirror spherically shaped to an accuracy of $\frac{1}{8}$ of the wavelength of light. With a focal length of 122 cm, and magnifying eyepieces, the angular magnification approaches 300X.

The field of view of a reflecting telescope is extremely limited, because off-axis aberrations such as coma and astigmatism are pronounced. For star surveys over a broad field of view, refracting telescopes are preferred, although Schmidt reflecting telescopes, which contain a special refracting device to compensate for off-axis aberrations, are also used.

8-10 BINOCULARS

The perception of depth comes from viewing with both eyes and cannot be achieved with a telescope. Opera glasses and inexpensive field glasses are constructed as a pair of Galilean telescopes, with achromatic objectives and diverging eyepieces. Because of their restricted field of view at high magnification, their usefulness is limited to angular magnifications in the range 2 to 4. Greater magnification and field of view can be achieved with a longer focal length and compound eyepieces. A *binocular* has a folded optical path, which greatly shortens its physical length (Fig. 8-29). A pair of prisms serving as mirrors are commonly arranged in each light path to provide two effects: they move the optical axes from the large separation distance at the objectives (needed to permit large diameter lenses) inward to the observer's interpupil separation at the eyepieces; and they invert the image of each objective so that it is upright (Fig. 8-29b).

The most common system of prisms uses a total of four reflections in each light path to redirect and invert the image. Two "Porro prisms" at right angles is the most common arrangement (Fig. 8-29a). Each is a 45°-45°-90° prism in which light is incident on each

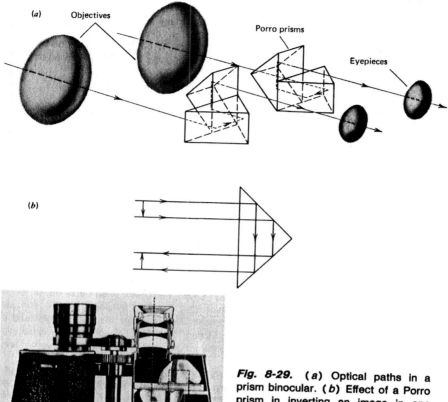

Fig. 8-29. (*a*) Optical paths in a prism binocular. (*b*) Effect of a Porro prism in inverting an image in one plane. (*c*) Typical prism binocular, with one tube cut away to show the arrangement of objective, prisms, and compound eyepiece (Courtesy Bausch and Lomb Optical Company.)

face at an angle exceeding the critical angle, and total internal reflection (Section 8-1) provides high reflectance and consequently maintains a bright image.

A magnification as high as 8 is possible without the need for a tripod to stabilize against inadvertant hand movement. Binoculars are described by a pair of numbers such as 7 × 50 or 8 × 30, where the first number gives the angular magnification and the second the objective diameter in millimeters. The quotient of the second divided by the first gives the diameter of the exit pupil in millimeters. When the exit pupil is larger than the pupil of the observer's eye, the image is perceived to be nearly as bright as without the binocular. But a much dimmer image is produced if the exit pupil is smaller than the observer's. For example, the exit pupil of a 6 × 30 binocular is $\frac{30}{6} = 5$ mm, which is more than sufficient for viewing in bright sunlight when the observer's pupil is constricted to a diameter of only a few millimeters. But at night with the pupil dilated to a diameter as large as 7 mm, only half of the pupil's area receives light from the 5 mm exit pupil of the binocular. By comparison a 7 × 50 binocular provides an exit pupil of $\frac{50}{7} = 7$ mm diameter, which matches the observer's. Twice as much light comes from the image, so it appears correspondingly brighter. This is why the 7 × 50 binocular is called a "night glass." During the Apollo spaceflights, the astronauts used high magnification 20 × 60 binoculars to observe features on the earth. The exit pupil of $\frac{60}{20} = 3$ mm was adequate for daylight viewing but was not particularly effective at night.

8-11 PROJECTION SYSTEMS

At this point we could logically proceed to discuss the optical systems in cameras; however, we shall defer that topic to the following chapter where it can be directly related to other important aspects of photography. Here we shall take up the subject of how a photographic slide can be projected so that an enlarged, real image is formed on a screen. Photographers use a similar procedure in the darkroom to project the image of a negative onto sensitized paper to make a print. Projectors of these types are designed to gather as much light as possible from a lamp, direct it through the slide or negative, and focus the image a comparatively large distance away. In a projector, a *condenser lens* of large diameter is mounted in front of the slide (Fig. 8-30) to collect light from the lamp. A mirror behind the lamp reflects additional light into the condenser. The elements of the condenser are designed to form an image of the lamp's filament near or at the projection lens. Then practically all of the rays passing through the slide are intercepted by the lens. This image of the filament does not appear on the screen, because it is so badly out of focus at that distance. Nevertheless by this clever idea the light leaving the slide can be refracted by the lens and focused on the screen. This is an efficient use of light, and it produces a bright image of the slide on the screen. An infrared absorbing filter is usually placed near the condensing lens to remove the infrared radiation and avoid overheating the film in the slide.

A similar optical system is found in overhead projectors that are widely used by lecturers. A transparency to be projected is laid flat on a glass plate, which is actually a Fresnel lens that acts as the condenser (Section 8-12). The projection lens, mounted directly over the transparency, contains a flat mirror oriented at 45° that directs the light toward the screen.

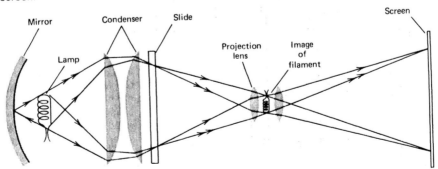

Fig. 8-30. Typical arrangement in a slide projector. The relative distance between lens and screen is more than 10 times larger than shown.

Projection Screens

A projection screen serves to reflect light from the projector back to the viewer. Its type of surface influences the image brightness, and uniformity of brightness across the screen. The *gain* of a screen is defined as the ratio of the screen's brightness compared with the brightness of a perfectly diffusely reflecting surface having 100% reflectance. A matte screen can be manufactured that diffuses light quite evenly in nearly all directions. A small amount of absorption at the surface reduces its gain from 1 to a value of about 0.85.

There are many applications in which a brighter image is desirable. When the audience is seated toward the center of the room, in front or in back of the projector, a semimatte screen having a somewhat glossy finish will serve this purpose because it directs more light back in the general direction of the projector. This increases the brightness in the desired direction. A screen gain of about 1.5 is possible without producing glare at the central portion of the screen (called a "hot spot") when seen by a viewer near the projector.

A screen with metallic paint on the surface has an even more restricted range of directions for the reflected light and achieves a screen gain as high as 2.5. However, viewers to one side will see the closer portion of the image as being brighter than the portion on the far side. A further twofold increase in gain can be provided by a *lenticular screen*, whose surface has small facets that are contoured to reflect light back closer to the projector.

The surface of a *beaded screen* is embedded with small plastic or glass spheres, which reflect a major portion of the light from their front and back surfaces into an even narrower cone of directions toward the projector, and the screen gain can be as high as 15. Beaded surfaces are also used on road signs and warning reflectors on bicycles. As a result of their high screen gain, their brightness decreases rapidly as the viewer moves away from the direction of the source of light.

■ 8-12 FRESNEL LENSES

When a lens of unusually large diameter is wanted for a situation where optical quality can be compromised, the *Fresnel lens* offers the advantages of light weight and thinness. Le Comte de Buffon in 1748 suggested that the weight of a plano-convex lens could be considerably reduced by recessing successive concentric rings while retaining their original curvature (Fig. 8-31a). Some light will be refracted in spurious directions by the steps, but this is inconsequential for many purposes. The first application of this idea appears to be due to Augustin Fresnel, famous for his research on interference, who designed such a lens for lighthouses. Here, as for signal lights and floodlights, the large size of the source requires a lens of a large diameter to gather most of the light. A conventional lens would be too thick and heavy to be supported. Cylindrical Fresnel lenses are also used to encircle lanterns where they direct light nearly horizontally for improved visibility.

Modern Fresnel lenses are molded of plastic or acrylic, with the width of each ring being only a fraction of a millimeter. Small versions are used with ground-glass screens in cameras, in overhead projectors, and in microfilm readers to increase

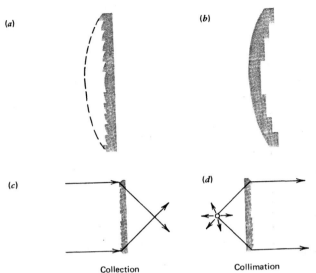

(a)   *(b)*

(c)   *(d)*

Collection   Collimation

Fig. 8-31. (*a*) Fresnel lens, with the dashed curve showing the curvature of the recessed segments; (*b*) stepped lens; (*c*) Fresnel lens focusing light from a distant source; (*d*) Fresnel lens producing a collimated beam from a source located at its focal point.

brightness in the outermost regions. The optical quality is acceptable for use as a window that provides a wide view for an observer standing some distance away. Image quality is usually limited by the sharpness of the steps, which reflect extraneous light (about 2%). Converging, diverging, cylindrical, and other types of lenses are commercially available completely free of spherical aberrations with effective apertures as large as f/0.6. Lenses 50 cm in diameter can be fabricated with a thickness of less than 1 mm. A *stepped lens* (Fig. 8-31*b*) has weight removed by cutting steps into the flat side, leaving the curved face smooth. This produces a thicker lens, which is sometimes awkward to mount.

Closely related is the *lenticular lens* formed on the front of automobile headlamps. The curved surface is composed of small rectangular segments, which are shaped to concentrate light preferentially to the side or ahead so that the desired pattern of illumination is provided.

■ 8-13 STAGE LIGHTING

The casual theatergoer may be unaware of the many special lighting effects that contribute significantly to our enjoyment. Certainly the choice of color for lights takes into account the laws of color addition and subtraction as they affect the color of curtains, stage setting, and costumes. But optical factors are also paramount. The naive approach of using direct front lighting provides an unacceptably flat appearance of actor and set, caused by the lack of shadows. Instead, a light angled downward at about 45° and directed from a vantage point in front at 45° to the side of center is found to render shadows on an actor's face that gives it character and shows expression. But using two such lights on either side to display him or her equally to both sides of the audience partially washes out the desired effect. This unfortunate situation can be largely compensated by relying on the contrast of colors when one light is given a red tint and the other blue. A green light is rarely used, because it provides a completely unnatural hue for the face and emphasizes skin blemishes.

With angled front lighting, shadows will appear behind the actor on the floor or backdrop unless overhead lights are mounted above the stage. These are chosen to *fill* areas that may be blocked by performers or scenery. In settings of natural appearance the color temperature of illumination is chosen to represent the suggested source: a white light (6000 K) for overcast days, slightly yellow (5000 K) for direct sunlight, and yellow-orange (3000 K) for incandescent lamps. Usually the color temperature for lights on one side is chosen higher than for the other, to represent the common mix of sunlight and reflected (or perhaps incandescent) light. To achieve this, *area lights* providing diffuse illumination and *spotlights* for direct illumination may be used. For front illumination, the brightest spotlight is usually an arc lamp. Such a source is essential if the beam of light is to be thrown a great distance. Very intense spotlights used in movie productions are sometimes called "Klieg" lights, a corruption of the name of its inventors Anton and John Kliegl.

Plano-Convex Spotlight

The simplest spotlight is a metal hood containing a high power source of light, which is made more effective by a reflector behind the source and a lens in front. The earliest version had a spherical reflector with the lamp positioned at its center (Fig. 8-32). Light emitted backwards from the lamp is returned forward to pass close by the filament, essentially doubling the forward intensity of the source. A plano-convex lens intercepts much of this and focuses it onto the stage. The lamp and reflector are mounted on a sliding carriage, so that the spread of the beam and size of its spot can be adjusted. Colored filters may be mounted in front of the lens. Stage lenses in the United States are commonly specified by two numbers: a 6-by-8-in. lens having a diameter of 6 in. and focal length of 8 in. The *plano-convex* or *"P-C"* spotlight is available in sizes ranging from a 250 W lamp with 5 in. diameter lens to a 2000 W lamp with 8 in. lens.

Most light is collected by the lens if the lamp has a specially arranged flat filament that emits light primarily in two opposite directions, toward the reflector and toward the lens. The lamp's bulb is clear to keep the source small for more effective focusing. Also the lens is best placed as close to the source as possible to intercept more of its light. Therefore, to project a narrow collimated beam, it must have a short focal length. The acute curvature of the lens unfortunately produces

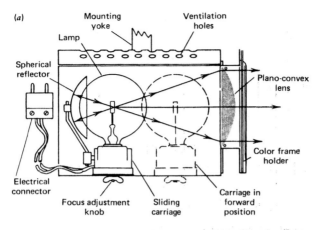

(a)

(b)

Fig. 8-32. Plano-convex spotlight: (*a*) details of the optical system, showing the lamp positioned with its filament at the center of curvature of the spherical mirror; (*b*) a modern example of a 1000 W tungsten-halogen spotlight with a Fresnel lens. (Courtesy of Kliegl Brothers, New York.)

a perceptible amount of chromatic aberration that colors the spot; another disadvantage is that the lens must be thick and consequently absorbs much of the infrared radiation from the lamp, the heat of which tends to crack the lens. In some cases a compound lens using two weaker elements can replace the thick lens. But more often a brighter beam can be obtained with a Fresnel lens, which is available with an especially short focal length. The beam from a Fresnel lens, because of refraction from the steps in the lens, is without a sharp border and therefore is particularly useful as overhead illumination where the spot from one light should merge smoothly into the next. The amount of light refracted to the side from a Fresnel lens can be reduced substantially if the steps of the lens are blackened.

Ellipsoidal-Reflector Spotlight

The most common instrument for lighting stages today is the *ellipsoidal-reflector spotlight* (Fig. 8-33). The ellipsoidal shape for the mirror affords two advantages: it reflects light that was emitted over a greater range of directions from the lamp;

(*a*)

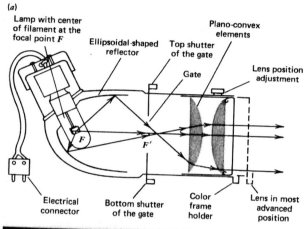

Fig. 8-33. Ellipsoidal-reflector spotlight. (*a*) Optical system showing the conjugate focal points *F* and *F'*. (*b*) Tungsten-halogen lamp version (Courtesy of Kliegl Brothers, New York.)

and it focuses that light through a different position within the hood. A geometrical property of an ellipsoidal shape is that light from a lamp placed at one focal point of the ellipsoid is reflected through the conjugate focal point. (The sphere is a special case where the two conjugate focal points coincide at the center of the sphere.) As the light converges on the conjugate focal point, it passes through an aperture known as the "gate." The lens subsequently intercepts this light and forms an image of the gate on the stage. The size of the spot is adjusted by closing or opening the gate, and the spot can be given special shapes by appropriately choosing the shape of this aperture. The sharp edge of the beam can be blurred by moving the lens to throw the image out of focus.

Because a spotlight is normally tipped at 45° and is serviced from above, a lamp designed for operation with its base on top is useful. Recently, tungsten-halogen lamps have become popular because of their longer life, and they can be operated in any orientation. Ellipsoidal-reflector spotlights as large as 3000 W with a 12 in. diameter lens are available. Fresnel or step lenses often are preferred over thick plano-convex or compound lenses because they are less subject to breakage under the extreme heat that is generated.

Colored Lights

Colored stage lighting is achieved by placing a colored filter in a special frame built into the spotlight. This is a spectacularly wasteful procedure, since the filter typically absorbs 70 to 80% of the light. The filters consist of dyes dissolved in glass, plastic, or gelatin. Several companies manufacture plastic-based color filters that are sold in large sheets that can be cut to fit the filter frame of any spotlight. These filters (e.g., Cinemoid ™, Dynamold ™, etc.) are waterproof and heat resistant.

Manufacturers of filters provide the transmission curve and total transmittance for each filter. The transmittance gives the percentage of luminous flux (*not* radiant flux) allowed to pass when white light is incident, as explained in Sections 2-4 and 6-4. Color Plate 5 shows the transmission data for some filters manufactured by the Rosco Corporation. Curves with narrow transmission bands such as Primary Red, Primary Blue, and Primary Green produce highly saturated colors, but transmit very little light. They are not useful if high levels of illumination are required, nor could the colors be successfully reproduced by the four-color printing used for the color plates of this book. For that reason only less saturated filters are illustrated. In selecting colored filters for stagelights the lighting expert seeks to achieve appropriate illumination levels while controlling the three-dimensional aspects of the stage (plasticity). This involves manipulation of both *key* (directed spotlights) and *fill* (background illumination). The result of superimposing light from spotlights with different filters can be predicted from the laws of additive color mixing developed in Chapter 2. Reflectance curves of objects in the set must also be considered. There is a time-honored principle saying "always project onto a subject something of its own color so that it will reflect that color back to you." The reason for this was made clear in Chapter 2.

When two spotlights producing equal illumination with two nearly complementary colors are superimposed, the spot is white. Thus two complementary spotlights aimed from opposite sides at about 45° produce white front lighting, but the different side lighting colors heighten the sense of depth. This technique is called "complementary tint cross-spotting" and can be done, for example, with the pale blue and light rose filters whose transmission curves are shown in Color Plate 5. Use of closely related tints tends to reduce the apparent depth of the stage, and use of a single saturated color produces an almost flat effect.

There are important differences between lighting for stage and for television or movie recording. Camera film and television cameras tolerate a narrower range of luminance than the human eye. Furthermore the quality of their color reproduction is completely determined by the SPD of the light, whereas the human eye compensates to an extent for imbalance toward one hue. Yet the basic principles for all three media remain the same. The foreground or area of prime interest is served by a key light, which is often a strongly directed source such as a spotlight or arc lamp. This throws a strong shadow to emphasize contrast with the background and perhaps give the effect of sunlight. The fill that illuminates the background complements this primary source. The fill is chosen to respect the limitations of the medium: in theater the background can be considerably darker than for cinema or television. For film, the luminance of the background can be as low as 1% of the keyed areas and still be recorded on the negative, but for television it is generally kept above 5%.

SUMMARY

All of the optical effects described in this chapter rest on Hero's law of reflection, Snell's law of refraction, or the more subtle influence of dispersion. Refraction of light at a boundary between media of differing refractive indices changes the direction of light rays. Indeed, rays approaching a medium of lower refractive index at an incident angle exceeding the critical angle will be reflected at the boundary as though it were a mirror. The utility of a lens comes from refraction at its two surfaces. Its strength is determined by how much curvature these surfaces have: a convex surface provides convergence of rays coming from a point on a distant subject; and a concave surface, divergence. The strength of a lens is indicated by the reciprocal of its focal length. A converging lens has a positive strength, and a diverging lens a negative one. Converging lenses serve to focus a real image at a distance of one focal length or more behind the lens, the exact distance depending on how far the subject is from the lens. Such lenses are applied in eyeglasses, cameras, and the objectives of microscopes, telescopes, and binoculars. They also are used as common magnifiers and eyepieces in microscopes and some telescopes. In these latter cases the subject is placed slightly less than one focal length from the lens so that the image is virtual and greatly magnified. Diverging lenses always produce a virtual image in front of the lens, and they find applications in Galilean telescopes to provide an upright image. Lenses of better quality are actually compound lenses containing two or more elements of differing refractive index and curvature chosen to minimize aberrations. Reflecting telescopes avoid chromatic aberration entirely and have other advantages when a large size is needed to gather as much light as possible. If image quality is not a primary concern—as in stagelights, lighthouses, and wide-angle viewing windows—a Fresnel lens offers the advantages of less weight and lower cost than a standard lens. In stagelights and other projection systems, a properly shaped reflector behind the lamp greatly enhances the beam intensity. The spread of the beam and sharpness of its edge can be adjusted by changing the distance between the source of light and the converging lens used to form the beam. ∎

QUESTIONS

8-1. Can total internal reflection occur as light comes to a medium of higher refractive index just as it does on arriving at a medium of lower index? Explain.

8-2. When a light ray in a bent lightguide having a refractive index of 1.7 meets the surface, what is the greatest incident angle that still enables the ray to be totally internally reflected?

8-3. A fish in a pond has a limited angle from the vertical within which it sees the world above the water's surface. Within what angle from the vertical do all subjects above the water appear? Predict this angle by using Snell's law.

8-4. What distinguishes a real image from a virtual image? What is meant by saying that a lens forms a real image of a subject?

8-5. When using a thin meniscus lens of a given focal length, does the image's position and size depend on whether the lens is oriented so that the concave or convex surface is closest to the subject? Why?

8-6. Under what condition will a "converging" lens cause light to diverge and produce no real image?

8-7. If a converging lens of a given shape is reproduced from a type of glass having a higher refractive index than the original lens, is the focal length increased or decreased? Explain your reasoning.

8-8. If a lens has a focal length of 50 mm, what diameter would you expect it to have if it is rated with an f-number of f/3.5?

8-9. The f-number of a camera is adjusted by changing the diameter of an opening in the lens to allow more or less light to enter. By what factor is the luminous flux increased when the f-number is decreased from f/5.6 to f/2.8?

8-10. Suppose that a subject is 30 cm in front of a lens having a focal length of 10 cm. Where is the image found? If the subject is 1 cm high, how high is the image? Work out your answers on a diagram by following the path of two rays leaving a common point on the subject in different directions.

8-11. Find the answers to question 10 by using the thin lens formula.

8-12. Use a ray diagram to show where the image lies when a subject is 5 cm from a converging lens whose focal length is 10 cm. Is the image real or virtual? How much larger is the image than the subject?

8-13. Use the thin lens formula to find the answers to question 12.

8-14. In close-up photography of flowers it is common to screw additional lenses onto the front of the principal lens of the camera. If one of these close-up lenses has a strength of +1 diopter and a second has +3 diopter, what is the total strength of the two lenses when attached simultaneously? What is the focal length of the two lenses together? Does the effect of using two lenses depend on which lens you attach first?

8-15. If a camera with a lens of 50 mm focal length cannot focus any subject lying closer than 30 cm, how much closer can the subject be brought if a lens of +3 diopter is attached to the front? How much larger is the image?

8-16. What is meant when a given lens is said to exhibit chromatic aberration? What is an achromatic lens?

8-17. When using a magnifying lens, how close should the subject be placed? What happens when the subject is moved further than the focal length away? What do you see?

8-18. Explain the principle of the compound microscope. What advantage is gained by using an objective with a large diameter? Why is the image inverted when seen through the eyepiece?

8-19. The Galilean and Keplerian telescopes differ in that one provides an upright image while the other's image is inverted, yet both may use the same type of objective. Explain how this difference comes about as a result of the type and position of the eyepiece.

8-20. Using ray diagrams explain why for the same objectives a Keplerian telescope with an eyepiece of the same diameter and focal length as a Galilean telescope has a larger field of view.

8-21. What is meant by the "exit pupil" of a microscope or telescope?

8-22. What advantage do reflecting telescopes have over refracting ones?

8-23. What is meant by the description "8 × 30" for a binocular?

8-24. What is the purpose of the condenser lens in a projector? Exactly what does it image, and where does it place the image?

8-25. Why are Fresnel lenses used in certain applications even though their shape always produces some blurring of the image?

8-26. Why is the lamp in many types of stage spotlights mounted on a carriage that can be slid closer or farther from the lens?

8-27. High quality spotlights often have an ellipsoidal reflector placed in back of the lamp. Why does the reflector have this special shape?

FURTHER READING

A. C. S. van Heel and C. H. F. Velzel, *What is Light* (McGraw-Hill, New York, 1968). A delightful introductory treatment of optics that avoids mathematics.

There are many texts covering the field of optics. Most have a moderate emphasis on mathematical analysis, for example, the following two books:

W. H. A. Fincham and M. H. Freeman, *Optics* (Butterworth, Reading, Massachusetts, 1974).

G. R. Fowles, *Introduction to Modern Optics* (Holt, Rinehart and Winston, New York, 1975).

C. J. Campbell, C. J. Koester, M. C. Rittler, and R. B. Tackaberry, *Physiological Optics* (Harper and Row, Hagerstown, Maryland, 1974). Excellent presentation of ray optics, including quantitative aspects, and a detailed discussion of the eye.

A. Stimson, *Photometry and Radiometry for Engineers* (Wiley-Interscience, New York, 1974). Detailed analysis of the photometric aspects of optical systems, including photographic applications.

W. O. Parker and H. K. Smith, *Scene Design and Stage Lighting*, Second Edition, (Holt, Rinehart and Winston, New York, 1968). Elementary introduction to the purposes and techniques of stage lighting, with descriptions of various lights.

F. Bentham, *The Art of Stage Lighting* (Sir Isaac Pitman and Sons, London, 1968). Rather more detailed explanation of optical techniques than the preceding reference.

From the *Scientific American*

A. Yariv, "Guided-Wave Optics" (January, 1979), p. 64.

W. S. Boyle, "Light-Wave Communications" (August, 1977), p. 40.

F. D. Smith, "How Images are Formed" (September, 1968), p. 97.

P. Baumeister and G. Pincus, "Optical Interference Coatings" (December, 1970), p. 58.

W. H. Price, "The Photographic Lens" (August, 1976), p. 72.

■ CHAPTER 9 ■

PHOTOGRAPHY

Photography, the permanent recording of images of our visual world, is one of the most widely known and appreciated applications of the scientific principles of light and color, which we have explored in the preceding chapters.

The photographic process and the technology that it requires can be divided into two major areas: First, there is the process of forming an image within the camera, a process based on the laws of optics discussed in Chapter 8; second there is the process of recording the image on light sensitive materials and "fixing" it so that it becomes permanent. Although the principles underlying the photographic process originated many centuries ago, development of the techniques that set the explosive growth of photography in motion occurred during a period of about 50 years in the middle of the nineteenth century.

■ 9-1 HISTORICAL PERSPECTIVE

The basic mechanism of forming an image within a camera has been known at least since the time of Aristotle. The word "camera" in Latin means "chamber" or room. Its modern connotations evolved from the earlier use of the camera obscura, a darkened room illuminated only by the light that enters through a small hole in one wall. On the opposite wall appears an inverted, dim image of the scene outside, which can be witnessed by persons whose eyes have adjusted to the darkness. This image-making ability of a small hole has been known for at least two thousand years: Arabic philosophers, using tents as camera obscuras, found that the image is sharpened but made dimmer when the hole is reduced in size, as we can now understand (Fig. 9-1).

A modern version of the camera obscura is the pin-hole camera, which provides pictures of acceptable quality with a hole of 0.05 mm ($\frac{1}{50}$ in.) diameter or smaller. Daniel Barbaro, a Venetian noble, is credited with discovering that the image in a camera obscura sharpens considerably if a simple convex lens is placed in the hole, provided the subject outside is a proper distance away. With a larger diameter lens and hole, a relatively bright image can be obtained. It would only be necessary to place a light-sensitive substance on the wall to record the image.

Light-Sensitive Crystals

The discovery that salts of silver blacken on exposure to light is attributed by some to the alchemist Albertus Magnus who lived in Bavaria during the thirteenth century and by others to the sixteenth century mineralogist Georg Fabricius. In 1727 Johann Heinrich Schulze discovered that silver nitrate precipitated on the bottom of a glass flask, when exposed to light through a paper stencil, retained a darkened image wherever it was exposed. In 1802 in England, Thomas Wedgewood and Sir Humphrey Davy described a method for reproducing drawings and silhouettes on paper or leather treated with silver nitrate or silver chloride. These images, like those of Schulze, were not permanent; the astronomer Sir John Herschel discov-

Fig. 9-1. The camera obscura provides an image because light from any one point on the subject is limited by the hole so that it strikes only a small region of the inside wall. If the hole is made smaller this region shrinks, making the image sharper but dimmer.

ered in 1819 that these images could be fixed with hyposulphites. (It was Herschel who coined the word "photography" in 1839.)

Credit for the first photograph made inside a camera goes to the French lithographer Joseph-Nicéphore Niepce who in 1826 employed a portable camera obscura to take an eight hour exposure on a pewter sheet coated with an asphalt solution. A later collaboration between Niepce and Louis Daguerre, a Parisian painter, impresario and scenery designer, sowed the ideas that later bore fruit when Daguerre, after Niepce's death, returned to experimenting with silver compounds. In 1837, he discovered a way to record direct positive images in the first practical photographic process. His light sensitive surface was a polished silver plate on which he had deposited iodine vapor to form silver iodide crystals on the surface. After exposure of about half an hour, a wash of mercury caused the unexposed silver iodide to form a black stable compound. The exposed silver iodide was then fixed with a salt solution to prevent it from turning dark. The daguerreotype process was made public by the French government in 1839 and quickly spread around the world. Its great success marked the beginning of the era of photography.

In 1838 Samuel Morse, inventor of telegraphy, visited Paris where he met Daguerre who introduced him to the daguerreotype process. On his return to New York, Morse began making daguerreotypes from the window of his office at New York University. His associate, John Draper of the Chemistry Department, placed his assistant in front of a daguerreotype camera on the roof of a university building.

The assistant's face was powdered with flour to increase his reflectance. The resulting 1839 photograph, which nevertheless required a half hour exposure in full sunlight, has been generally considered the first photograph made of a person, and it opened the age of portrait photography. However, there is some evidence that the first portrait was actually made by another New Yorker, Alexander Walcott. Draper also recorded in 1840 the first photographic image of the moon that showed features of its surface.

Antoine Jerome Balard, the discoverer of the element bromine, showed in 1826 that one of its compounds, silver bromide, also darkens on exposure to light. Subsequently it was found that the sensitivity to light of mixtures of silver iodide and either silver bromide or chloride (called collectively, *silver halides*) far exceeded that of pure silver iodide or chloride. By 1841, daguerreotype plates made with mixed silver halides allowed photographs to be obtained with exposures of less than one second.

A major advance in photographic optics was achieved by Josef Petzval, professor of mathematics at the University of Vienna, who designed a compound lens admitting 15 times more luminous flux than Daguerre's simple lens, which had a small diameter in order to limit aberrations. Subsequently built by the German optical firm headed by Peter von Voightländer, it enjoyed immediate commercial success and remained the most favored lens for more than half a century. A flexible bellows connecting the lens to the body of the camera permitted the lens to be moved back and forth to achieve a focus on the light-sensitive daguerreotype plate at the back.

A contemporary of Daguerre, the Englishman Fox Talbot invented "film" by coating paper with crystals of silver chloride. After exposure this produced a negative, because illuminated regions slowly turned dark during the exposure. Then, after fixing, the negative could produce a positive when pressed against similarly sensitized paper and exposed to sunlight. Any number of identical "contact prints" could be reproduced in this way. Fox Talbot fixed his silver chloride images with sodium thiosulphate, which he had acquired from Herschel. This fixing solution still is widely used and known (incorrectly) to photographers as "hypo" for the less effective sodium hyposulphate introduced earlier by Herschel.

Developing an Image

Talbot's early photographs on paper treated with silver chloride and silver bromide were made directly in the camera. In 1840 he discovered that if after a much shorter exposure that produced no visible image the paper was treated chemically, a complete image slowly appeared. This procedure of chemically developing a *latent image* reduced the required exposure times drastically, making Talbot's process competitive in speed with Daguerre's.

Talbot's paper film was eventually displaced by emulsions on glass plates that gave far superior pictures. The glass plate technique was invented in 1847 by Claude Niepce, a cousin of Nicéphore Niepce. The "Niepcotype" was coated with silver salts suspended in a layer of albumen (egg white). In 1850, the albumen was

Fig. 9-2. Nineteenth century photographer using a wet-plate bellows camera. The required portable darkroom is set up in a tent. (Courtesy of George Gilbert.)

replaced by the newly discovered collodion. Frederick Scott Archer in England and Gustave LeGray in France independently introduced the wet plate collodion process. Scott Archer coated a silver nitrate solution onto a film of collodion liquid, which adheres smoothly to a glass plate serving as support. Wet plates had to be prepared and developed immediately before the emulsion dried. The disadvantage of having to carry a darkroom with the camera, illustrated in Figure 9-2, was overcome by the *dry-plate process*. Here tiny silver bromide crystals are suspended throughout a layer of fast-drying gelatin covering a glass plate. Introduction of the dry plate process by the English physician Richard Maddox in 1873 led to its general adoption, and improved versions continue to be used for special purposes, particularly in scientific work. The gelatin increased the light sensitivity of silver bromide crystals by a factor of 50, making it possible to photograph a moving subject without also having to move the camera.

In 1884 the American George Eastman opened photography to the amateur with the forerunner of modern *roll film*. Light-sensitive crystals were suspended in a flexible emulsion coated onto paper; following development the emulsion was stripped off and attached to a glass plate for drying. Before the turn of the century Eastman abandoned the tedious stripping procedure when he introduced a flexible transparent base on which the emulsion was coated. In 1888 Eastman produced

TABLE 9-1 Historical Evolution of Film Speeds

Year	Process	Exposure Time (sec) Daylight at f/16 Aperture	Sensitivity (Approximate ASA Film Speed)
1839	Daguerreotype	2400	0.0004
1856	Wet collodion	30	0.03
1880	Gelatin dry plate	5	0.2
1880	Eastman's first roll film	1/2	2
1931	Kodak Verichrome	1/25	25
1938	Kodak Plus-X	1/125	125
1953	Kodak Tri-X	1/400	400
1965	Kodak Royal X Pan	1/1250	1250
1966	Kodak 2485 High-speed recording film	1/4800	4800
1981	Polaroid Ultra-high speed instrument film	1/20,000	20,000

Based in part on "Technology the Enabler" by Martin L. Scott, *Image,* Vol. *24* (December 1981) and on "Film Speeds 1930–1965" by David A. Gibson (unpublished). Film speeds have been measured in different ways and have changed with time. They also depend on the development process (speeds can be significantly increased—"pushed"—during development). In this table, speeds are taken as the inverse of the exposure time in seconds at f/16 in sunlight as currently recommended by the manufacturer.

the first hand-held camera to use roll film, with a fixed lens providing good focus for everything beyond two meters. The film was placed close to the lens which reduced the size of the image, thereby concentrating the light on the film with the advantage of quicker exposures. Projection systems permitted large positive prints to be made on sensitized paper from the magnified image of the negative.

Substantial improvements in film performance have come in this century with refinements of a discovery in 1873 by H. W. Vogel in Germany. He observed that combining a dye with the light sensitive emulsion generally enhances light sensitivity and extends the range of sensitivity over a broader span of the spectrum. And in 1947 Edwin Land of the Polaroid Corporation in the United States introduced a 60 second development film containing development chemicals within the film itself. Table 9.1 indicates the way film speeds have increased from the Daguerreotype to modern high speed films.

Color Photography

Interest in color photography grew apace with the advances in monochrome technology just outlined. James Clerk Maxwell demonstrated the first color photograph in 1861 before London's Royal Society. It was an awkward arrangement in which he had taken three monochrome photographs of a tartan ribbon successively through blue, green, and red filters. By the laws of additive color mixing that he had just investigated, he knew that a wide gamut of colors could be reproduced if three positives illuminated by similar blue, green, and red lamps were projected in register onto a white wall. Some years later a puzzle arose as to why Maxwell's demonstration worked, because it was discovered that the wet collodion plates on which the images developed were insensitive to red light. It would seem impossible to have a faithful color rendition, contrary to contemporary accounts. The answer was found in 1961 by Ralph Evans and his colleagues at the Eastman Kodak Research Laboratories. Careful measurements showed that the red dyes in Maxwell's tartan ribbon also reflected ultraviolet light in a portion of the spectrum that fortunately coincides with a region of high transmittance of the ferric thiocyanate solution used as a red filter. It was actually ultraviolet light that exposed the red negative!

In the early 1900s August and Louis Lumières found a reliable means for recording a color photograph on a single plate. In effect, three color filters for exposure and projection were an integral part of the plate. They dyed equal portions of fine-grain starch blue, green, and red; and after mixing them together spread the grains in a fine layer one grain thick over the light sensitive emulsion. After exposure in which light passed through the grains before hitting the emulsion, the emulsion was processed to yield a positive transparency with the colored grains remaining in place. When backed by white paper, the transparency at each point reflects light that has passed through the same colored grain of starch. Thus partitive color mixing of the primaries as in a fine mosaic (Section 2-3) reproduces the colors of the original scene. This *Autochrome process* became the most popular color technique by 1910; and even today the soft, luminous quality of the Lumières' photographs is a delight to behold.

The first simple means for making color photographs was invented in 1935 by Leopold Godowsky, Jr., and Leopold Mannes. First working in New York City as musicians while experimenting in their spare time, and later with improved support carrying out their research at the Eastman Kodak Laboratories, they evolved the process known as Kodachrome. This is a multilayer color reversal film, which we shall describe later in this chapter. Essentially it requires two development processes, one to render a negative in the emulsion and a second that includes the addition of dyes to convert the negative to a color positive. The positive can be projected onto a screen for large-size viewing. The most recent major advance in color technology came in 1962 when Edwin Land introduced Polacolor film, a multilayer rapid development film with dyes included within the film itself. Although the many types of color film now available can reproduce an extremely wide gamut of colors, they do so in slightly different ways, because of differing dyes, emulsion thickness, surface texture, and backing. This affords the photographer a varied means of self-expression.

In Sections 9-7 to 9-9 we shall return to the topic of color photography and consider the steps by which color information is captured in color films. We should note that all color photography processes represent a compromise in the film's ability to reproduce the full color range of the scene being photographed. The films we shall discuss are all designed to produce a single emulsion containing all the color information. Although these films are the most popular for general use, they all suffer from problems of color range and color balance. In commercial work, the more difficult and expensive *color separation process* is widely used in which three separate negatives are produced representing the blue, green, and red images. These are subsequently used to make positive prints through the dye transfer printing process which is related to the four-color printing process that we described in Section 2-5. Since the color separation negatives are processed separately, there is considerably more room for correction of color balance with this process, and far superior color rendition is possible.

■ 9-2 TAKING A PICTURE

Before taking a picture, the photographer adjusts the *focus, exposure time,* and *aperture setting* on the camera. As we shall see, there is a great deal of freedom in setting these adjustments that will determine which elements of the scene are in sharp or blurred focus, which range of tones will be emphasized, the color rendition, etc. It is the interplay of these choices that is the key to photography as an art.

How to Focus

Focus is achieved by the mechanism that moves the lens back and forth to establish its correct distance from the film, providing a sharp image on it (Fig. 9-3). The photographer selects which elements of the scene will be in focus when making this adjustment. For a lens of given focal length, this distance is related to the distance of the subject from the front of the lens, as discussed in Section 8-2. The farther the subject, the closer the lens must be to the film. When the subject is a great distance away, the lens is one focal length from the film.

Obtaining a Proper Exposure

The two other adjustments of *aperture setting* and *exposure time* are interrelated. The aperture is an adjustable opening to admit more or less light. An arrangement of thin overlapping metal leaves called the diaphragm is positioned behind the lens. When the rim of this element is rotated, the aperture expands as the leaves move apart, allowing more luminous flux to enter. On the other hand, exposure time is controlled by the shutter. It is a complicated device that opens and closes at a preset rate, allowing light to strike the film when open. During exposure the film responds to the accumulated amount of light (or number of photons) that it absorbs. Therefore the response is determined by the *product* of the illuminance (E_v) as governed by the aperture setting and the exposure time (t) fixed by the shutter. The product of illuminance and time is called the *exposure* (H_0):

$$H_0 = t E_v \qquad (9\text{-}1)$$

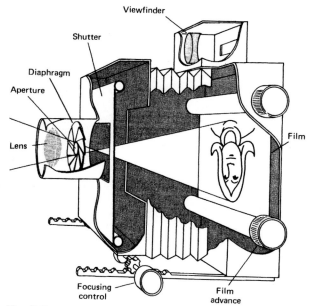

Fig. 9-3. Elements of a bellows-type camera: the lens is moved by the gear-and-track system, which shifts the whole front of the camera; a flexible bellows ensures that the interior is light-tight. A simplified form of shutter is shown, consisting of a rectangular hole in a ribbon that is pulled past the diaphragm.

Because illuminance E_v varies from one place to another across the film, so will the exposure H_0. Let us disregard this feature for the moment and just consider the exposure of the major area of interest. Since the product of E_v and t appears in Eq. 9-1, a long exposure time can compensate for situations of low illumination, when little light reaches the camera. Similarly short exposure times are possible under brightly lit conditions. The possibility of trading off exposure time and illuminance is called the *reciprocity* property of a film. It holds for a modest range of times and values of illuminance, yielding the same results for the final picture. Reciprocity fails under extreme conditions—either very short exposure times (less than about $\frac{1}{1000}$ sec) or very long times (longer than 1 sec). For these extreme conditions the aperture should be opened somewhat more to expose the film properly.

Shutter Setting

Two types of shutter mechanisms are common today, the *leaf shutter* and the *focal plane shutter*. Figure 9-4 depicts a leaf shutter positioned within a compound lens, commonly found in inexpensive cameras. As shown by the sequence of illustrations, when the shutter release is pushed, the leaves of the shutter swing open, remain open for a moment, and close.

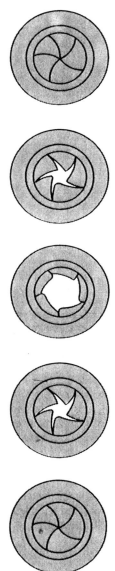

Fig. 9-4. Sequence of drawings showing how the metal blades of a leaf shutter open and then shut during an exposure.

The focal plane shutter found in more expensive cameras can operate somewhat faster than $\frac{1}{500}$ sec, which is the limit of the leaf shutter's speed. Instead of the two movements of opening and closing, an open slit slides in one direction past the film. Mounted immediately in front of the film, nearly at the focal plane, this focal plane shutter allows the photographer to view the scene directly through the lens of the camera by a system of mirrors. This is a major asset when special lenses are used for closeup or telephoto work.

The shutter speed is generally selected according to the amount of movement in the scene. To record high-speed action in sports photography without blur, a short exposure time of perhaps $\frac{1}{500}$ sec or less is usually essential. A timing device in most cameras allows a choice of the exposure time, commonly from 1 sec to $\frac{1}{1000}$ sec in steps of approximately a factor of 2. The sequence is designed to be compatible with the indicated aperture settings, which we describe next.

Aperture Setting

Aperture size is indicated on a scale of numbers called "f-stops." The selected f-stop indicates the f-number of the camera's optical system. Recall from Section 8-2 that the f-number of a lens is defined as the ratio of its focal length f to the diameter of the aperture. A large aperture is thus indicated by a small f-number. A common sequence of f-stops is f/1.4, f/2, f/2.8, f/4, f/5.6, f/8, f/11, f/16, f/22, and f/32. Figure 9-5 illustrates the aperture sizes that correspond to various f-stops. Turning the aperture control ring on the lens adjusts the aperture formed by the set of metal leaves inside. For this lens, f/2.8 admits the most light, providing the greatest illuminance on the film. In the colloquial language of photographers, it is the "fastest" setting because by admitting more light it allows a faster shutter speed to achieve proper exposure (Eq. 9-1) than any of the other apertures. An important characteristic of a lens is its smallest f-number. With a large diameter lens the f-number may be as low as f/1.2.

The numerical sequence of f-stops on a lens may seem curious at first but is chosen for a good reason. Neighboring values are in the ratio of about 1.4, which approximates the square root of 2. Thus, decreasing the f-stop by one setting causes the diameter of the aperture to increase by $\sqrt{2}$. The area admitting light, proportional to the *square* of the diameter, is consequently doubled. To keep the film exposure unchanged, the exposure time should then be halved (Fig. 9-6). This explains why exposure times progress by a factor of 2 but f-stops by the square root of 2. The same exposure is retained when the f-stop and exposure time are simultaneously increased or decreased by one setting (Fig. 9-6).

■ 9-3 PHOTOGRAPHIC OPTICS

A modern camera has a compound lens with many elements (Fig. 9-7). This minimizes chromatic aberration as well as the other aberrations that would result if a simple lens were used, as described in Section 8-5. Antireflection coatings permit

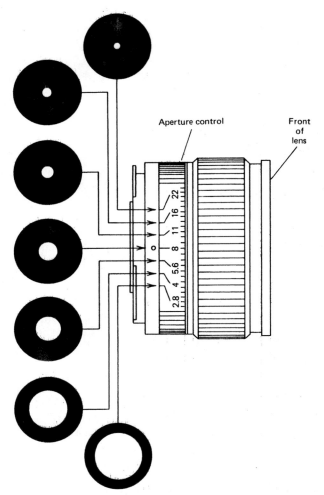

Fig. 9-5. f-stop settings on the aperture control ring of a lens. The corresponding apertures are shown on the left. The area of each differs from that of its neighbors by a factor
of 2.

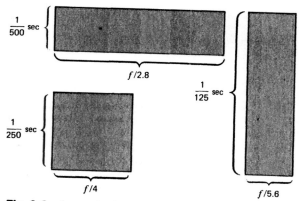

Fig. 9-6. Increasing both the exposure time and f-number by one setting maintains the same exposure. The height of each rectangle represents exposure time and the width represents aperture area; therefore the area of each block represents exposure.

a half dozen or more elements to be used in a high quality lens while maintaining high transmittance and minimizing multiple reflections (lens flare) that would misdirect light onto the film. The principal optical properties of a compound lens are determined by its focal length.

Field of View

The focal length of the lens and size of the film determine how much of a scene will be recorded. A useful way to indicate what is meant by "how much" is the total angular view that can just span the diagonal distance between opposite corners of the photograph. This is called the *field of view*, and for convenience it is specified for the lens focused at infinity. A strip of the popular "35 mm" film size has a total width of 35 mm, but because space must be devoted to sprocket holes running along each edge so it can be pulled through the camera, the photograph dimension in this direction—which is usually the height of the photograph when the camera is held in its horizontal position—is only 24 mm. The other dimension has been set at 36 mm to provide an aspect ratio of 2:3, a pleasing ratio. The diagonal distance across the photograph is therefore 43 mm.

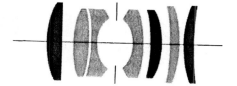

Fig. 9-7. Typical arrangement of elements in a modern large-aperture (small f-number) photographic lens.

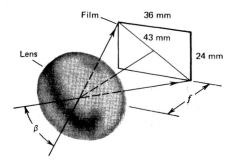

Fig. 9-8. Angular field of view β with the lens focused for subjects a great distance away. The angle β is given by the relation $\tan(\beta/2) = 0.043/2f$, where f is the focal length of the lens in meters.

Figure 9-8 indicates how the field of view is simply determined by those light rays from the scene that pass straight through the center of the lens toward the opposite corners of the frame. A lens having a long focal length is positioned far from the film and has a correspondingly small field of view. This expresses the fact demonstrated earlier (Eq. 8-9) that the magnification is proportional to the focal length of the lens; thus a smaller portion of the scene is included in the photograph when a longer focal length is used. This is illustrated in Fig. 9-9 for lenses with focal lengths of 50, 100, and 200 mm. The magnification doubles each time the focal length is doubled.

The curve giving the field of view of a 35 mm camera for lenses of various focal lengths is shown in Fig. 9-10. Marked on the curve for special emphasis are popular focal lengths, ranging from a 28 mm lens for "wide angle" shots to a 200 mm "telephoto" lens for magnification of subjects at a long distance. A 200 mm lens has about the longest focal length that permits the camera to be held in the hand for taking a photograph with an exposure time as long as $\frac{1}{60}$ second. With higher magnification the slight inadvertent movements during this time would cause appreciable blurring of the image unless the camera is braced mechanically. Lenses with a focal length as long as 1000 mm are useful for wildlife photography when supported by a tripod.

For most popular photography a focal length of 50 mm provides an acceptable field of view. If the lens diameter is sufficiently large so that the aperture can be opened to a diameter of 25 mm, the lens has a speed of f/2 for effective light gathering. The *zoom lens,* first introduced in the early 1930s, has a variable focal length with elements arranged to maintain focus as the magnification (and corresponding field of view) is adjusted (Fig. 9-11). Modern zoom lenses for amateur use can be varied by a factor of 3 in focal length, say from 70 to 210 mm. A special zoom lens for television cameras incorporates 20 elements and has a 20:1 range for its focal length. The advantage of variable focal length of a zoom lens must be balanced against its greater weight and smaller aperture compared to lenses of fixed focal length.

Fig. 9-9. Views of Notre Dame Cathedral in Paris photographed from the same vantage point with lenses of the following focal lengths: (*a*) 50 mm, (*b*) 100 mm, (*c*) 200 mm. Doubling the focal length reduces the field of view by a factor of two.

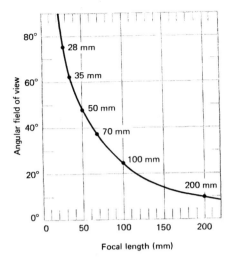

Fig. 9-10. Field of view of a 35 mm camera for lenses of various focal lengths.

Example 9-1   Tele-Extender Lenses

To achieve greater magnification than a given lens normally provides, it may be possible to mount a small, inexpensive "tele-extender" unit between the lens and camera body. This unit is a diverging compound lens that increases the overall focal length by a fixed factor. Units of $2\times$, $3\times$, or $4\times$ are common. We can ask how the f-stop numbers on the original lens should be interpreted when a $2\times$ tele-extender is used, say, with a lens of 200 mm focal length. As the new focal length is now doubled (400 mm), the aperture diameter at each setting is unaffected, but the f-number at each setting is actually twice the indicated value.

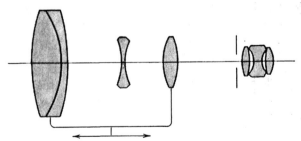

Fig. 9-11. A seven element zoom lens with an aperture of f/1.9 designed for 8 mm motion-picture cameras. The first and third components move forward or back together, to provide a continuous range in focal lengths from 10 to 30 mm, while also keeping the image focused on the film.

The f-stop indicated by f/2.8 is actually f/5.6. In effect, the increase in linear magnification of the image by 2× spreads the same luminous flux over 4 times the film area, and the illuminance is reduced by a factor of 4.

Depth of Field

The range of distances from the lens for which subjects remain "essentially" in focus at the film is called the *depth of field*. When the subject is too far away or too close, blurring of the image becomes noticeable. This is shown by the "out of focus" foreground and background in Fig. 9-12a compared with the greater depth of field in Fig. 9-12b. Then light from each point on the subject is spread over a small circular region of the film called the *blur circle* (or *circle of confusion*). In Fig. 9-13a, the blur circle reaches the just noticeable size when the subject is too far (at a distance called the *far field* denoted by s') or too close (the *near field* denoted by s''). A subject is at the far field when its image is focused so far in front of the film that the spreading light from a point of this image produces a blur circle of just noticeable size at the film (i). Similarly, a subject is at the near field when its image

Fig. 9-12. (*a*) The shallow depth of field in the scene on the left places the neck of the pot in focus but not its front surface or background. This is due to a large aperture setting (f/1.4, with an exposure time of 1/125 second); (*b*) the greater depth of field on the right is acheived with a small aperture setting (f/16, with an exposure time of 1 second).

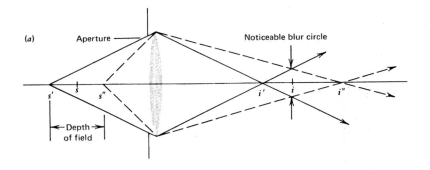

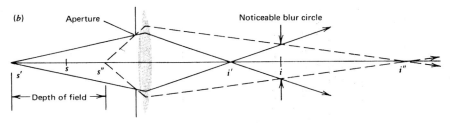

Fig. 9-13. (*a*) With an aperture of large diameter, the images at *i'* and *i"* must be close to the film at *i* to keep light from a point on the subject at *s'* or at *s"* from exposing more than the just noticeable blur circle at the position of the film. (*b*) With a smaller aperture, the distance between s' and s" can be increased.

is focused so far behind the film (*i"*) that the light converging to a point on the image also produces a blur circle of just noticeable size at the film. The distance between far and near fields is called the depth of field. .

The geometry in Fig. 9-13 shows that a smaller aperture confines the light admitted into the camera so that rays from a given point in the scene converge to their focus within a narrower cone of angles (Fig. 9-13*b*). This increases the depth of field, because even though *i'* and *i"* are further from the film than in Fig. 9-13*a*, light coming away from *i'* and the light heading toward *i"* will still not expose a region that exceeds the noticeable blur circle. Therefore for a given lens, depth of field increases as the f-number increases.

Photographers exploit the freedom of choice of aperture size to put foreground or background subjects either in focus or out of focus, thereby helping to establish a desired effect. Although a small aperture size enhances depth of field, it also reduces the luminous flux allowed into the camera. Exposure time must be increased accordingly for proper exposure. The pair of photographs in Fig. 9-12 also illustrate reciprocity of the film: to gain greater depth of field the f-number and exposure time were both increased by six stops.

■ 9-4 TYPES OF CAMERAS

The hundreds of models of cameras available today can be divided into three main types according to how the optics provides the view to the photographer called the *view*, *reflex*, and *viewfinder* cameras . The *view camera* (Fig. 9-14) is the oldest kind and is still popular for architectural and still-life photography. Built like an accordian, the front of the camera which contains the lens, aperture, and shutter can be moved back and forth for focusing. A very large piece of film can be exposed, an advantage for recording details when photographing a large group of people. With the film removed, the photographer views the image that appears on a glass plate at the position where the film will be exposed. The plate is chemically etched or mechanically ground on one side to provide diffuse transmission so that the entire image can be seen by the eye scanning its surface. The image can also be examined on this *viewing screen* with a magnifying glass to inspect for sharpness of focus everywhere. Unfortunately the image is inverted on the screen, which is disconcerting at first to the novice photographer.

Other advantages of a view camera come from being able to shift or tilt the lens relative to the plane of the film. The bellows permits the front of the camera which holds the lens to be tipped foward or backward to improve depth of focus: thus when the top is tipped forward, distant subjects near the top of the scene and nearby subjects near the bottom can both be recorded in focus. Similarly the distorted image from an upward view of a tall building, whereby it appears to bend backwards because of the converging vertical lines, can be eliminated with a view camera. One simply tilts the back so that it is parallel to the building face, leaving the lens pointing upwards to project the image onto the film, and then the sides of the building will appear straight and parallel.

The *reflex camera* in Fig. 9-15 like the view camera provides an image on a viewing screen, but to permit the film to remain in place while framing and focusing the scene, a mirror drops down to reflect the light onto the screen. The two-dimensional image seen on the screen more closely corresponds to the photograph that will be recorded than the scene observed directly, although the use of a mirror reverses the image right to left. When an exposure is made the mirror swings up out of the way momentarily as the shutter opens. The mirror is positioned at equal distances from the screen and film, so that when the image appears focused on the screen it will also be focused on the film. Although this camera is smaller than a view camera, it still permits a comparatively large film size (5 × 5 cm), with an accompanying good resolution.

The past decades have seen another version of the reflex camera rise to popularity for its versatility and comparatively low cost: it has become known as the *single-lens reflex* (or "SLR"). As shown in Fig. 9-16, for viewing the scene a mirror drops down behind the lens to direct the admitted light to a five-sided prism that reinverts the image. The scene is virtually identical to that presented to the film, no matter what lens is attached. When an exposure is made, the mirror swings up out of the way momentarily before the focal plane shutter of this type of camera

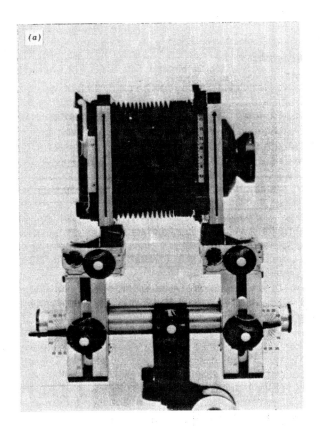

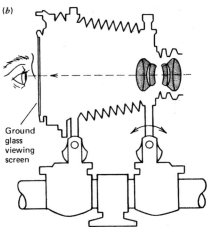

Ground
glass
viewing
screen

Fig. 9-14. (*a*) View camera (courtesy of Sinar Bron, Inc.); (*b*) The lens focuses the image on a glass viewing screen at the back where the film will be placed.

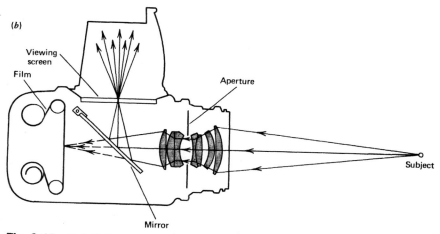

Fig. 9-15. (*a*) Reflex camera (courtesy of the Victor Hasselblad Corporation); (*b*) diagram showing how a mirror reflects light to the viewing screen.

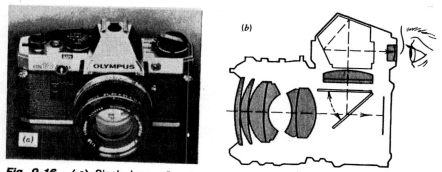

Fig. 9-16. (*a*) Single lens reflex (courtesy of Olympus); (*b*) a mirror, five-sided prism, and eyepiece serve to present the image as it will appear on the film.

opens. A reflex camera's main disadvantages are the vibration and noise caused by the moving mirror and the fact that during the moment of exposure the photographer cannot see the subject being photographed.

The *viewfinder camera* is the simplest of the three types of cameras. The optical system used in viewing the scene is separate from that used in focusing the image on the film (Fig. 9-17). Because the two optical axes do not coincide the viewfinder

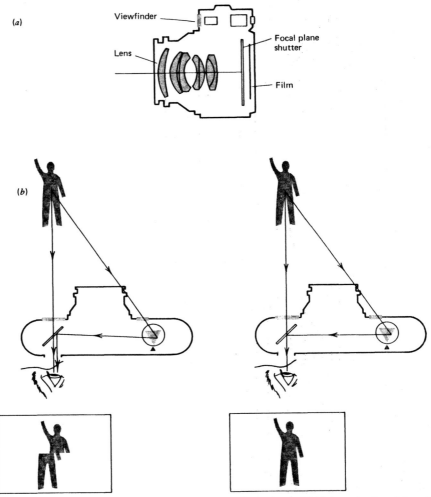

Fig. 9-17. (*a*) Viewfinder camera with a focal plane shutter; (*b*) focusing is achieved with a rangefinder that properly superimposes two images when the prism is rotated to the correct angle for the subject distance. The camera lens, coupled to the prism, has then been set at the proper distance from the film.

camera may be unsatisfactory for closeup work when the different positions of lens and viewfinder do not include exactly the same view. This is known as *parallax*. Focusing is achieved by using a *rangefinder system* that superimposes in the viewfinder two images of the same subject as seen from different angles. As shown in Fig. 9-17b the angle of the beam of light intercepted by the reflecting prism depends on the distance of the subject. Therefore with the prism properly linked to rotate as the lens is moved away from the film (usually done in a screwlike fashion by rotating the lens), when the images in the viewfinder line up, the lens has the image focused on the film.

■ 9-5 MONOCHROME FILMS

Serious photographers must understand their films just as artists know their palettes. The film's sensitivity to light, its fineness of grain, its contrast, and for color film its color balance are all important considerations that affect the nature of the image.

The light sensitive component of modern film is a layer of tiny crystals of silver bromide (Fig. 9-18a). The crystals are suspended in gelatin made from animal hides and bone, and the combination is called the *emulsion* (Fig. 9-19a). The emulsion is supported by a flexible base made from a cellulose product. Formerly, nitrocellulose prepared by treating cotton with nitric acid served the purpose; but this material was highly flammable and has since been replaced by slower-burning cellulose acetate, made by treating cotton with acetic acid. For scientific purposes such as astronomical photography, a glass plate serves as the base to maintain flatness for accurate measurements. With any film, during exposure some light penetrates through the emulsion and would be partially reflected at the back surface of the base (Fig. 9-19b). This reflected light could expose neighboring regions of the emulsion and blur the image, an effect called *halation*. To avoid blurring, an antihalation coating is applied to the base to absorb the transmitted light.

Latent Image

Once the emulsion is exposed to light the silver bromide crystals are said to be "sensitized" and the film retains a *latent image*. The details of how this occurs are not completely understood. In any case, it is clear that one important feature of silver halide crystals is a continual migration of many silver ions through the spaces in the crystalline array of atoms. When a photon is absorbed by an atom in the crystal, an electron acquiring much of this energy becomes liberated and begins to migrate, usually stopping when it reaches an imperfection. This most likely occurs at a surface of the crystal (Fig. 9-20). Its negative charge attracts one of the migrating positive silver ions, which on being neutralized by the electron becomes an atom trapped at the impurity. If two or three photoelectrons and ions migrate to the impurity before the first one leaves, their mutual interaction causes them to bind together in place and form what is called a *development center*. Only

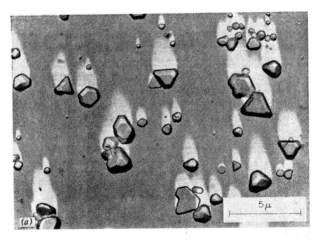

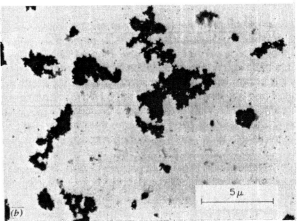

Fig. 9-18. (*a*) Photosensitive crystals of silver bromide in a film emulsion as recorded by an electron microscope technique; (*b*) silver grains in the developed negative from the same kind of film. (Courtesy of Eastman Kodak Company)

1% of the photons absorbed by a crystal produce development centers, because of the need for these special conditions.

The reciprocity property of a film—that exposure is determined by the product of illuminance and exposure time—rests on the fact that under normal conditions the number of development centers that form is proportional to the number of photons that are absorbed by the emulsion. However, reciprocity fails if the exposure time is either too long or too short. It fails and compensation must be made when the illuminance is extremely low, because then photons are absorbed at a slow rate and there are correspondingly fewer free electrons migrating through a given

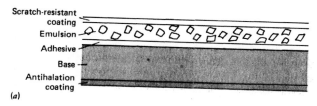

Scratch-resistant coating
Emulsion
Adhesive
Base
Antihalation coating

(a)

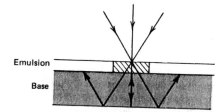

Emulsion

Base

(b)

Fig. 9-19. (*a*) Monochrome (black-and-white) film is typically only 0.13 mm thick, and most of that is the base. The thickness of the emulsion is much exaggerated in the figure. (*b*) Halation occurs if light is reflected back into the emulsion and exposes adjacent areas.

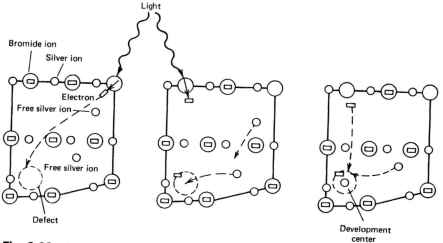

Light

Bromide ion

Silver ion

Electron

Free silver ion

Free silver ion

Defect

Development center

Fig. 9-20. Three stages in the creation of a development center: when an electron liberated by absorption of a photon becomes trapped at a defect, when joined by a migrating silver ion to form a neutral atom, and when one to three other electrons and ions also encounter the defect before the first atom migrates away.

silver bromide crystal. It is thus more likely that a silver atom trapped at a defect will escape before a sufficient number of other electrons and ions arrive to stabilize the complex. Reciprocity also fails when the exposure time is too short, because then there is insufficient time for the necessary number of electrons and ions to reach a given defect in their migrations. This is why a greater exposure (about one-half to one stop) must be given with most films when the exposure time exceeds 1 second or is shorter than $\frac{1}{1000}$ second.

Development

After exposure the film is developed by immersing it in the appropriate developing solution of which there are many varieties. Silver atoms in an exposed crystal are released by the action of the solution, and are attracted to a nearby development center where they produce a tiny grain of metallic silver. Each grain consists of fine, intertwined filaments of silver, looking much like steel wool. Multiple reflections of light from the many surfaces enhance the weak absorptive effect of a single reflection, so that as discussed in Section 7-1 the silver grain appears black. Formation of a grain is a remarkable transformation, because as many as 10^9 atoms now cluster together where previously there were only 3 or 4! The concentration of grains is correspondingly greater in regions of greater exposure. The chemistry of a simple photographic development process is represented as:

$$AgBr + H_2O + R \rightarrow Ag + HBr + ROH \qquad (9\text{-}2)$$

where the symbol R represents the reducing agent in the developer. The film is then immersed in a "stop bath," usually acetic acid, to stop the action of the developer. Finally it is placed in a bath of sodium thiosulfate ($Na_2S_2O_3$), colloquially known as *hypo*. This "fixes" the film by converting the remaining unexposed silver bromide into a soluble salt that dissolves, leaving these regions transparent. Finally the film is washed to remove any remaining soluble products of the developing and fixing stages. Then it is dried. This sequence of steps leaves a "negative" image on the film: when held before a lamp, light shines through in regions where it had *not* been exposed, the light elsewhere being blocked by the silver grains.

A close examination of the negative reveals that regions which should have had uniform exposure actually display a texture, known as "graininess." This is particularly evident in films prepared with especially high sensitivity to light (Fig. 9-21). Manufacturers know that one way to increase the sensitivity, or "speed," of a film is to increase the size of the silver halide crystals, since the number of photons striking a given crystal in a short time determines whether it will produce a development center. Larger crystals will intercept the required number of photons more quickly. But they also produce a more uneven distribution of silver grains, since development centers are further apart. Individual grains are too small to be seen; the unevenness that we see as "graininess" comes from this uneven distribution of many grains. Many films of high sensitivity are thus described as "coarse-

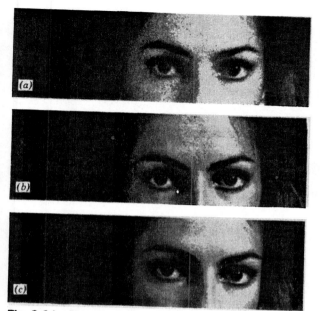

Fig. 9-21. The graininess of standard films varies with their sensitivities: (*a*) portrait taken with high-speed Royal-X Pan film; (*b*) one with intermediate speed Tri-X film; and (*c*) one with low speed Panatomic-X film. (© Eastman Kodak Company, 1980.)

grained,'' and those of low sensitivity ''fine-grained.'' The graininess can also be affected by the choice of development process.

Dye Sensitizers

Untreated silver bromide responds only to short wavelengths, from ultraviolet through cyan (see curve *A* in Fig. 9-22). In 1873 H. W. Vogel discovered that including an appropriate organic dye in the emulsion extended the wavelength range through the green. Apparently the dye molecules become attached to the crystals in such a way that the energy they absorb can be transferred to the electrons in the crystals. Emulsions of the type that respond to the green portion of the spectrum are known as *orthochromatic* (curve *B* in Fig. 9-22). Still other dyes extend the range through the red as illustrated by the curves *C* and *D* for film that is called *panchromatic*.

The most useful sensitizing dyes are cyanines. An example is the violet-blue dye pinacyanol whose molecule is shown in Fig. 9-23*a*. This molecule can electrically attach to bromine or another halide ion through the electron-attracting nitrogen bond. This dye molecule has both conjugated bonds along the central chain of carbon atoms and a resonant structure in the rings at both ends. As explained in Section 5-5 these features spread the electron quantum states over a large portion of the molecule and they have correspondingly low excitation energies. Therefore the molecule absorbs in the long as well as intermediate wavelength portions of the visible spectrum. Pinacyanol has been used to sensitize panchromatic films.

Still other organic dyes extend the range of a film into the infrared (curve *E* in Fig. 9-22). Another cyanine molecule *kryptocyanine* (Fig. 9-23*b*) is effective for absorbing both red and infrared radiation. With appropriate sensitizers, film can respond to wavelengths as long as 1200 nm. Other chemicals such as silver sulfide may be added to the emulsion to increase the number of imperfections at the surface of the silver bromide crystals and thereby enhance the film's sensitivity to light.

For some purposes only infrared recording is of interest, and for this purpose the film can be rendered insensitive in the visible by incorporating an appropriate dye filter within the upper layer of the emulsion. Two types of infrared photography are currently practiced. In the more popular type an infrared source is placed to

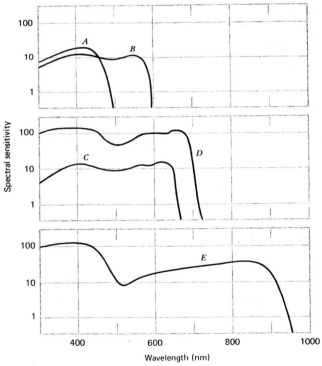

Fig. 9-22. Spectral sensitivity of several monochrome films: *A*, blue sensitive; *B*, orthochromatic; *C*, panchromatic; *D*, high-speed panchromatic with extended red sensitizing; and *E*, high-speed infrared. (Adapted from Carroll et al., *Introduction to Photographic Theory: The Silver Halide Process*, Wiley, New York)

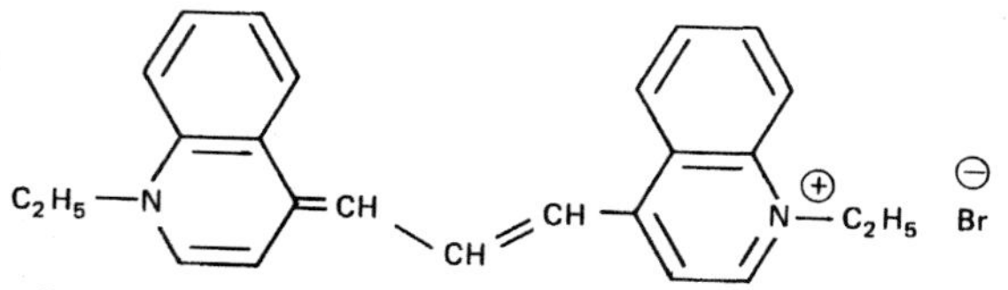

Fig. 9-23. (*a*) Molecular structure of the violet-blue dye pinacyanol. The bond assignments leave one of the nitrogen atoms with a net positive charge, which allows that end of the molecule to be attracted to a bromine ion (Br^-). (*b*) Molecular structure of kryptocyanine, which absorbs strongly in the red and infrared.

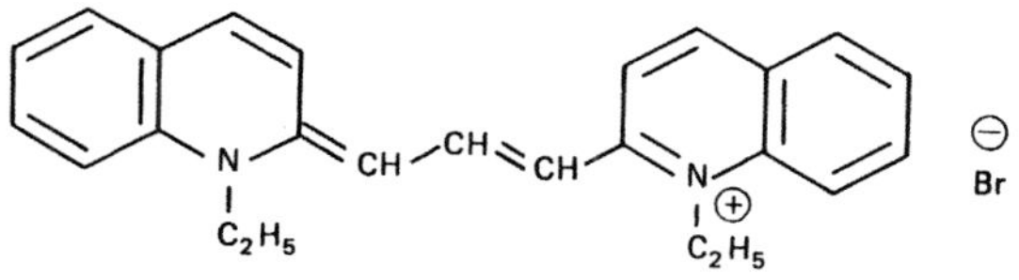

illuminate the subject and the photograph is recorded by *reflected* infrared radiation. But in most scientific work it is the infrared radiation *emitted* by the subject that exposes the film. This records information about the relative temperature of subjects in the field of view, since it is sensitive to the thermal radiation they emit (see Section 7-2). Subjects warmer than 800 K (about 500°C) are incandescent and easily photographed. Subjects in the range 250° to 500°C will produce an image on infrared film if a sufficiently long exposure time is allowed. But infrared photography of colder subjects (such as the skin of a person) is virtually impossible without taking elaborate precautions, because these subjects emit so little radiant flux within the range of sensitivity of films.

Positive Prints

A monochrome "positive" print of a negative can be produced by a similar process to that used to form the negative, but in this case the silver halide emulsion is supported by a highly reflecting white paper. An enlarger with an optical system similar to that of a slide projector (Fig. 8-30) may be used in a darkroom to project a magnified image of the negative onto the light sensitive side of the paper. Then following developing, fixing, washing, and drying, light areas of the print correspond to regions protected from exposure by dark areas of the negative. This "negative of a negative" is therefore a positive.

Unlike film, ordinary printing paper is not sensitized to long wavelengths of light and only responds to blue light. This allows red or yellow "safelights" to be used for illuminating a darkroom during the printing process without danger of fogging the print. "Variable contrast" paper has an expanded range of sensitivity so that the contrast can be modified by varying the SPD of the light leaving the enlarger.

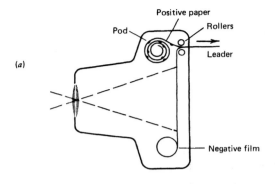

(a)

(b)

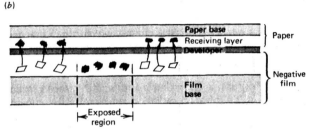

Fig. 9-24. (*a*) Arrangement of film and paper in a Polaroid Land camera. (*b*) Sandwich of paper and negative film with the intervening layer of liquid developing chemical that releases silver atoms from the unexposed regions of the negative.

Film must be developed in total darkness, although some types may be examined with either a red or green safelight during the final stages of development.

Rapid Development Film

In 1947 Edwin Land of the Polaroid Corporation announced a major innovation by which a high quality positive print could be turned out a few seconds after exposure, without resort to a darkroom. A package of Polaroid Land film contains both the negative film and positive paper attached to a common leader (Fig. 9-24). After exposure the two are sandwiched together as they are pulled out of the camera and pass through rollers that break a small pod attached to the inner side of the paper. Chemicals released from the pod are spread between negative and paper and transfer the image from the negative to the paper. They do this by releasing silver atoms from crystals in regions of the negative that were *not* exposed, whereupon the atoms migrate directly across to the paper surface and form silver grains to produce dark regions. This forms a positive image on the paper. With the negative film and positive paper pressed together, held less than 5×10^{-4} cm apart, the silver atoms have no chance to migrate sideways an appreciable distance before contacting the paper and becoming trapped in the receiving layer. This eliminates the opportunity for the silver grains to develop into clumps that would be seen as "graininess." Even for high-speed negative film, a fine-grain positive is achieved. After 10 seconds or so the positive can be pulled away from the negative to reveal a finished photograph.

■ 9-6 EXPOSURE METERS

The eye is notoriously poor at judging levels of illumination because it adapts to changing light conditions. More reliable judgement is obtained with a light sensitive meter indicating the amount of light falling on it. Early devices employed a selenium cell that provides a voltage indicated by the deflection of the needle in a voltmeter. More sensitivity is achieved now with a photoconductive sensor whose electrical resistance decreases with increasing illuminance, such as a cadmium sulfide or gallium arsenide cell (Fig. 9-25). The electrical resistance of the cell decreases when it absorbs light; thus when connected in a circuit with a battery, the flow of current increases with illuminance, and this is registered by a current meter.

To indicate the proper settings of aperture and exposure time for a given film, the scale on the meter must be adjusted for the appropriate sensitivity of the film being used. The sensitivity of a film is denoted by the value of its *film speed*. Two systems are popular: the DIN system (Deutsche Industrie Norm) in Europe and the ASA system (American Standards Association) in the United States. In both, more sensitive films are characterized by higher numbers. An ASA film speed of 32 is considered relatively insensitive to light (a "slow" film) and 400 or more is quite sensitive ("fast" film). How the manufacturer determines the speed of a given film is described in Advanced Topic 9B.

There are two types of exposure meters: *reflected light* meters, which register the amount reflected toward the camera and *incident light* meters, which are used

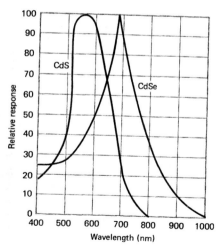

Fig. 9-25. Spectral sensitivity of representative cadmium sulfide (CdS) and cadmium selenide (CdSe) cells. When used in conjunction with filters, the shape of the sensitivity curve for CdS can be made quite similar to the sensitivity curve of the eye.

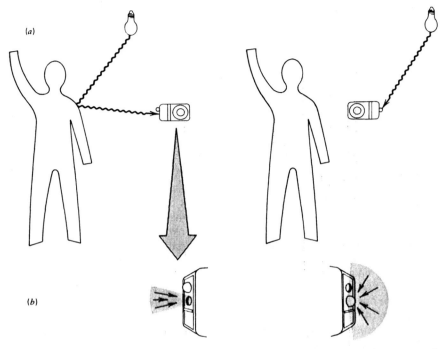

Fig. 9-26. (*a*) A reflected-light meter *(left)* is aimed at the subject and measures the light reflected from its surface; the incident-light meter *(right)* is held close to the subject to measure the light incident on it. (*b*) The opening in a reflected-light meter accepts light from a narrow range of directions to respond to light from the subject and not the background; the incident-light meter receives light from a much broader range of directions.

to indicate the amount falling on the subject (Fig. 9-26). Cameras with built-in meters are of the first type.

Reflected Light Meter

Reflected light meters seem the simplest to use. The photosensitive cell receives light from a narrow range of directions that is determined by an internal system of multiple lenses and internal louvers. The field of view can be as small as 20°, and some meters built into cameras have an even narrower field of view. In all cases the cell simply responds to the total received flux and recommends an exposure value based on the assumption that the scene is an average one. This can surprise the unwary photographer in unusual circumstances. For example, directing the meter toward a white sheet of paper will cause it to recommend an anomalously low exposure that renders the sheet of paper in the final positive print an average gray. Similarly, if the meter is pointed toward a dark paper instead, the meter recommends a high exposure that would also render the paper as gray. Gauging exposure by directing the meter toward a standard gray surface (with a reflectance of about 18%) is one way to avoid these problems. This is the typical reflectance of green leaves and other objects in an outdoor scene. It is good practice when outdoors to incline a hand-held meter downward by about 30° below the horizon to exclude skylight. With a meter built into the camera it is best to expose for the subject of prime interest.

For a subject of nearly uniform luminance, a reflected light meter gives the same reading whether held at the camera position or close to the subject. As explained in Advanced Topic 8A, the reason is that the luminous flux received by the meter from each small area of the subject *decreases* as the square of the distance from the subject, but the total area of the subject encompassed by the meter's field of view *increases* as the square of the distance. The increase of one cancels the decrease of the other. Thus when photographing a scene in which the luminance of the subject differs greatly from its surroundings, the meter can be brought close to the subject to judge the proper exposure.

Incident Light Meter

An incident light meter is preferred by many professional photographers. For measurements it is held near the subject and is aimed in the general direction of the camera. Because it has a wide field of view—almost 180°—it can monitor most of the incident illumination that will be reflected toward the camera. A translucent hemisphere of etched glass or plastic positioned over the photosensitive cell refracts the light toward it. Figure 9-27 shows how this can be slid in front of a cell to convert a meter from reading reflected light to incident light. To use either type of meter, the photographer sets the meter's calibration (knob at the center in this example) for the film's ASA number ("400" in this instance). Then, when the meter registers the light that is incident on it, the needle of its meter indicates the amount. With incident light meters the calibration is arranged for a subject of average reflectance. The photographer must compensate for the actual reflectance if the subject has an unusually high or low reflectance. Figure 9-27 shows a reading of about 7.3 on the "light scale," an arbitrarily numbered scale that the manufacturer chose to use. Then the photographer rotates a dial (with the names "aperture" and "speed" printed on it) as shown to align the arrow with the number corresponding to the light scale reading. This places the choices of proper aperture settings and corresponding shutter speeds in register. In this example the scene could be shot at f/8 in $\frac{1}{60}$ sec, or equally well at f/4 in $\frac{1}{250}$ sec.

An incident light meter avoids a major shortcoming of the reflected light meter. Reflected light meters add contributions from various regions of the scene *arithmetically*. But film responds *logarithmically* to exposure. Comparatively speaking the reflected light meter will favor surfaces of high luminance, allowing those of average and low luminance to be underexposed. If the meter is highly directional, this problem could be overcome by taking readings of the brightest and darkest

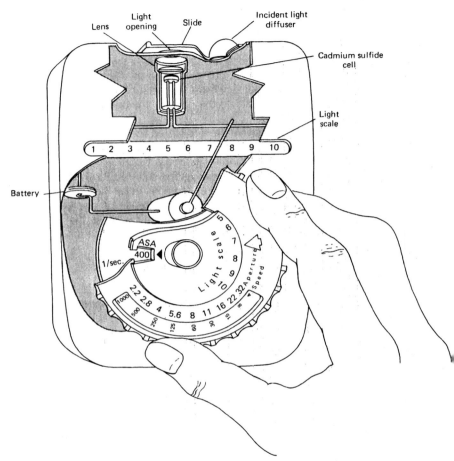

Fig. 9-27. Light exposure meter set for a film having an ASA speed of 400, with an illumination level of 7.3 indicated by its meter. The light enters directly through an opening restricting its field of view to a narrow range of angles for use as a reflected light meter. When used as an incident light meter the light diffuser is slid over the opening to increase its field of view.

surfaces and then calculating the geometrical mean (square root of their product). For example, a scene represented by four surfaces of equal area having steps in luminance of 1, 4, 16, and 64 when simultaneously monitored by a reflected light meter would be characterized by the arithmetical average: $(1 + 4 + 16 + 64)/4 = 21$. However, if the extreme values of luminance (1 and 64) are monitored separately, the calculated geometrical average would be $\sqrt{1 \times 64} = 8$. The difference between 21 and 8 in light entering the lens is a factor of 2.6, so they are more than one full f-stop apart. This is too great a difference to tolerate in careful work.

Transmittance of Lenses

Built-in exposure meters that register light admitted by the camera lens take into account the transmittance of the lens. As shown in Fig. 4-27 the glass from which lenses are made reflects about 4% of the incident light at each surface. This percentage is substantially reduced by a lens coating whose refractive index is the geometric mean (square root of the product) of the indices for air and glass, as will be shown in Section 11-9. This minimizes the overall reflectance of the two surfaces between the air and coating and between the coating and glass. Further reduction of the reflectance is obtained by having the coating made one-quarter wavelength thick for the center of the visible spectrum. Then destructive interference occurs in this spectral region between light reflected by the two surfaces. Transparent magnesium fluoride is widely used as a lens coating, but it still allows as much as 1 to 2% reflectance at each element of a lens. Further reduction in reflectance, as well as broadening of the spectral region in which destructive interference is effective, can be achieved by adding more coatings of intermediate values of refractive index. These are commonly called "multicoatings."

As mentioned earlier, light reflected within a compound lens, known as lens "flare," causes unwanted exposure over the film. Light is also reflected and scattered from dust and fingerprints, and for this reason lenses should be kept clean.

■ 9-7 COLOR NEGATIVE FILM

Color prints made from color negative film are made in much the same way as monochrome. Once exposed, the film is developed to produce a "color negative" with bright areas rendered dark and colors represented by their complements. A color positive is then produced by passing white light through the negative and focusing the image onto an emulsion-coated paper. When developed this "negative of a negative" provides the desired positive. The details of how this is accomplished illustrate the principles of subtractive color mixing.

Color negative films such as Kodak's Kodacolor and Vericolor have a multilayer emulsion with one or more filters sandwiched between them. Color Plate 27 shows a cross section of an exposed and developed film where yellow, magenta, and cyan dyes are found in layers that were sensitive to the blue, green, and red portions of the spectrum. The individual layers are actually very thin, for example, the yellow layer is only $\frac{1}{70}$ the thickness of a human hair. Color Plate 28 shows how the various layers are affected by light and by the subsequent development. During exposure light first reaches a layer of silver bromide crystals dispersed in gelatin that is sensitive only to the blue component. Then it passes through a layer serving as a yellow filter, which transmits only the green through red portion of the spectrum. The transmitted light then exposes the next emulsion layer, which contains a dye sensitizer to extend its sensitivity into the green. The remaining light exposes the third emulsion layer, which has a different dye sensitizer to render a response to the red. In this way three separate layers register latent images of the blue, green, and red spectral components of the incident light.

Developing the film leaves silver grains where development centers had formed. But each layer of emulsion also contains a chemical known as a *dye coupler,* which reacts with the developer in areas where the grains are formed. The product of this secondary reaction is a dye. Thus the dye-forming reaction leaves a concentration of this dye in proportion to the number of silver grains. The color of the dye in each layer is the complement of the color to which the layer responds. This is illustrated in Color Plate 28. The film is then immersed in a second chemical that removes or "bleaches" away the silver grains and yellow filter, leaving only the layers of dyes to affect the passage of light. When illuminated by white light a negative image is seen, formed by complementary colors of the original subject. Actually, the overall color has a strong orange component to compensate for distortions that will occur in producing a print.

The paper used to make such a positive has a multilayer emulsion similar to the films. When the paper is exposed by projecting an image of the negative onto it, development centers form precisely where they had not appeared in the negative (Color Plate 28). However the emulsions do not contain particularly effective sensitizers for long wavelengths, so that a safelight can be used in the darkroom. This is also the reason why the negative must provide more light at long wavelengths than would come from the true complementary colors. Then development and bleaching produces a positive print. Each layer of emulsion contains a dye of the complementary color to which the corresponding layer in the film responded, but it is found where light had *not* been absorbed in the film. A careful study of Color Plate 28 should convince you that when white light is reflected from the print after passing through the layers of dye, the colors are a close rendition of the original scene.

■ 9-8 COLOR REVERSAL FILM

The first practical color film was Kodachrome, invented by Godowsky and Mannes in 1935. It is called a color "reversal" film because after exposure it undergoes two development stages, the first of which produces a negative; the second stage produces a positive while erasing the negative. A color slide is made from this positive merely by mounting it in a cardboard or plastic frame.

The sandwich of three layers of emulsion and yellow filter in Kodachrome film is similar to that of a color negative film, but the emulsion contains no dye couplers. Once exposed the film is first developed in a way that leaves the unexposed areas still sensitive to light. The "reversal" aspect of the process then comes into play when the remaining emulsion is exposed to light (or more commonly this is done chemically). First, the film is exposed to red light that renders the lowest layer developable. The film then is developed in a solution containing a dye coupler, so that as metallic silver grains form so does a transparent cyan dye in the same regions. Next the film is exposed to blue light and developed with a dye coupler that leaves a yellow dye where the top layer originally had not been exposed. Finally the film is developed with a magenta coupler to deposit that color in the middle layer. Then the metallic silver grains and the yellow filter are bleached out by a solution, and the film is fixed, washed, and dried. The result is a positive picture when white light is passed through.

The Kodachrome process requires a complicated development sequence, which must be controlled precisely. Kodak's Ektachrome film and Agfa's Agfacolor have dye couplers included in the emulsion, making for simpler development that can be carried out in a home darkroom. Because in current color photography three light-sensitive layers must simultaneously be given proper exposure when taking a photograph, there is less latitude for error when choosing the exposure, and development must similarly be done more carefully.

■ 9-9 RAPID DEVELOPMENT COLOR FILM

In 1963 the Polaroid Corporation introduced a color film that could be developed within seconds. This Polacolor film, consisting of eight layers and a base, contains developer and dyes (Fig. 9-28). Three layers of emulsion, each sensitive to one of the three primary regions of the spectrum, have adjacent layers in which dye molecules of the complementary color are coupled to developer molecules. Once exposed, the film upon being pulled from the camera is pressed against a sheet that will become the positive. One of the layers of the positive contains a pod of alkaline chemical that is broken when the sandwich of negative and positive is pulled between rollers, and this chemical triggers the development. Exposed regions of the film absorb the developer and dye from the adjacent layer as silver grains form. Elsewhere this leaves the dye free to migrate through layers of negative to the first layer of the positive (the "mordant" layer) where they are retained. As the emulsion is only 0.05 mm thick there is very little sideways migration to blur the picture. Fixing takes place when a small portion of the alkaline solution reaches the acidic material in the positive. The resulting chemical reaction produces water, which washes away any remaining alkali, thus terminating the process. The mixture of dyes residing on the positive when it is pulled away from the negative gives a faithful rendition of the scene by the laws of color subtraction.

SX-70 Film

The more recent SX-70 film has an upper protective plastic layer and three positive layers (all four transparent) coating a nine-layer negative that is similar to that in Polacolor (Fig. 9-28). After exposure a motor spins a roller that ejects the film and breaks a pod containing a complex alkaline processing fluid (Fig. 9-29). The pod contains chemicals to initiate and control development of the negative, some titanium dioxide particles and two dyes, one orange and the other blue, which are

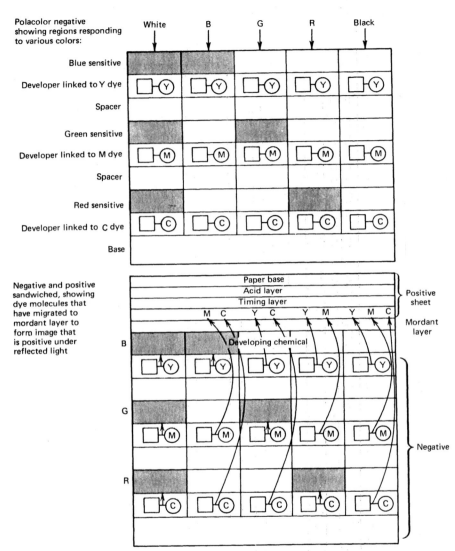

Fig. 9-28. Polacolor process of the Polaroid Corporation showing the exposed negative (top) and the sandwiched negative and positive (below), with dyes being transferred from the unexposed portions of the negative to the mordant layer of the positive. When the positive is stripped from the negative, these dyes form the image.

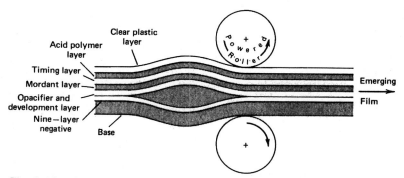

Fig. 9-29. Chemical pod of SX-70 film being broken on passing between rollers in the camera. The opacifier and development fluids are then spread evenly between the layers of negative and the mordant layer, which will become the positive.

highly reflecting when in an alkaline environment. They provide a protective shield over the negative to keep it from inadvertently being exposed as it emerges from the camera. The fine titanium dioxide particles scatter light and thereby decrease the transmittance through the dye layer as light is made to travel a greater distance through it. Even in bright sunlight the underlying emulsion is effectively protected. When the concentration of alkali in this opacifying layer decreases as the solution is absorbed into the negative, these dyes change color and become transparent. Prior to this the negative has been developed as in the Polacolor process and the released picture dyes penetrate upward through the opacifying layer to form a colored image in the mordant layer above. Meanwhile the alkali also penetrates the mordant layer and more slowly through the "timing layer." On reaching the acid layer a little water forms and the alkalinity of the processing fluid is rapidly reduced. The opacifying dyes become completely colorless so that the picture can be seen, but the white titanium dioxide remains to form a highly reflecting background for the picture, giving it unusual brightness. All this occurs in less than a minute after exposure.

■ 9-10 COLOR BALANCE

A film records according to the SPD of the incident light it receives. The eye does not respond in unbiased fashion because it adapts to varying light conditions. We do not notice the excess red component in a room illuminated by incandescent lamps with a low color temperature, nor do we notice the excess green component under most fluorescent lamps, contributed by their strong line spectrum (Section 7-5). But color films do record these characteristics and, unless compensating filters are placed before the lens, the resulting photographs do not appear "true to life." With sources emitting strong line spectra such as sodium or mercury vapor lamps, even a filter cannot compensate. When viewing the print, other objects such as the white border provide a reference that prevents the viewer from compensating for the color imbalance, although he would have compensated when viewing the original scene. However, improper color balance of a projected slide viewed in a dark room is often not apparent because compensation can occur.

A "daylight" film has its relative sensitivity to the blue, green, and red components of the spectrum balanced to provide a true rendition of the scene for illumination of a color temperature of 5600 K. As illustrated in Color Plate 29, a photograph taken of a scene illuminated with incandescent light appears distinctly reddish unless taken with a compensating blue filter. A film balanced for a "tungsten" source of light is designed for illumination from an incandescent source (usually for a color temperature of about 3200 K) and so is less sensitive in the red portion of the spectrum. Thus the developed photograph will not show the excess red of the illumination. Correspondingly, if a tungsten film is used outdoors it will provide photographs with a bluish cast unless taken with a compensating red filter.

Color temperature as used in photography refers to the SPD that characterizes an incandescent source, as discussed in Section 7-2. Although a particular type of fluorescent lamp may to the eye have a color temperature of 5600 K, because of contributions from its line spectrum it may produce results on film that differ considerably from what is produced by a tungsten filament of the same color temperature.

SUMMARY

The photographic process has two major components: forming an image within the camera, and recording the image on light sensitive material to form a permanent record. Photography began in 1837 with the invention of the daguerreotype—a silver plate treated with iodine to form light sensitive silver iodide. Subsequently, silver salts suspended in an emulsion were coated onto glass plates and, starting in 1884, onto rolls of celluloid to form roll film. Modern films still employ silver salts in gelatin emulsions, but their sensitivity and spectral response have been greatly extended by the use of sensitizers.

Color photography is nearly as old as black and white. By 1861, color separation was used to record three separate monochrome images representing the red, green and blue content of the scene. The original color could later be reconstituted by addition, a process still in widespread commercial use. The first practical process for recording color in a single film was Kodachrome, invented in 1935. This was the first of the multilayer films in which a stack of three emulsions with different spectral responses record different wavelength ranges. Dyes added during processing produce the finished color transparency. Kodachrome, and later films such as Ektachrome, undergo two stages of processing resulting in a positive transparency that can be viewed by projecting it on a screen. Color negative films such as Ektacolor are processed to produce a color negative in which the colors are complementary to the original scene. The negative is used to produce a positive paper print. In all color films, the spectral sensitivity must be balanced to match the SPD of the illumination, particularly for daylight versus incandescent lighting.

The first cameras used a pinhole or a small simple lens to form the image within the camera. In the mid-nineteenth century the first compound lenses were introduced, greatly reducing the required exposure time. Modern photographic lenses are complicated multi-element optical systems that provide extremely high light-gathering power while also correcting most aberrations. In addition to the lens, a camera also contains a shutter to control the duration of exposure, an aperture to control the amount of light admitted during exposure, and a mechanism for adjusting the focus. These elements are combined in various ways in different types of cameras such as the viewfinder, the view, the reflex, and the single-lens reflex cameras. Most of the modern single-lens reflex cameras contain built-in light meters, frequently coupled to the shutter or aperture to provide automatic exposure control. ∎

QUESTIONS

9-1. Explain how a pin-hole camera can form an image.

9-2. The autochrome process in color photography relied on additive color laws whereas the modern color transparency is based on subtractive laws. Why is this so?

9-3. "Exposure" has a precise meaning in photography. What is it, and in what units is exposure measured?

9-4. If the f-stop of a lens is increased by a factor of 2, by how much is the area of the aperture decreased?

9-5. If the f-stop of a lens is reduced from f/16 to f/11, by how much should the shutter speed be increased or decreased to maintain the same exposure?

9-6. Why does the depth of field for a camera increase when operating with an aperture of a higher f-number?

9-7. With the reflex camera and view camera the image is seen on a ground glass. Why might a photographer wish to view the image on a ground glass? Why is a ground glass used for viewing instead of a polished plate glass?

9-8. What is a "latent image" in an exposed photographic film?

9-9. What is meant by the term "development center"?

9-10. Why do films have an "antihalation" layer behind the emulsion?

9-11. Explain why sensitizers are included in the emulsion of a film.

9-12. Why can "rapid development" films achieve a finer grain when developed than negative films, even while providing high speed?

9-13. Which way should a reflected light meter be directed when taking a reading? An incident light meter?

9-14. What differences in color are apparent between a developed color negative film and a color reversal transparency of the same scene?

9-15. Why does a color reversal film have less exposure latitude than a monochrome film?

9-16. Why is a light scattering layer incorporated in Polaroid's SX-70 type color film?

9-17. To what are the three dyes in Polaroid's Polacolor film attached in the undeveloped film?

9-18. Explain what can produce color imbalance when exposing a film.

FURTHER READING

Editors of Time-Life Books, Life Library of Photography, includes volumes on *Color, The Camera, Light and Film,* and *Photography as a Tool.* Lavishly illustrated examples of the photographic art that should inspire amateur and professional alike. Clear explanations of photographic principles are given at the introductory level.

L. Larmore, *Introduction to Photographic Principles,* Second Edition (Dover, New York, 1965) (paperback). Fundamentals of the science on which photography is based, for the serious photographer.

A. Stimson, *Photometry and Radiometry for Engineers* (Wiley, New York, 1970). How camera settings affect exposure is included in this general text on how light is measured. Contains a rather complete characterization of light sources used in photography. At a technical level.

J. M. Eder, *History of Photography* (Dover, New York, 1978) (paperback). Entertaining account of the challenges overcome by early photographers.

H. E. Edgerton and J. R. Killian, Jr., *Moments of Vision, The Stroboscopic Revolution in Photography* (MIT Press, Cambridge, Massachusetts, 1979). Fascinating examples of "stop" motion photography by its inventor, but very little is described about the details of the technique.

B. Coe, *The Birth of Photography* (Taplinger, New York, 1977). A delightful account of the formative years of photography (1800–1900), with numerous early photographs.

R. W. G. Hunt, *The Reproduction of Color,* Third Edition (Fountain Press, London, 1975). Authoritative explanation of color reproduction techniques in photography, television, and printing.

From the *Scientific American*

W. H. Price, "The Photographic Lens" (August, 1976) p. 72.

R. Clark Jones, "How Images are Detected" (September, 1968), p. 111.

R. M. Evans, "Maxwell's Color Photograph" (November, 1961), p. 118.

CHAPTER 10

THE EYE AND VISION

The human eye is an incredibly complex optical instrument in which the stimulus—a constantly changing pattern of light—is transformed into a response—a pattern of electrical signals that travel along the optic nerve to the brain where they produce the perception we call "seeing." The operation of the eye involves optics, photochemistry, neurobiology, and electrophysiology. We shall consider each of these aspects of the eye's activity in this chapter, and will show how the concepts we discuss can explain some of the more puzzling aspects of vision that do not fit within a framework of the theories of colorimetry and photometry of Chapters 3 and 6.

■ 10-1 OPTICAL SYSTEM

The optical system of the eye is frequently compared to a camera, a useful analogy (possibly originated by Leonardo da Vinci), which we can follow in Fig. 10-1. The iris, which functions like the diaphragm of a camera, changes the size of its opening, or pupil, to control the amount of light entering the eye. The lens forms an inverted image of the subject on the retina, which is located in the position corresponding to that of the film in a camera.

Not until the early seventeenth century was the purpose of the lens understood. The correct explanation was offered by the mathematical physicist Johannes Kepler (1571–1630), working in Prague. Noted for formulating the laws describing the orbits of the planets, Kepler apparently acquired an interest in optics during the solar eclipse of 1600. Using the law of refraction recently formulated by the Dutch optician Willebrord Snell, Kepler worked out mathematically the directions taken by rays when they leave a point on a subject and are intercepted by a glass sphere. He found that the rays should converge behind the sphere at another point that depends on the position of the point on the subject. Arguing that the lens of the eye acts in the same way, Kepler realized that rays from different parts of a subject will be focused by the lens at different positions spread across the retina. In 1604 Kepler wrote: "Vision, I say, occurs when the image of the whole hemisphere of the external world . . . is projected onto the . . . concave retina."

We now appreciate that the major focusing element is the transparent cornea of the eye, and the lens serves for fine adjustments of the focus. The image on the retina is inverted—upside down and right to left. To account for how we perceive an upright image Kepler was forced to conclude that optical concepts do not apply beyond the retina, and some other mechanism—which he realized was yet to be explained—conveys an impression of the scene to the brain via the optic nerve. To Kepler goes the credit for explaining the essence of the optics of the eye and the retinal theory of vision.

The actual formation of an inverted image on the retina was demonstrated in 1630 by the French scientist-philosopher-mathematician René Descartes, who scraped the opaque backing from the eye of a freshly slaughtered ox and observed the inverted image directly.

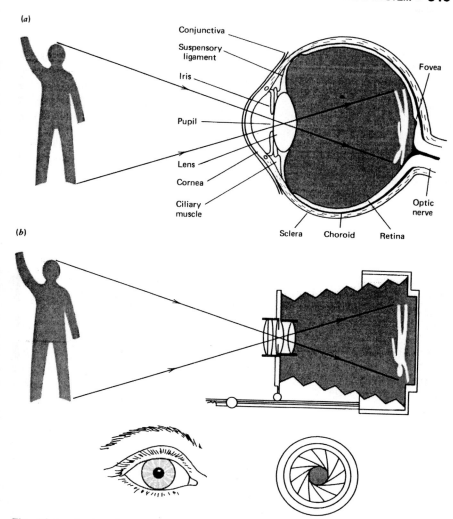

(a)

Conjunctiva
Suspensory ligament
Iris
Pupil
Lens
Cornea
Ciliary muscle
Sclera
Choroid
Retina
Fovea
Optic nerve

(b)

Fig. 10-1. Analogy between the (a) eye and (b) camera.

Accommodation

One obvious problem that comes up immediately in attempting to understand the optics of the eye is how the image on the retina can remain in focus if the distance between the eye and the subject being viewed changes. In Chapter 8 our discussion of ray optics showed that as the distance from a subject to a lens decreases, the image is focused further behind the lens. In a camera, to compensate for this change, the lens is moved further ahead of the film. In the eye, however, the distance from the lens to the retina is fixed by the anatomical structure of the eye. How then can the image remain focused on the retina if the subject's distance changes? Only by a variation of the strength of the lens, that is, by changing its focal length.

Figure 10-1*a* shows that the lens is supported by *suspensory ligaments* that in turn are attached to a muscular system, the *ciliary muscle.* The lens is not solid, but has the consistency of a stiff gel, and its natural elasticity causes it to bulge. This tendency is countered by tension exerted by the suspensory ligaments, so that when the eye is relaxed, these ligaments stretch the lens and reduce its curvature. The normal eye is then focused for infinity (for a subject more than 10 meters away). As the subject comes closer, the ciliary muscles contract so as to slacken the suspensory ligament, allowing the curvature of the lens to increase. This decreases the focal length in such a way that the image remains focused on the retina. This process is called *accommodation.*

In Section 8-2 we noted that the strength of a lens is the reciprocal of its focal length in meters. The cornea alone has a strength of about 43 diopters, and the lens of the relaxed eye about 19 diopters. The total strength is then 62 diopters, corresponding to a focal length of slightly more than 16 mm. When the ciliary muscles exert tension and allow the lens to bulge when accommodating to a nearby subject, the lens strength increases to about 33 diopters. The lens and cornea then provide a total strength of 76 diopters, or a focal length of 13 mm.

This "auto-focus" system is limited by the elasticity of the lens, which decreases markedly with increasing age and thus will not bulge as much. Young children can easily focus on subjects as close as 5 to 10 cm from the eye. For young adults, the closest comfortable viewing distance is about 25 cm. For people over 40, the elasticity of the lens often becomes so reduced that focusing at normal reading distance becomes impossible, and reading glasses are required. This condition is known as farsightedness, or *presbyopia* (from the Greek words for "old eye").

Corrective Lenses

Eyeglasses compensate for the eye's inability to focus on nearby subjects by providing a few diopters additional strength. If a distortion of the lens causes the image always to be focused behind the retina ("farsighted"), a converging lens may be needed all the time, and either eyeglasses or contact lenses will correct the problem (Fig. 10-2*a*). On the other hand, if the relaxed eye focuses the image in front of the retina ("nearsighted") a diverging lens is required for correction (Fig. 10-

(a)

(b)

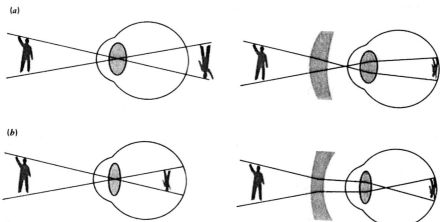

Fig. 10-2. (*a*) An eyeglass for a farsighted eye has a converging lens. (*b*) An eyeglass to correct a nearsighted eye has a diverging lens.

2*b*). If there is also a loss of accommodation because of loss of lens flexibility, portions of the surface of a correcting lens are ground with different curvature so that there is a corresponding difference in strength when the wearer looks straight ahead toward a distant subject or downward at a nearby one. A lens with two focal lengths is a *bifocal* and one with three, a *trifocal*. A new type of lens has a shape that effectively provides a continuously variable focal length from top to bottom.

Even without a corrective lens a nearsighted person can see a nearby subject better than a distant one. This is because as a subject is brought closer, the image is focused further behind the lens. From Fig. 10-2 we conclude that it lies closer to the retina and therefore is seen in sharper focus. Similarly a farsighted person sees a more distant subject with better clarity than a nearby subject.

Often there is departure from sphericity in the cornea or lens as well, leading to the aberration called astigmatism, which was described in Section 8-5. To correct for astigmatism, a cylindrical lens is required. Frequently, eyeglasses correct both for spherical and astigmatic errors, which requires a rather sophisticated procedure for grinding the lens. The immense number of possible combinations of spherical and cylindrical curvatures requires a very large assortment of lenses to be maintained in a good opticians' supply house. The prescription for a lens specifies its strength in diopters. A positive strength denotes a converging lens, and a negative strength, a diverging one. For correcting astigmatism the orientation of the axis of cylindrical curvature must be specified in addition to its strength.

How much visual impairments blur the image on the retina depends on the size of the pupil: at low light levels when the iris is wide open, very slight defocusing causes severe blurring; conversely, blurring can be virtually eliminated—even for subjects within only an inch of the eye—by looking through a very small pinhole. This effect, first discussed by Scheiner in 1619, was recently utilized commercially in a set of "lensless eyeglasses" consisting of two thin black disks, each containing a tiny pinhole! Forming an image with small apertures is also the basis of the pinhole camera and its predecessor, the camera obscura, discussed in Chapter 9.

■ 10-2 PERSPECTIVE

Before continuing with our analysis of the functioning of the eye, we shall see how the concepts of the previous section help to explain the ideas of perspective—the technique for representing a three-dimensional scene on the two-dimensional surface of a painting. In Fig. 10-3 we illustrate schematically how the images of subjects are formed on the retina. The size of the image is determined entirely by the visual angle—the angle θ between the rays reaching the eye from the extreme edges of the subject. In the illustration, if $\theta_1 = \theta_2$ the distant tree and the nearby bird each subtend the same visual angle so that the sizes of the images are identical. This basic concept, that the visual size of a subject depends on both its physical size and distance from the eye through the visual angle, was clearly spelled out in Euclid's *Optics* around 300 B.C. This may seem surprising since Euclid—and indeed many writers before Kepler and Leonardo da Vinci—believed that vision was an emission process by which light rays were emitted from the eye rather than being received by it. However, Euclid in addition assumed that the rays travel out-

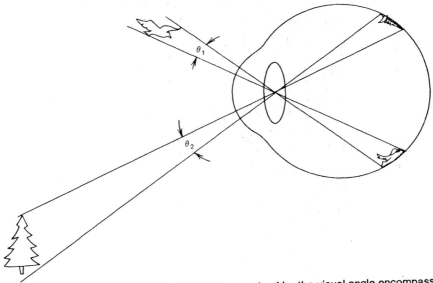

Fig. 10-3. The size of the retinal image is determined by the visual angle encompassing the subject.

Fig. 10-4. The principle of linear perspective showing how rays entering the eye from corners of a cube project onto a two-dimensional surface. The spectator is depicted holding his hand to his eye presumably because in earlier illustrations of this period strings were used to determine the direction of rays. (From Brook Taylor, 1811 edition of the 1719 treatise *New Principles of Linear Perspective* as it appeared in M. H. Pirenne, *Optics, Painting and Photography*, Cambridge University Press, 1970)

wards in straight lines; therefore the geometrical conclusions he reached were correct.

The application of Euclid's geometrical analysis to painting developed during the Renaissance into the theory of linear perspective, largely through the work of Leonardo da Vinci. Figure 10-4, from a 1719 treatise on perspective by Brook Taylor, illustrates the basic ideas: the eye (Point *O*) is at the center of projection and receives light rays from each point on the subject, for example, from the points labeled *A* through *E*. Placed in front of the subject is a stand with a transparent "Leonardo window" (*FGHI*) representing the flat surface of the painting to be executed. Each ray entering the eye from the subject passes through the window and defines a point (*abcde*) on the two-dimensional surface corresponding to the original three-dimensional object. This projection onto the plane surface defines the correct form of a painting, which, when viewed from the original center of projection (*O*), will appear to reproduce the original three-dimensional scene.

The illusion of depth in a painting is relatively easy to achieve if the painting is viewed with one eye from the proper viewing position, the center of projection. This occurs, for example, in the "peep-show" where a painting is located within a box and viewed through a small hole in the front panel. In practice the perception of a three-dimensional scene in a painting is hindered both by differences in viewing positions and by viewing with both eyes, which produces an awareness of the *surface* of the painting as a plane surface. Techniques of painting do exist which overcome much of this difficulty and create a startling impression of three dimensionality. They are referred to as *trompe-l'oeil* (fool the eye) and often rely on a heavy use of shadow to achieve a sense of depth. There are some striking large-scale paintings that achieve the effect of three dimensionality with great intensity, most notably the painting of the ascendance of Saint Ignazio in the Church of St. Ignazio in Rome, painted in the late seventeenth century by Fra Andrea Pozzo.

■ 10-3 THE RETINA

The inner surface of the eye on which the image of the external scene is formed is called the retina. Within the retina, there are roughly 130 million specialized photoreceptor cells, the rods and cones, that respond to the pattern of illuminance of the image and its spectral power distribution (SPD). These photoreceptors are connected, through a complicated structure of nerve cells within the retina, to about one million nerve fibers, which leave the eye through the optic nerve.

Referring to Fig. 10-1a, we note on the optical axis of the eye a small indentation in the retina called the *fovea*. Although only about the size of a pinhead, the fovea is the crucial region where photoreceptors provide high visual acuity as well as most color vision. To "look over" a subject consists of scanning the eyes so that the image passes over the fovea. The photoreceptors of the fovea are almost entirely cones—about 30,000 of them—that are connected to nearly the same number of fibers in the optic nerve. The 120 million rods and six million cones distributed throughout the remainder of the retina share the remaining fibers.

The area where the optic nerve makes its connection to the retina contains no photoreceptors and therefore is a blind spot in the field of vision. Most people are unaware of their blind spots since the two eyes have them in different parts of the visual field, and the eyes "fill in" for each other. To observe the blind spot, close your right eye and stare at your right thumbnail held in front of you at arm's length. Now move your left thumb gradually away from the right. When the left thumb is about 15 cm (6 in.) from the right thumb it will disappear!

Returning to the anatomy of the retina, let us consider a cross section of the human retina magnified approximately 150 times (Fig. 10-5). The bottom half of the figure shows the outer cover of the eye, the sclera. The inner surface of the retina that faces the front of the eye is at the top. Note that light reaching the retina must pass through several layers of transparent neural tissue before reaching the photoreceptors near the midpoint.

Eyes of diurnal animals such as man have a layer of black pigment at the back of the retina to absorb light that bypasses the photoreceptors. It reduces reflection that would lead to blurring of the image. But nocturnal animals, such as cats, lions, and deer, that must see at night, have a different structure to use light more efficiently, at the expense of clarity. The cells at the back of the retina reflect the light back past the photoreceptors to give them a second chance. Such animals seen in the strong beam of automobile headlights have eyes that reflect as a yellow glow. Even in humans, some reflected light can be seen if a bright light is shown

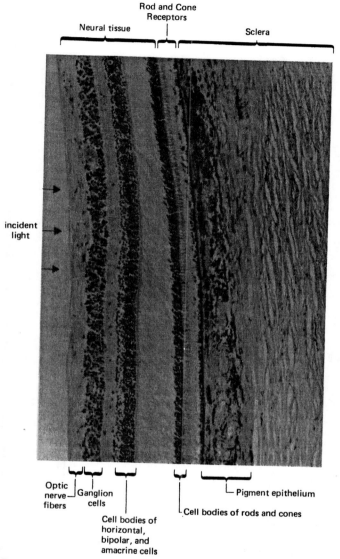

Fig. 10-5. Cross section of the human retina. (Courtesy Norman C. Charles, M. D., 1982)

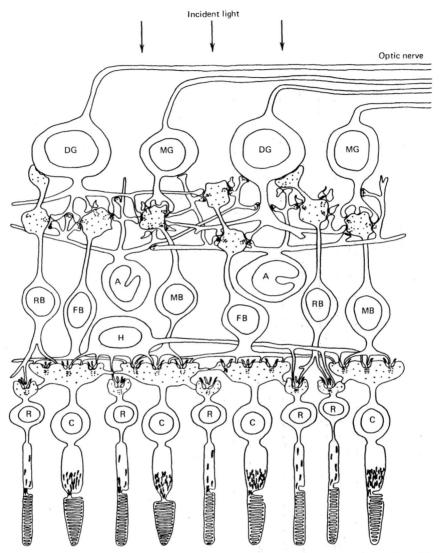

Fig. 10-6. Rods (R) and cones (C) connect to several types of bipolar cells: midget bipolars (MB), rod bipolars (RB), and flat bipolars (FB). Communication along the retina is provided by horizontal (H) and amacrine cells (A). These and the bipolars send signals to the midget ganglion (MG) and diffuse ganglion (DG) cells, whose axons comprise the fibers of the optic nerve.

directly into the eye along its optical axis. The lens of the eye helps to redirect the weakly reflected light back toward its source. Photographers using a flash lamp know this as "red eye" and are aware that it can be avoided by positioning the flash lamp further to the side of the camera.

The receptors—rods and cones—are tightly packed in the retina. In the fovea, the cones form a mosaic with a distance between adjacent cones of only about 2 μm, barely larger than the wavelength of light! This region of high-acuity vision within the fovea is called the *macula*. A schematic diagram of the retinal structure is shown in Fig. 10-6, where the differences in the shapes of the rods (R) and cones (C) can be seen. Light arrives from the top in this figure, traversing the layers of nerve cells to reach the rods and cones.

Within these receptors are many visible layers of membranes (Fig. 10-6). These membranes contain the visual pigments that absorb light and then undergo a complex series of chemical changes resulting in the production of an electrical signal within the cell body.

The visual pigments of the rods and cones are all very similar, consisting of a *retinene* joined at both ends to one of the retinal proteins called *opsins*, which are associated with the membrane. A source of retinene is vitamin A (Fig. 10-7a), commonly found in foods such as the carrot. In rods the retinene is neoretinene-b (Fig. 10-7b) and the protein is scotopsin. The combination is a molecule called *rhodopsin*, known also by the less formal name *visual purple*.

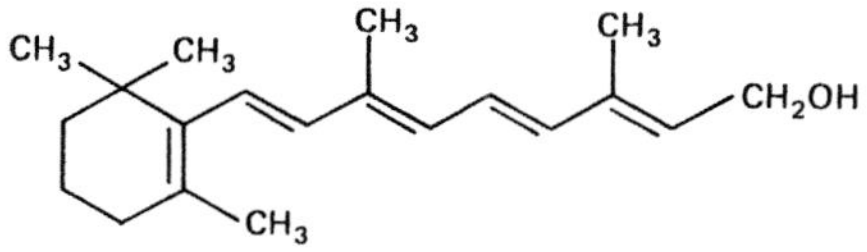

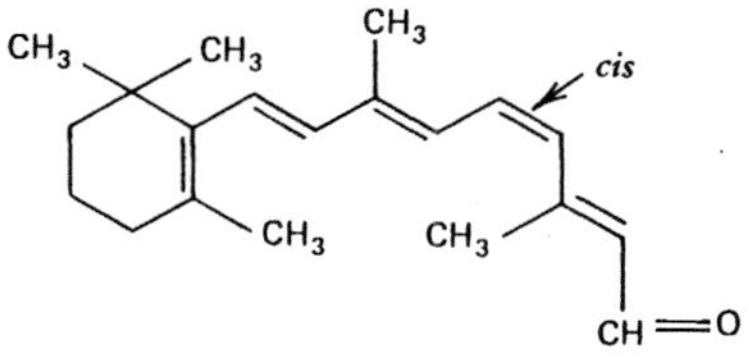

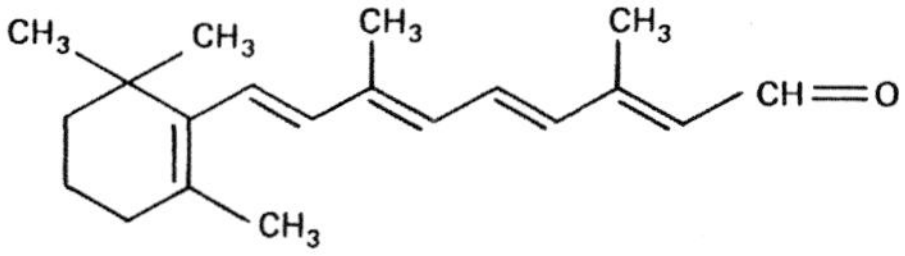

Fig. 10-7. The molecules of (*a*) vitamin A; (*b*) *cis*-neoretinene-b found in rods of the dark-adapted retina; and (*c*) *trans*-retinene produced by the illumination of the retina.

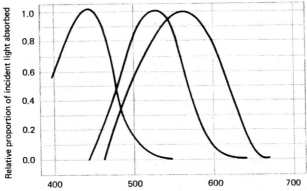

Fig. 10-8. Relative sensitivity of the three types of cones in the human retina as deduced from differences in the absorption spectra between bleached and unbleached cones. The heights of the curves have been adjusted to have a maximum value of 1.0. (From G. Wald and P. K. Brown, "Symposia on Quantitative Biology," The Cold Spring Harbor Laboratory of Quantitative Biology, New York, 1965)

Cone vision involves different retinene-protein combinations called *iodopsin*, which absorb at longer wavelengths than rhodopsin. There are three types of cones, each containing a different pigment formed of retinene joined to one of three different opsins. The three pigments have different absorption spectra (Fig. 10-8), so that the responses of the three types of cones depend differently on the spectral power distribution (SPD) of the light reaching them. As originally proposed in 1801 by Young, this difference is the physiological basis of color perception.

A rod or cone contains several million molecules of visual pigment. When one absorbs a photon, part of the energy causes the bonds of the retinine to rearrange, and it changes from the *cis* form in Fig. 10-7b to the all-*trans* form in Fig. 10-7c. In the process the straightened end breaks loose from the opsin protein attached to the membrane and a nerve pulse is initiated. If no other photon is absorbed, a sequence of adjustments takes place in the opsin over a period of about one second, and the retinene would be released. If a great many retinene molecules are released from the membranes, the pigment is said to be "bleached." The action of an enzyme eventually reconverts the *trans*-retinene to its *cis* form and joins it to an opsin to regenerate the pigment. The retina's sensitivity to light is reduced when the pigment is bleached, because the free retinene cannot absorb a photon and generate a nerve impulse. This is why it seems darker when you walk from the bright sunlight into a movie house than after you have been inside for a few minutes.

Rods are extremely sensitive since each is capable of responding to a single absorbed photon. Furthermore, neural interconnections serve to enhance the signal in the ganglion cells when several nearby rods are simultaneously excited. Rhodopsin exhibits its maximum absorption at 507 nm, corresponding to the wavelength where the eye's scotopic sensitivity is also greatest (Fig. 6-1). The sensitivity of a single cone is about five times lower than that of a rod. Consequently at low light levels the cones are inoperative and one sees entirely by rod vision with no color discrimination possible. Conversely, at high light levels the rods are inoperative and vision is entirely mediated by the cones. We shall discuss this distinction further in Section 10-8.

■ 10-4 CONTRAST: LATERAL INHIBITION

In our discussions of vision until now, we have followed a "local" point of view in which the psychological perception of the scene in a small region of the visual field is assumed to be determined by the SPD of the light incident on the corresponding region of the retina. This assumed simple connection between the stimulus and resulting perception is grossly inadequate, however, and formed one of the main obstacles to general acceptance of the physical ideas of light and color from Newton's time until the last century. Color interactions, for example, were well known by artists of the sixteenth century, who recognized that the perceived color of an area of the canvas could be seriously modified by strong colors in adjacent areas. We shall consider the effects of some of these interactions in the next section. However, a purely local theory turns out to be inadequate even for describing color-free aspects of vision.

For example, imagine that we stand in a room whose ceiling, walls, and floor are painted white. The brightness of these surfaces will appear unchanged if the illuminance is increased or decreased, once the transitory period is over during which the illumination was changed. In special situations such as this, our perception of brightness does *not* increase with the illuminance on the retina. This is called *brightness adaptation.*

Part of our compensation for changing illuminance comes from the action of the iris of the eye, which regulates the entering luminous flux to some extent by changing its diameter in response to changing illumination. However, this is not sufficient to account for the entire effect, because the maximum change in admitted flux is about a factor of 16, whereas the range of intensities for which the eye compensates is much greater.

An additional effect comes from adjustment in the sensitivity of the retina itself to the level of illumination. As explained in the preceding section, because the number of rhodopsin and iodopsin molecules fixed to membranes decreases with increasing illuminance, the retina's sensitivity to a given change in illuminance is reduced. Nevertheless the neural signal from a given rod or cone would still increase with increasing illuminance; thus the explanation for brightness adaptation must be found at a subsequent stage in neural processing.

Lateral Brightness Adaptation

A related visual effect is the modification of the sensitivity of one area of the retina by the level of illuminance on an adjacent area. This produces a variety of interesting contrast effects sometimes called *lateral brightness adaptation.* In Fig. 10-9, for example, if two identical gray cards are viewed with one in the center of a large black background and the other in the center of a large white background, then the gray card in the black background will appear brighter than the other one. This effect indicates that the white background tends to reduce the sensitivity of the visual system not only at those points where its image is formed but also in the adjacent area where the image of the gray card falls.

An important clue to the origin of brightness and lateral brightness adaptation is another effect of contrast known as *Mach bands* (Fig. 10-10). These were named after the physicist Ernst Mach who first investigated these phenomena in the 1860s. Each gray strip in the upper portion of the figure has uniform reflectance, so that the luminance (stimulus) is uniform over the entire strip. But the perceived brightness is not at all uniform, each strip appearing to have a bright edge along its left side and a dark edge along its right as indicated in the figure. Another example occurs on the home television screen. With the set turned off (but the room lights on) the screen of most sets will appear gray-green. But when turned on and displaying a picture some areas will appear jet black—for example the dark hair of an actor. Clearly, that area of the screen cannot be darker with the set on than with the set off. The perception of blackness is a localized effect of lateral brightness adaptation, the lowering of the sensitivity of an area of the retina caused by the strong stimulation of adjacent areas receiving brighter portions of the image.

Are these effects "psychological" or "physiological"? The question is, of course, without meaning until the terms are defined. We shall define as "physio-

Fig. 10-9. The gray card on a black background appears brighter than an identical gray card on a white background.

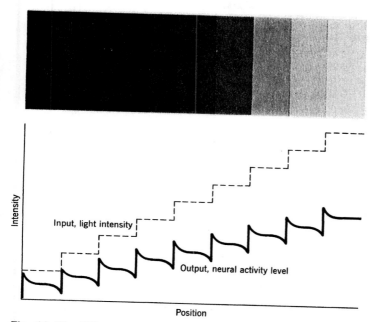

Fig. 10-10. Differences in brightness are enhanced at the edges of adjacent rectangles, each having a uniform but different luminance. (From Buss, *Psychology: Behavior in Perspective*, Second Edition, Wiley, New York, 1978.)

logical'' any process that can be observed to take place at some point along the visual pathway before it disappears into inaccessible regions of the brain. Conversely, by "psychological" we mean a process that occurs too far along in the information processing system to be accessible to measurement with external probes. The line between physiological and psychological processes constantly shifts as experimental methods and instrumentation improve. The domain of physiological effects grows as we learn more about the detailed processes of the brain.

Contrast effects were generally considered to be psychological (the viewer "compensates" for the brightness of the surroundings) before the 1930s. At that time, H. Keffer Hartline initiated a program of research on the eye of the horseshoe crab *limulus* that has profoundly modified our ideas about vision in general and contrast in particular.

Neural Signals

If we return to Fig. 10-6, we can next consider what happens when a receptor (rod or cone) responds to the incident light. The photoreceptor cells and the other cells above them in the figure represent a kind of switchboard that is an extension of the brain. Each nerve cell or neuron consists of receptive areas (dendrites), a cell body, a nerve fiber (the axon), and a terminal junction or synapse at which the axon of one neuron terminates on the dendrites of another. When a photoreceptor is stimulated, it transmits an electrical signal to nearby bipolar cells and horizontal cells. Signals are passed along until they reach the ganglion cells. From this point information is transmitted over the much longer distance to the brain by the same process as found elsewhere in the nervous system: when a neuron is stimulated, it sends out a short electrical pulse ("action potential"), about 0.1 volt in amplitude and 0.001 second long, that travels along the axon until it reaches the "target" neuron.

The size of the pulses is essentially independent of the strength of the stimulus. It is the *rate* that increases as the strength of the stimulus increases. There is considerable evidence that the rate at which many of the ganglion cells produce pulses varies as the logarithm of the illuminance on nearby photoreceptors. Thus the number of pulses per second increases by a constant amount each time the light intensity is doubled, over a very large range of intensities. This logarithmic response accounts in part for the very large range of light intensity to which our visual system can respond while not making excessive demands on the signal handling capability of the neural system. It also may explain how neural signals are coded so that the information they provide the brain would make perceptually equal steps in brightness correspond to constant multiples (rather than constant differences) of intensity.

Once a neuron has been triggered, the signal that it produces travels along the axon until it reaches the synapse at the target neuron. At that point it may trigger the target neuron to react, sending a signal along its axon, so that a signal is relayed along from neuron to neuron until reaching the brain. The dendrites extending from a cell body contain a large number of synapses and a target neuron receives inputs from many different photoreceptors. Not all synapses are "excitatory." There are also "inhibitory" synapses. If a signal arrives at an inhibitory synapse, then it is less likely that this cell will fire if it receives another signal at an excitatory synapse. Thus target neurons generate output signals that depend on the combination of excitatory and inhibitory inputs they receive. The processing of information that takes place along interconnected neural pathways is extremely complex. As shown in Fig. 10-6, the photoreceptor cells connect to a large number of other neurons within the retina, and the interconnections between them are so complex that they have still not been fully analyzed.

Lateral Inhibition

Nevertheless, the basic physiological mechanism underlying brightness adaptation phenomena can be observed as follows. In Hartline's research on the horseshoe crab a tiny electrical probe (microelectrode) was inserted into a single nerve fiber of the optic nerve. A point light source was then moved around in the visual field until pulses were observed in the probe. The response to this strictly local stimulus could now be studied by changing the intensity of the light source.

Next, a second light source was turned on close to the first, and it was observed that the pulse rate at the probe *decreased.* This observation established the fact that some of the cross connections in the neural network of the retina are inhibitory so that stimulation of one photoreceptor lowers the effective sensitivity of its neighbors. This mechanism—called *lateral inhibition*— is easily seen to explain brightness adaptation phenomena. The strong stimulation of one area of the retina produces inhibitory pulses in neurons carrying signals from adjacent areas which lowers the pulse rate from those areas traveling along the optic nerve. Also, if a large area is strongly illuminated, each small section will inhibit the areas near it: although the signals coming from individual photoreceptors increase with increasing illumination, the inhibition introduced by the subsequent neural processing leaves the perceived brightness unchanged (brightness adaptation).

This very condensed discussion of lateral inhibition illustrates the sense in which brightness adaptation, lateral brightness adaptation, and Mach bands may be considered as physiological phenomena, since the effect of lateral inhibition is identified as present in the pattern of neural pulses in the optic nerve before it reaches the brain.

Pathways to the Brain

Finally, we note briefly that the partially processed visual information leaving the retina via the optic nerve does not proceed directly to the brain. As shown in Fig. 10-11, fibers of the optic nerve divide at a region known as the *optic chiasm* (named for the Greek letter chi (χ) representing a crossing), where the signals from the right half of each retina (corresponding to the left half of the visual field) proceed to the right side of the brain, and information from the left half of each retina to the left side of the brain. The fibers of the optic nerve finally terminate at the *lateral geniculate nucleus,* whose function is still not properly understood.

The separation of the right and left halves of the visual field at the optic chiasm was discovered by Isaac Newton in 1666 during the dissection of an animal, probably a sheep. A drawing in his notebook of that year closely resembles Fig. 10-11. New neural fibers that originate within the lateral geniculate nucleus carry the visual signals along the final path to termination on cells of the visual cortex. This thin gray surface layer of the brain is the vision center and is where much of the complex signal processing basic to perception goes on.

■ 10-5 INDUCED COLOR

Johann Wolfgang Von Goethe published *Theory of Colours* in 1810. This book by the acclaimed author of *Faust* and other fictional masterpieces caused a tremendous sensation in Europe with its violent attack on Newton and all of scientific color theory. To some extent, Goethe was motivated by a persistent aesthetic distaste for Newton's view of white light as a mixture of colors which can be separated into

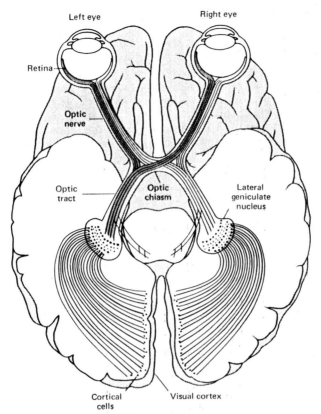

Fig. 10-11. Visual pathways from retina to cortex as viewed from above.

its true constituents by a prism, and he proposed to return to Aristotle's position that white light is pure and develops color as a sort of contamination. This aspect of Goethe's treatise is entertaining, but of no objective merit.

On the other hand, much of the *Theory of Colours* deals with various color contrast phenomena which Goethe had carefully demonstrated and clearly described. For example:

> 65. Let a short, lighted candle be placed at twilight on a sheet of white paper. Between it and the declining daylight let a pencil be placed upright, so that its shadow thrown by the candle may be lighted, but not overcome, by the weak daylight: the shadow will appear of the most beautiful blue.

This is an example of colored shadows, one of a group of dramatic color contrast phenomena beyond the scope of the local theory of Newton and his followers.

At the present time, there is no fully established explanation for color contrast phenomena. However, a plausible explanation can be constructed from a mild extrapolation of the lateral inhibition mechanism discussed in the preceding section. It is only necessary to suppose that the inhibitory connections are mainly between cones of the same type. It is then apparent that if one area of the retina is strongly stimulated by red light, then the sensitivity of the red-sensitive cones in neighboring areas of the retina would be reduced.

In Goethe's experiment, the yellow light of the candle preferentially stimulates the red-sensitive and green-sensitive cones. The shadow area is illuminated by the white daylight. But the sensitivity of the red- and green-sensitive cones in this region has been reduced by lateral inhibition from the surrounding yellow region. Consequently, the blue-sensitive cones are more sensitive than the others so that the shadow appears blue.

The phenomenon of colored shadows is easily demonstrated with two matched slide projectors. If a colored filter is placed in front of one, and an object—your hand, for example—is placed between the projector with the filter and the screen, then the shadow (which is illuminated by the other projector) will take on a hue complementary to that of the filter.

The classical theory of colorimetry presented in Chapter 3 need not be abandoned, but must be used with care when different hues are simultaneously present in the visual field. The reader should recall that in psychophysical experiments where color matches between a standard and a sample color are made, the two colors are displayed with a black surround that effectively eliminates any possibility of color interactions. The appearance of a colored area close to a large area of another color cannot readily be predicted by any colorimetry system, and is one of the problems that a working artist or decorator must solve on the basis of experience.

■ 10-6 SUCCESSIVE CONTRAST

In the preceding sections we have discussed a group of visual effects known collectively as *simultaneous contrast phenomena*. In these effects, the visual intensity or chromaticity of part of the visual field is modified by the intensity or chromaticity of adjacent areas. There is another class of visual effects known as *successive contrast phenomena* in which the intensity or chromaticity of an area of the visual field is modified by the preceding stimulus. An example is shown in Color Plate 30. If you stare at the black fixation point in the center of the picture for 20 seconds and then shift your view to the fixation point in the blank square, you will see an afterimage in colors that are complementary to those in the picture. (Blinking helps to make this appear.)

Complementary afterimages are easily understood as a kind of fatigue effect. Strong stimulation of a photoreceptor tends to bleach the light sensitive pigment and lower the sensitivity. Thus, when viewing a red object, the red sensitive photoreceptors will be temporarily fatigued. When a white surface is viewed subsequently, the red sensitive detectors respond less strongly than the others; thus the impression is of cyan. Note that this is a purely local effect and would occur even if there were no mechanism of lateral inhibition. In the past, surgeons working at a strongly illuminated operating table were disturbed when such complementary afterimages appeared if they temporarily looked away from the table towards the white clothing of their colleagues. The afterimages were often cyan, complementary to the color of blood. The clothing worn in modern operating rooms is usually light green or blue, which nearly eliminates this problem.

■ 10-7 FLICKER

When the eye is stimulated by a very brief flash of light, the neural activity does not die away immediately, but persists for about 0.1 second. Consequently, if one sees a series of flashes sufficiently closely spaced in time, the perception is of continuous light. A useful application of this principle is found in the stroboscope or strobe light, a device producing very short flashes of light at intervals that can be adjusted by rotating a dial. If a strobe light is operated at less than five flashes per second, you perceive a series of distinct "snapshots" of the visual field. A person appears to move in a series of jerks. At higher rates, the scene appears to flicker, and the flicker gradually disappears as the flash rate increases. The rate at which all trace of flicker vanishes—called the *critical fusion frequency*—depends on the nature and intensity of the pulses, but is generally about 40 flashes per second.

If a strobe light illuminates an electric fan, the rate can be adjusted so that the movement of the fan appears to stop. Each flash is so short that the fan does not move appreciably during the flash, and with proper adjustment the blades will have turned through one or more full rotations by the next flash and thus appear not to have moved at all. Nevertheless, because the flashes occur more often than 40 times per second, the perception is of a continuously seen motionless object.

Movies and television both depend on this *persistence* of visual perception to produce the illusion of continuous motion in response to a sequence of distinct pictures. However, their mode of presentation must also contend with brightness adaptation. Movies are normally made at 24 frames per second. During the $\frac{1}{24}$ second during which the frame appears, the eye undergoes significant adaptation, and during the subsequent dark moment when the light is blocked from the film and the next frame is moved before the lens, the eye begins to readapt. Thus it regains high sensitivity for the initial viewing of the next frame. A movie shown at only 24 frames would appear to flicker because of this successive adaptation to the bright scene and intervening periods of darkness. An ingenious technique avoids this problem: during projection, the film is advanced through the projector at 24 frames per second, but each frame is shown twice. Thus the viewer sees 48 images per second, which exceeds the critical fusion frequency where adaptation effects become imperceptible.

A similar problem occurs in television where the entire frame is swept 30 times per second. As will be described in greater detail in Chapter 13, in order to avoid flicker the picture is displayed by first sweeping out every other line (1, 3, 5, etc.) in one pass, and then sweeping out the interleaving lines (2, 4, 6, etc.) on the following pass. Thus the entire screen is covered 60 times per second to avoid flicker.

■ 10-8 DAY AND NIGHT VISION: RODS AND CONES

The distribution of rods and cones covering the human retina is depicted in Fig. 10-12: the cones are concentrated in the tiny rod-free fovea, while the rods dominate outside of the fovea. The sensitivities of these two systems differ in two important ways, which we have already described in Section 6-1: (1) The sensitivity of the rod system is about 1000 times greater than that of the cones and accounts

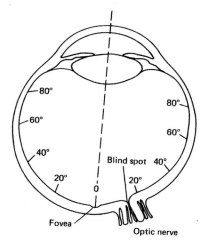

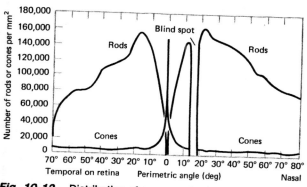

Fig. 10-12. Distribution of cones and rods in the human retina. (From M. H. Pirenne, *Vision and the Eye*, Associated Book Publishers, London, 1967.)

for scotopic (dark-adapted) vision; (2) the peak sensitivity of the rods is at about 507 nm while the cones' sensitivity peaks near 555 nm. Note that the cone sensitivity depicted by the photopic curve V_λ in Fig. 6-1 is actually the sum of the sensitivities of the three different classes of cones.

At night, when the illumination level is too low for the cones to respond, vision is mediated exclusively by the rods. Since there is only one type of rod, night vision is in black and white, and any sense of color comes either from memory or from a psychological judgment based on relative brightness. Because the scotopic sensitivity curve peaks at shorter wavelengths than the photopic curve, blue objects appear brighter than red ones at night, whereas the opposite is generally true in day vision. This *Purkinje shift* makes it extremely difficult to identify colors at night. The spectral sensitivity of night vision matches almost perfectly the absorption spectrum of rhodopsin.

At the high illumination levels characteristic of day vision, the rods are completely saturated, so they do not respond to changes in the illumination, and vision is mediated by the cones. Consequently, in order to see a subject clearly one turns the eye so that the image falls directly on the fovea where the cone density is greatest. But at night this strategy fails completely. Since the cones are inoperative, the fovea is effectively blind. Therefore, to see a subject distinctly at night, one should "look off" slightly to one side so that the image falls on a rod-containing region away from the fovea.

■ 10-9 PHYSIOLOGY OR PSYCHOLOGY?

In discussing visual phenomena, there is always some confusion over the distinction between the physiological and psychological aspects of perception. As we noted in Section 10-4, modern physiological studies of the retina have shown that many contrast phenomena previously thought to be psychological are actually physiological in that the level of neural activity already contains the new features of the perception, such as Mach bands.

More recently, it has been shown that there is also a physiological basis for shape recognition. Studies of monkeys and cats show that certain cells in the accessible outer layers of the brain's visual cortex receive signals from many fibers of the optic nerve, and are found to "fire" preferentially only when simultaneously stimulated by signals originating in specific spatially related areas of the visual field. Thus there are "line detectors" and "corner detectors," among others, which fire only when a bright line or bright bent line is imaged on the retina. This is evidence that certain features of a scene are singled out for emphasis in the processes of forming a perception. The survival value of this ability as an evolutionary development is obvious. Indeed, there is some recent evidence for cells that are "face detectors," responding when essential features of a face appear in the visual field. The response of such cells is the culmination of many preceding stages of neural processing that are yet to be understood. Nor in fact do we know whether signals appearing in such cells imply that the individual is then conscious of perceiving a face.

Fig. 10-13. An ambiguity in perception.

Perception of Form

In the early 1900s psychologists in Europe generally became aware that perception is not the simple sum of information provided by our sensory systems. A triangle may be recognized if represented by three lines, as three dots depicting its vertices, or as a solid, three-sided patch of color. The form ("gestalt") is independent of the elements that describe it. This school of gestalt psychology argued that it is impossible to infer what a person would perceive from knowing the image on the retina. Figure 10-13 illustrates the ambiguity. Sometimes we see a white vase on a dark background, and at other times two dark faces. What determines our perception: the color interpreted as foreground? the color forming a contiguous area? the form best matching a form in memory? or the form best associated with a recent event? This question is remarkably difficult to answer on a scientific basis. While gestalt psychology and more recently developed schools of thought provide descriptions of what cognitive processes might be taking place in the brain, it has been impossible to verify these theories by experiment. Thus questions at this fundamental level continue to remain a frontier of modern psychology.

■ 10-10 COLOR BLINDNESS AND COLOR VISION

The physiological basis of color vision, first postulated by Young in 1801, lies in the different spectral sensitivities of the three types of cones. Normal vision, or *trichromacy,* occurs in about 97% of the population, while the rest suffer from one of a group of color deficiencies collectively referred to as "color blindness." Young was a physician, and his studies of color vision were stimulated by an interest in color blindness.

Color blindness occurs when genetic factors or disease have eliminated one or more types of cones. If only one type is present, then the individual is a *monoch-*

romat and sees only in black and white. Monochromacy is quite rare, occurring in about 0.003 percent of the population. More frequently, two types of cones are present, which results in dichromacy. About 5 percent of the male population and 0.4 percent of the female population are *dichromats*. The most common type of dichromacy is known as deuteranomaly, or red-green blindness. Deuteranopes lack green-sensitive cones and therefore respond only to the blue and red content of light. They cannot distinguish among red, yellow, and yellow-green since all of these colors stimulate only the red-sensitive cones. The confusion is most serious in this spectral region because of the closeness of the absorption maxima of red-sensitive and green-sensitive cones (Fig. 10-8). ■

> As it is almost impossible to conceive each sensitive point of the retina to contain an infinite number of particles, each capable of vibrating in perfect unison with every possible undulation, it becomes necessary to suppose the number limited, for instance, to the three principal colors, red, yellow (sic) and blue, and that each of the particles is capable of being put in motion more or less forcibly by undulations, differing less or more from perfect unison. Each sensitive filament of the nerve may consist of three portions, one for each principal color.
>
> Thomas Young, 1802

10-11 OPPONENT PROCESS COLOR THEORY

The trichromatic color theory, introduced by Young and made quantitative by Helmholtz, with the inclusion of lateral inhibition seems to provide a reasonably complete model for explaining most color perception phenomena. Nevertheless, the actual functioning of the retina may be still more complex. Recent experimental evidence has provided support for a different explanation of color vision originally proposed in 1878 by Ewald Herring. He sought an answer to why no light allows us to experience yellowish-blue. An additive mixture of yellow and blue appears either as an unsaturated yellow or unsaturated blue or gray, as described by the additive color mixing rules. Similarly we know of no color that could be called reddish-green. Additive mixtures of red and green could be called yellowish-red, orange, yellow, or yellowish-green, but never reddish-green. Thus there seem to be four "pure" colors—blue, green, yellow, and red. All other colors tend to be described as mixtures of pairs of these colors. How can this be explained by trichromatic theory, which is based on only three color receptors in the retina?

Herring suggested that there are two pairs of color receptors and that members of a pair act in opposition to each other: a blue-sensitive receptor and a yellow-sensitive one acting in opposition, as well as a green-sensitive one and red-sensitive one acting in opposition. Although this possibility has been ruled out by the measurements summarized in Fig. 10-8, modern versions of Herring's theory suppose that this opposition is a neural process carried out by the network in the retina or at higher levels in the brain, as indicated schematically in Fig. 10-14. Thus, if only green and red photoreceptors provide signals of equal strength, they cancel each other in the green-red "Herring receptor" and

Photoreceptors

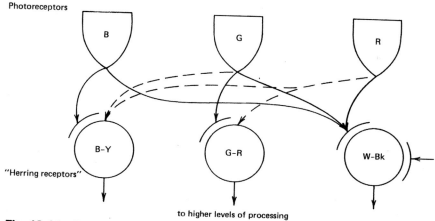

"Herring receptors"

to higher levels of processing

Fig. 10-14.   Schematic of a modern version of Herring's opponent color theory depicting excitatory connections (solid lines) and inhibitory connections (dashed lines). Beside the two Herring receptors dealing with aspects of chromaticity is a third concerned with brightness.

they just produce a yellow signal from the blue-yellow "Herring receptor." From an analogous argument, the other perceivable color combinations can be deduced: blue-greens, blue-reds, yellow-greens, and yellow-reds are associated with signals of various relative strengths coming from the range of signals produced by the blue-yellow and the green-red "Herring receptors." This way of intercomparing photoreceptor signals provides a logical framework for describing many color phenomena. Current research is exploring the implications of this theory, and the results suggest that the details of the processes of color vision are far from being completely understood.

10-12   RETINEX THEORY OF COLOR VISION

Our earlier investigations of color mixing and colorimetry in Chapters 2 and 3 led to the conclusion that the color of an isolated surface seen without reference to its shape or texture is determined by its spectral power distribution. On the other hand, we know from experience that in commonplace situations the perceived color of a subject is largely insensitive to the SPD of the light illuminating it. A tablecloth appearing white under sunlight still appears white under candlelight where there is an excess of red. This phenomenon whereby color is little changed under varying conditions of illumination is called *color constancy*. Its occurrence is a hint of a fundamental aspect of color vision that so far has remained unexplained.

Edwin Land, the inventor of the sheet Polaroid and rapid development film, has proposed what he calls the *retinex theory* for color vision. The name was coined from a combination of "retina" and "cortex" to emphasize that higher level physiological processes in the neural system are an essential component in how color perception is formed. His theory is based on many experiments that he and his collaborators have performed and is an outgrowth of many concepts developed earlier by psychologists.

In essence the retinex theory says that signals sent along neural pathways from photoreceptors responding to the illumination on each region of the retina are intercompared in such a way that characteristics of the reflectance curve for each region are deduced. Thus our perception of the color of each region is formed on the basis of its reflectance curve and not its SPD. Because the reflectance curve is a property of the subject being viewed and not the illumination, subjects retain their color identity under a wide variety of lighting conditions.

SUMMARY

The human eye is both an optical instrument and an information processing system. The cornea and lens form an image on the retina, with the iris controlling the amount of light admitted. The effective focal length of this compound lens is adjusted by varying the curvature of the flexible lens to accommodate subjects at varying distances from the eye. Within the retina there are two classes of light-sensitive elements called photoreceptors: the rods, which are the most sensitive, mediate night (scotopic) vision at low light levels; the less sensitive cones mediate daylight (photopic) vision. There are three types of cones with sensitivities that peak at short, intermediate, and long wavelengths of the visible spectrum. It is this differential sensitivity that provides color vision at daylight illumination levels. Night vision, in contrast, is essentially black and white.

The electrical signals generated in the photoreceptors are processed in several layers of nerve cells within the retina before they proceed via the optic nerve through additional processing centers and eventually to the visual cortex at the back of the brain. Studies of the electrical signals traveling along individual fibers of the optic nerve reveal that important visual effects such as lateral brightness adaptation—reduction of the apparent brightness in one region of the visual field by strongly illuminated nearby areas—is already coded into the signals leaving the eye. This is a result of a physiological process called lateral inhibition, which lowers the sensitivity of photoreceptors in areas of the retina that are adjacent to brightly illuminated areas. The related induced color phenomena, which have historically been a major obstacle to the development of color theory, can also be explained qualitatively on the basis of lateral inhibition, which lowers the sensitivity of an area of the retina to a color that is present in adjacent areas. Successive contrast effects, on the contrary, are strictly local effects caused by fatigue or bleaching of photoreceptors, which temporarily lowers their sensitivity. Many aspects of color vision are still not well understood, and newer theories such as the opponent process theory and Land's retinex theory have been proposed in an effort to achieve a more complete understanding of this phenomenon.

QUESTIONS

10-1. What is the major refracting element of the eye? What part of the eye provides accommodation?

10-2. Explain why a person needing eyeglasses can improve vision by looking through a small hole, perhaps one formed by curling a finger toward the palm.

10-3. What type of lens provides correction for nearsightedness?

10-4. Linear perspective in art gives the impression of depth, but it does not mimic the total perception of depth when we observe subjects directly. Explain why.

10-5. From your experience, indicate what clues in a visual scene help to convey the relative distances of subjects.

10-6. What is the difference between the fovea and the blind spot of the eye?

10-7. What happens to the retinene of the visual pigment in a photoreceptor of the retina when it absorbs light?

10-8. Our perception of the brightness of a surface is judged according to the intensity of light coming from it. Yet the neural signal sent from the eye to the brain does not become bigger if the intensity increases; instead it shows an increase in "frequency." Explain what is meant by this "frequency."

10-9. Describe the path taken by neural signals once they leave the retina. How is the visual field divided between the hemispheres of the brain?

10-10. If you wish to detect a faint light at a known location at night, in which direction should you look?

10-11. Give some examples where the color of a surface is not properly predicted by its position on the C.I.E. chromaticity diagram.

10-12. What is the reason why Mach bands appear on both sides of a border between a light and dark surface?

10-13. Explain what is meant by "visual persistence."

10-14. Why should each frame of a movie not be shown for a period of $\frac{1}{24}$ second if the film is advanced at a rate of 24 frames per second?

10-15. What is meant when a person is said to be a monochromat? A dichromat?

10-16. What is a possible explanation for the afterimage having complementary colors after staring at a picture and then quickly shifting our gaze onto a white wall?

10-17. What evidence suggests that an opponent process is involved in our perception of color?

10-18. How does Land's retinex theory account for the phenomenon of color constancy?

FURTHER READING

There are many introductory texts describing the optical, anatomical, and physiological aspects of the visual system and psychological aspects of perception. The first two books in the following list make these topics particularly accessible.

R. M. Boynton, *Human Color Vision* (Holt, Rinehart and Winston, New York, 1979).
T. N. Cornsweet, *Visual Perception* (Academic Press, New York, 1970).

H. Davson, *The Eye,* Volumes 2 and 4 (Academic Press, New York, 1962). A more technical discussion of the physiology of the eye.

S. Polyak, *The Vertebrate Visual System* (University of Chicago Press, Chicago, 1957). A monumental presentation of the physiology of the eye and visual centers of the brain, with a fascinating historical review of vision research.

L. Kaufman, *Perception: The World Transformed* (Oxford University Press, New York, 1979). An elementary introduction to physiological and psychological aspects of perception by various senses, including vision.

L. Kaufman, *Sight and Mind: An Introduction to Visual Perception* (Oxford University Press, New York, 1974). An advanced text on a wealth of psychological aspects of visual perception.

R. L. Gregory and E. H. Gombrich, Eds., *Illusion in Nature and Art* (Charles Scribner's Sons, New York, 1973) (paperback). A fascinating presentation of illustrations that should appeal to those interested in understanding the basis of visual perception.

J. Albers, *Interaction of Color* (Yale University Press, New Haven, Connecticut, 1963). Many examples of the interplay of colors.

J. W. Von Goethe, *Theory of Colours* (1810. English Translation by C. L. Eastlake, republished by M.I.T. Press, 1970). Of historical interest.

M. H. Pirenne, *Optics, Painting and Photography* (Cambridge University Press, 1970). A full discussion of perspective accompanied by excellent illustrations.

S. Y. Edgerton, Jr. *The Renaissance Rediscovery of Linear Perspective* (Basic Books, New York,)

From Journals

P. L. Pease, "Resource Letter CCV-1: Color and Color Vision," *American Journal of Physics,* Vol. 48 (1980), p. 907. A useful annotated bibliography of the literature on the visual system and color vision, including books, journals, original research articles, and teaching aids.

From the *Scientific American*

G. Wald, "Eye and Camera" (August, 1950), p. 32.

E. L. Thomas, "Movements of the Eye" (August, 1968), p. 88.

A. T. Bahill and L. Stark, "The Trajectories of Saccadic Eye Movements" (January, 1979), p. 108.

M. F. Land, "Animal Eyes with Mirror Optics" (December, 1977), p. 126.

U. Neisser, "The Processes of Vision" (September, 1968), p. 204.

J. D. Pettigrew, "The Neurophysiology of Binocular Vision" (August, 1972), p. 84.

G. Oster, "The Chemical Effects of Light" (September, 1968), p. 158.

F. S. Werblin, "The Control of Sensitivity of the Retina" (January, 1973), p. 70.

R. W. Young, "Visual Cells" (October, 1970), p. 80.

R. W. Sperry, "The Eye and the Brain" (May, 1956), p. 48.

D. H. Hubel and T. N. Wiesel, "Brain Mechanisms of Vision" (September, 1979), p. 150.

E. F. MacNichol, Jr., "Three-pigment Color Vision" (December, 1964), p. 48.

W. A. H. Rushton, "Visual Pigments and Color Blindness" (March, 1975), p. 64.

E. Land, "The Retinex Theory of Color Vision" (December, 1977), p. 108.

A. L. Gilchrist, "The Perception of Surface Blacks and Whites" (March, 1979), p. 112.

J. Beck, "The Perception of Surface Colors" (August, 1975), p. 62.

C. S. Brindley, "Afterimages" (October, 1963), p. 84.

O. E. Favreau and M. C. Corballis, "Negative After-Effects in Visual Perception" (December, 1976), p. 126.

CHAPTER 11

COLOR IN ART

In this chapter we shall investigate the applications of color science and ray optics to the materials used in the visual arts. Our purpose is to understand how these materials "work," that is, how they produce the perceived qualities that the artist uses.

Although the contemporary artist can readily obtain tubes of premixed paint, painters prepared their own materials until very recently, and the grinding and mixing of pigments and paints was as crucial a part of the artist's training as preparation of the canvas, brushwork, and setting of the palette. The development of new materials and techniques has frequently been associated with major developments in art, so an understanding of these technical developments is an important component of a full appreciation and understanding of art.

We shall be interested, first of all, in the fundamental problem of color—where it comes from, and how it is changed in the process of mixing. Subsequently we shall explore the more complex issue of how paints function optically.

■ 11-1 SOURCES OF COLOR

There are two major categories of colorant materials, *dyes* and *paint pigments.* Dyes are materials that are usually dissolved in a solvent: the individual molecules of the dye become separated when mixed within the solvent. Individual dye molecules absorb light selectively in some parts of the spectrum while allowing light at other wavelengths to pass unaltered. Dyes dissolved in clear plastic (or gelatin) become the plastic filter materials used in stage lighting, photography, and color film; dyes dissolved in water (or other solvents) are used in coloring fabrics and paper.

Paint pigments are usually powders that can be suspended in a medium such as oil or acrylic to form the paint. Individual pieces of the powder retain their identity within the medium. Pigments also absorb selectively in certain regions of the spectrum when light passes through the powder, but their action is more complex than that of dyes since they occur not as individual molecules but as small pieces of solid matter. Reflection from the surfaces of these pigment particles influences the color.

Between the soluble dyes and the paint pigments there is another category of coloring agents, the *lakes,* which are pigments prepared from dyes. They are particles whose color is determined by dyes applied to them.

The rules of a color mixing for dyes are relatively simple: if the transmittance curves for two dyes are known separately, then the mixture of the two will have a transmittance curve that can be found by simply multiplying values on the two individual curves together, wavelength by wavelength. For paints, predicting the results of mixing is far more complicated, as we shall see later.

■ 11-2 DYES

Prehistoric humans colored materials by rubbing mineral pigments directly into them. Reddish hematite (an oxide of iron, Fe_2O_3), bluish manganese oxide, and white chalk or ground shells are examples. With the end of the ice age came the opportunity to apply vegetal colors by dipping the material into solutions made from crushed berries, roots, and leaves. Neither the pigments nor these vegetal dyes are "fast," meaning that the color fades with rubbing, washing, or exposure to sunlight. In some cases better fastness could be attained with a binder of egg albumen, saliva, or various resins from trees. Dyes have been used in coloring fabrics for clothing and decorative objects at least since 3000 B.C. All of the earliest dyes were natural, derived from either plants or animals, and the methods for their preparation were little changed until the mid-nineteenth century. A fluid obtained from the sea snail *Murex* and other shellfish was the basis of the dye Tyrian Purple used by the Phoenicians, Greeks, and Romans (Fig. 11-1a). It was the Royal Purple of ancient courts and the purple of Byzantium. Indigo (Fig. 11-1b), extracted from Indian plants now known as Indigoferae, and saffron yellow, from the dried stigmas of the crocus, are among the other important natural dyes.

Synthetic Dyes

The tremendous range of synthetic dyes available in the twentieth century can be traced to the discovery of the first aniline dye, aniline purple, in 1856 by the 17-year-old student William Perkin. In an attempt to synthsize quinine for the treatment of malaria, Perkin while working under August Hofmann at London's Royal College of Chemistry studied the properties of coal tar, a black substance obtained by heating coal and condensing the vapor given off. The black sludge he obtained looked unpromising, but he proceeded further to study its properties by adding some alcohol and discovered that the solution turned purple. When he found that this could dye silk and was reasonably colorfast, Perkin borrowed his father's savings and developed a commercial process to manufacture it.

The French called this new color "mauvre" after the purple of the mallow flower, and its popularity spread around the world within a decade. Hofmann later developed a range of brighter, bluish-red dyes, which he called "Rosanilines." Both

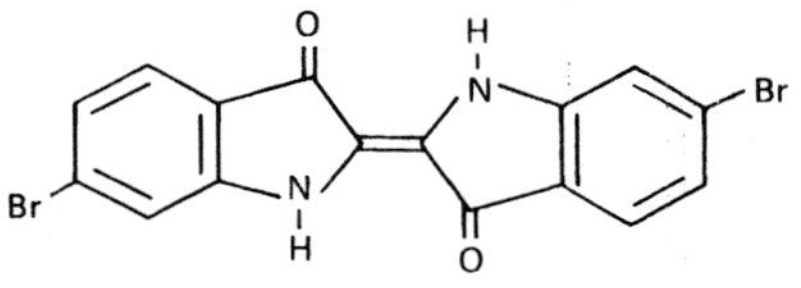

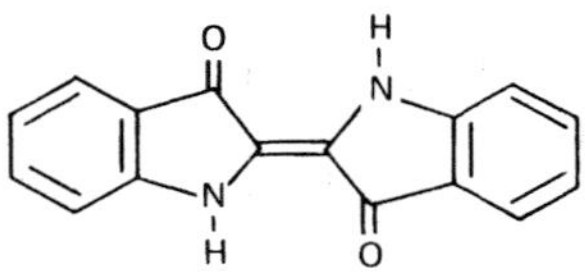

Fig. 11-1. (a) Tyrian purple and (b) indigo molecules have the characteristic structure of one type of colorant: carbon rings with particular groups of atoms attached to the side of the rings. A carbon atom is found at each intersection of straight lines, the lines representing electron bonds between the atoms. Side groups that are particularly effective in causing the molecule to absorb in the visible portion of the spectrum either donate an electron to the adjacent ring or accept one from it. Oxygen serves as an electron donor while bromine and the NH group are acceptors.

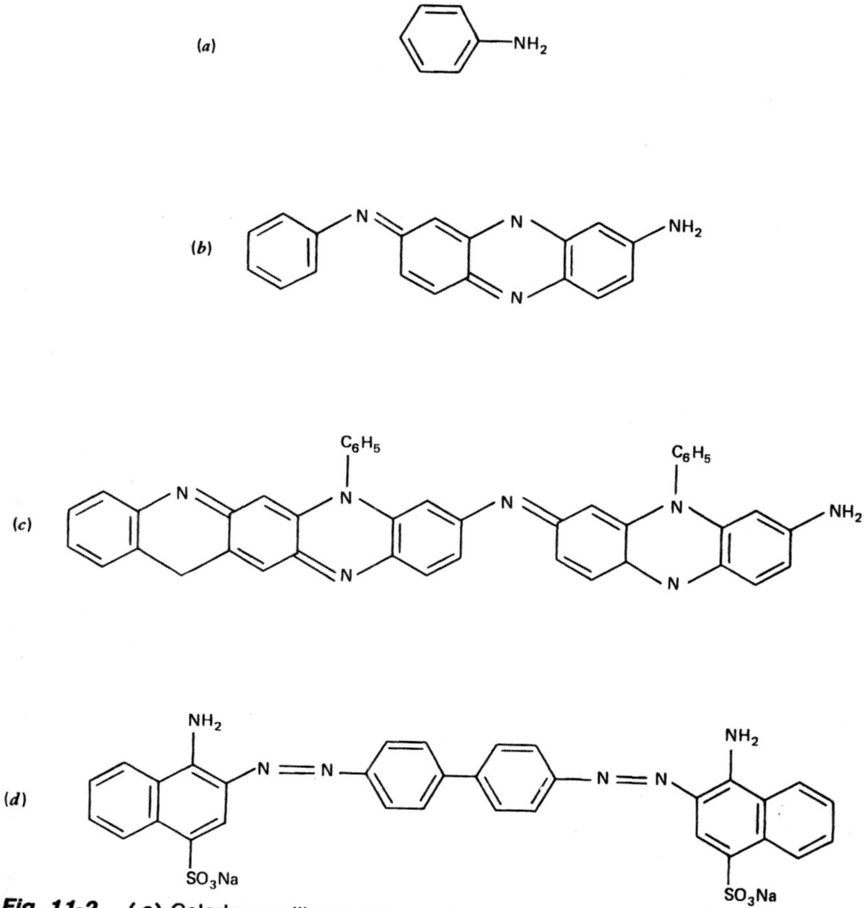

(a)

(b)

(c)

(d)

Fig. 11-2. (a) Colorless aniline that forms the basic unit for aniline dyes. (b) Perkin's aniline purple. (c) A possible structure for aniline black. (d) Congo red.

Perkin's and Hofmann's dyes are produced from the simple ring-shaped molecule aniline ($C_6H_5NH_2$) shown in Fig. 11-2a. This forms a colorless, oily liquid that can be produced commercially by the reduction of nitrobenzene. Chemical treatment of aniline can produce Perkin's aniline purple (Fig. 11-2b) and the Rosanilines, as well as a wide variety of colored dyes that are soluble in water. The family includes aniline black (Fig. 11-2c), malachite green, and Congo red (Fig. 11-2d).

Some aniline dyes such as the blacks are not particularly stable under light and tend to fade to a characteristic mossy green. This deficiency prompted the development of a more colorfast family of dyes—the *azo dyes.* They were first synthesized in 1863 and are commonly used in dyeing cotton and other cellulose fibers.

The azo family comprises an exceptionally wide range of colors. Chemically they are distinguished by a pair of nitrogen atoms joined by a double bond that serves as an electron acceptor, with benzene rings attached to each nitrogen. Their bold, saturated colors have been used to dramatic effect in tapestries when reproducing the designs of such artists as Fernand Léger.

These and other dyes developed by organic chemists have now largely replaced both the natural and the early synthetic dyes for textiles, filters, and film. Natural dyes continue to be used extensively for coloring food and drink. A list of some 8000 dyes classified according to chemical structure and their properties and applications can be found in the *Colour Index* of the Society of Dyers and Colourants in England.

When dyeing fabric, paper, leather, or fur the nature of the material as well as the colorant must be taken into account. The basic structural unit of these materials consists of long chains of atoms that link together to form a much bigger molecule, or *polymer* ("many parts"). Electrical forces may cross-link one polymer to another. Many polymers more or less aligned and bound together comprise a "fiber." For textiles several fibers may be twisted together to form yarn. Friction between the fibers endows yarn with additional strength.

The four most commonly used natural fibers are wool, silk, cotton, and flax. The first two come from animals, and the latter from plants. Cotton has a structure of long, cylindrical cells composed mainly of cellulose. Flax, one of the oldest fibers known and the material from which linen is made, is a composite fiber obtained from the stem of the cultivated flax plant. Its irregular fiber consists of both cellulose and lignin. (Linseed oil used in oil painting is obtained by grinding, steaming, and pressing the seed of the plant.) Wool is the hairlike covering of a variety of animals, but fibers from sheep and goats are favored. This material consists of a protein (keratin) that is rich in the sulfur-bearing amino acid cystine. Its outer surface is comprised of overlapping scales that give a rough texture. Silk by contrast is a smooth, slender protein fiber spun by the silkworm. Its surface is prized for its glossy appearance.

A dye molecule must provide the desired color and also adhere to the material. One important characteristic of the material is the electrical balance between the distribution of electrons and the arrangement of positive nuclei of its constituent atoms. The molecules of wool and silk contain side groups where there is a surplus of electrons, resulting in a net negative charge there. They also have groups with a deficiency of electrons and a corresponding net positive charge. Such *polar molecules* can form strong electrical bonds with dyes that also have polar groups.

Fibers with polar molecules can be colored by a *direct dye process* simply by immersing them in a solution of the dye. Most water-soluble dyes dissociate in solution, separating into a part that is positively charged and a remainder that is negatively charged. When a fiber composed of polar molecules is immersed in such a solution, strong electrical forces attract these ions of the colorant toward locations on the molecules having the opposite charge, where they remain when the water is drawn off. Martius yellow (Fig. 5-16c) is a polar dye that can be applied in water solution directly to fibers of wool and silk.

Some artificial fibers have a good electrical balance and are *nonpolar.* Polypropylene and nylon are two examples. If a water-soluble color is applied directly to such materials, it does not adhere strongly and quickly washes away. Cotton, linen, and rayon, which are cellulose-based fibers lie midway between the extremes of polar and nonpolar molecules. They form weak bonds with dyes and also tend to fade unless the dye is properly chosen.

Synthetic nonpolar molecules such as nylon can have their affinity for direct dyes enhanced by chemically incorporating polar groups at the ends of the molecules. Cotton also may be directly dyed with appropriate colorants. Congo red (Fig. 11-2*d*) has polar side groups that allow it to be applied directly to cellulose-based fibers of this type.

When it is not possible to dye a nonpolar or weakly polar fiber directly, it may nevertheless be possible to apply a colorant that is insoluble in water but soluble in the material itself. A *dispersed dye* is applied as finely divided particles suspended commonly in a soap solution and, when the particles make contact with the fiber, they break down and the molecules become dispersed in the fiber. Acetate dyes are an example of this process.

A few dyes can actually react chemically with a fiber to form a permanent bond. The natural vegetable dye henna is such a *reactive dye,* and it combines with the protein keratin of hair when used as a colorant. A family of synthetic dyes with the commercial name Procions that bond to cellulose fibers were developed in the 1950s. These reactive dyes encouraged the introduction of color into men's shirts and remain one of the most popular colorants of the dyeing trade.

However, there are relatively few dyes that chemically combine with the polymers of a fiber. It is more common to use a process that provides a group of atoms to establish a chemical link, or *mordant,* to the fiber. One process used for centuries relies on aluminum hydroxide $Al(OH)_2$, or more commonly now $KAl(OH)_3$, to attach Turkey red to wool or cotton. This coloring agent was traditionally obtained from the root of the madder plant, and was an extremely popular dye, even though it must be applied in a complex sequence of steps. Thus it was noteworthy when in 1868 the dye molecule alizarin, shown in Fig. 11-3*a,* which is the colorant of Turkey red, was first synthesized. Within 10 years the extensive fields of cultivated madder in Europe were turned to other crops as the less expensive synthetic colorant became commercially available. At about this time a similar dye called fuchsine after the fuchsia plant was produced in France. This reddish blue color was then named "magenta" after the Italian village near the site of a tragic battle fought in the Franco-Prussian war shortly after the dye was introduced. Mordants commonly used for alizarin and similar dyes rely on a metal complex to bind the dye molecule to side groups of the fiber molecule. Figure 11-3*b* shows a mordant based on the chromium ion. This metal is particularly effective for attaching azo dyes to wool.

Some dyes can be formed right in the fiber—the color does not appear until the material is given the proper treatment. For example, Tyrian purple (Royal Purple) was obtained as a colored juice from the sea snail *Murex,* but the colorant molecule illustrated in Fig. 11-1*a* is not formed until the material is oxidized by exposure to oxygen of the air and sunlight. To enhance colorfastness during washing, *vat*

(a)

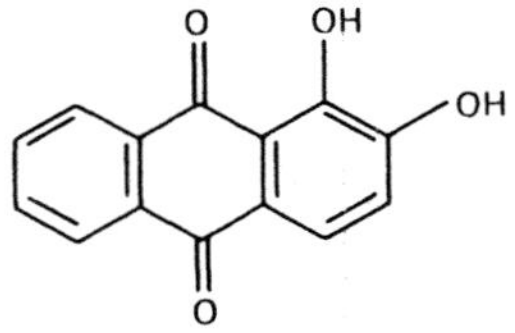

1, 2-dihydroxyanthraquinone
(alizarin)

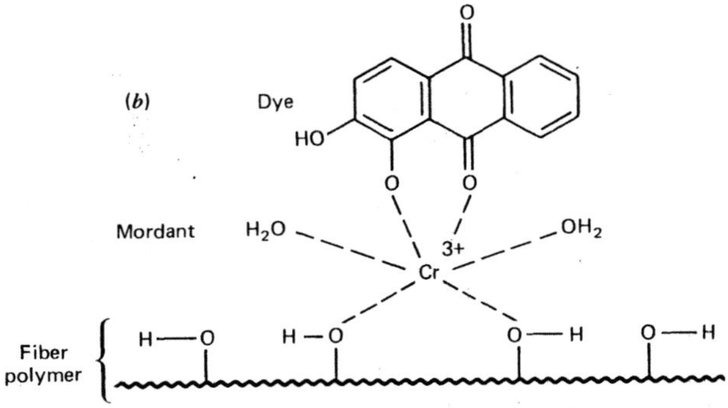

Fig. 11-3. (*a*) Alizarin, the colorant of Turkey red. (*b*) Alizarin attached to wool by a chromium-based mordant.

dyes, as this type is called, may be incorporated in the fiber in a form that is insoluble in water. This is often used for cotton and rayon, where the dye is first chemically reduced in vats to a soluble form having a different color and is then applied to the fiber. Finally the material is oxidized, which in modern commercial processes is accomplished by an acid solution in another vat. Thus the dye is actually formed in the last stage. Indigo (Fig. 11-1*b*) is insoluble in water and is applied in this way.

Dye in the modern denim blue jean is synthesized indigo. Remarkably, it is the way indigo fades that is prized by many people today. This is an advantage to the manufacturers, since a colorfast indigo for cotton fiber is costly to produce. The distinctive appearance of faded denim comes from the way the cloth is made. It is woven with an indigo warp (the threads showing at the surface) and a white weft (threads running at right angles, which can be seen on the underside). As the warp wears and fades, the white shows through and a subtle pattern develops that would be virtually impossible to duplicate by using a dye alone.

Light promotes fading of many dyes because a colorant molecule once excited by absorbing a photon may be more chemically reactive and combine more quickly on contact with oxygen from the air. Water absorbed onto the fabric accelerates fading especially for dyed cotton because ultraviolet absorption also promotes the oxidation of water (H_2O) to form hydrogen peroxide (H_2O_2), which in turn reacts with the colorant. This is why precious documents, drawings, and paintings may be sealed in an atmosphere of dry, inert helium or nitrogen. A transparent case made of the material with the trade mark Plexiglas (especially the type UF-3) helps to preserve such displayed objects because it effectively absorbs ultraviolet.

■ 11-3 PIGMENTS

A dye is absorbed by the material to be colored, whereas a pigment is applied to the surface. Pigments are usually highly insoluble substances. Just as with dyes, the pigments used in making paints can be classified as natural or synthetic. The natural pigments are nearly all minerals found in the earth. We shall discuss a few examples of them in this section.

The earliest natural pigments, found in prehistoric cave paintings, were the yellow, brown, and red earth colors. These materials are found on the modern artist's palette as yellow and red ochre, raw and burnt sienna, umber, Venetian red, and Indian red, which are all forms of clay containing compounds of iron. In sites of the Upper Paleolithic period in France, dating from about 40,000 B.C., caches of iron oxide powder were found that had been heat-treated to modify its color. This marks the earliest known example of pyrotechnology from which have emerged techniques for producing a rich variety of pigments and glazes for pottery. For example, partially oxidized iron known as *magnetite* (Fe_3O_4) is black, but material with a higher proportion of oxygen known as *hematite* (Fe_2O_3) is red. The hydrated oxides of iron in which hydrogen is incorporated are yellow, or, in certain compounds such as terre verte, green.

Ultramarine, a beautiful deep blue pigment, was extracted from the mineral lapis lazuli. Cinnabar, a red sulphide of mercury found in Spanish mines during Roman times, was the only bright red pigment known to the ancients. The mineral malachite when ground produces a green pigment, which was very popular in medieval painting. Cadmium yellow is made from the mineral greenockite, or cadmium sulfide, and occurs in many gradations from pale yellow through orange depending on the presence of impurities. Pigments derived from plants and insects are generally less stable than the minerals, although madder and indigo are reasonably stable.

During the Middle Ages many synthesized pigments were added to the artist's repertoire, greatly expanding the available range of colors. Vermilion red, originally made from natural mercury sulfide, has long been prepared by reacting sulfur and mercury. Various artificial blue pigments (copper blue, blue bice) were made from the green pigment verdigris by mixing and grinding with lime (calcium oxide) and sal ammoniac (ammonium chloride) to form blue copper-ammonia compounds. Verdigris itself is a synthetic pigment that was made by hanging plates of copper over hot vinegar in a sealed pot until a crust of green formed on the copper. Color Plate 31 shows a fourteenth century manuscript from Florence displaying the use

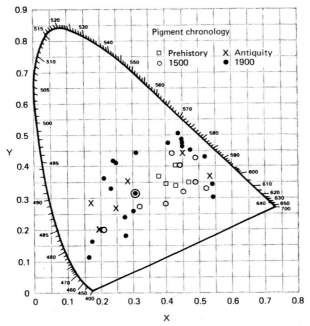

Fig. 11-4. Expansion of the artist's palette from prehistory through 1900. (From *Application of Science in Examination of Works of Art*, Museum of Fine Arts, Boston, 1965.)

of blue ultramarine from lapis lazuli, gold leaf produced by hammering gold particles together, vermilion from the reaction of mercury and sulfur, and verdigris from the corrosion of copper.

The gamut of colors available to artists during different periods of history has been explored by Stephen Rees Jones in "The History of the Artist's Palette in Terms of Chromaticity." Figure 11-4 from his article uses the C.I.E. diagram as a color map in the spirit of Section 3-4 to show the range of chromaticities available in prehistory, antiquity, the Early Renaissance, and the year 1900.

Directly Applying Pigments

As in prehistory, some pigments are simply rubbed onto a surface to color it. A pencil consists of a core of compressed graphite that breaks up into fine particles when rubbed on a rough surface. Graphite, being pure carbon, the major constituent of coal, provides an effective black because the small particles reflect light between each other many times before it can reflect back to the viewer. Lamp black, made from incompletely burned oil or fat, and charcoal black, obtained by heating vine clippings, are almost pure carbon as well.

Some pigments are applied to a surface as a suspension of fine particles in a liquid carrier, and the particles remain weakly attached when the liquid evaporates. Watercolors are examples, but the water carrier also contains a glue to help attach the pigment. India ink is another example, and here the pigment consists of fine carbon particles, usually produced by heating wood in a closed kiln, suspended in a mixture of water with some shellac and a little borax, the latter serving to make the shellac miscible in the water. Once dry, the shellac makes the ink waterproof. We defer a detailed discussion of watercolors and other types of paints to a later section.

■ 11-4 LAKES

There is one more important category of colorant materials known as *lakes.* Lakes are usually made from tiny grains of translucent colorless alumina (aluminum oxide), produced by heating soda (sodium carbonate) and alum (potassium aluminum sulfate), which are then colored by the addition of a dye. The name "lake" derives from the red lac insect formerly used as the source of a dye.

In fabric dyeing, lakes may be used to establish a mordant for materials where the dye by itself would not permanently adhere to the fiber. The alumina particles suspended in water adhere to the fabric and react with the dye to form a fast color.

In paintings, lakes prepared with many of the natural dyes discussed in Section 11-2 have been used for centuries. The introduction of lakes as pigments during medieval times greatly enriched coloration. Grain lake, for example, was widely used to obtain rich, saturated transparent reds. The red dye called grain is extracted from berrylike bumps that form on oak trees when stung by an insect. It has been in use for 2000 years and its name derives from the Latin word grana, which means a berry.

■ 11-5 COLOR SPECIFICATION

There is evidence that in their early stages of development, all languages have very few color names. After black and white, red usually appears first, then yellow, green and blue. As David MacAdam reports in *Color Essays,* the first color name in the bible occurs only after 16 chapters of Genesis (Esau was born all red and hairy) while blue, purple, and scarlet are first mentioned in Exodus 26:30. Gladstone (a Prime Minister during the reign of Queen Victoria) noted that "although Homer used light with greater splendor and effect than any other poet, yet the color adjectives and color descriptions of the poems were not only imperfect but highly ambiguous and confused." Similar confusion persisted in later civilizations. From the time of Pliny (first century A.D.) through the medieval period, for example, there was constant confusion between descriptions of the orange lead pigment *minium,* and the red sulphide of mercury cinnabar, which is *vermilion.* It is therefore generally difficult to decide what colors correspond to descriptions given in written accounts before the twentieth century. Even when paintings or other examples survive, it is never certain that the colors have not changed with time.

Until the nineteenth century, artists prepared their own paints, and the colors depended both on the source of pigments and on the method of preparation. Each school or studio had its own recipes and color designations, and there was little interest in objective color standards that today are considered essential. During the nineteenth century with the growth of industry, however, there arose a need for an objective system of color specification for fabrics and for the newly available pre-mixed oil paints that allowed landscape painters to work out-of-doors rather than in the studio where paint grinding and mixing equipment was available.

As we have seen in Chapters 2 and 3, the basis of objective color specification had already been established by Newton in the seventeenth century. Newton's ideas were generally rejected by artists, however, partly because they were aesthetically unacceptable. Goethe, in his book *Theory of Colours* strongly attacked Newtonian color theory:

> A great mathematician was possessed with an entirely false notion on the physical origin of colours; yet, owing to his great authority as a geometer, the mistakes which he committed as an experimentalist long became sanctioned in the eyes of a world ever fettered in prejudices.

We should realize that Newtonian color theory was far from complete in Goethe's time. The connection between additive and subtractive color mixing was still not understood and was not to be fully explained until the work of Helmholtz in the late nineteenth century. Moreover, as a strictly local theory, it completely failed to account for the various contrast phenomena that were already well known to artists even before Newton. Thus, it is not surprising that artists generally preferred to develop their own color theories based largely on aesthetic rather than objective principles, and many such theories were proposed at least since the time of Leonardo da Vinci.

The art of the tapestry-weaver is based upon the principle of mixing colors, and on the principle of their simultaneous contrast.

Michel-Eugène Chevreul
Director of Color
Gobelin Tapestry Factory, 1854

One of the most interesting artistic color theories was formulated by the British painter Joseph Turner early in the nineteenth century. Turner began with the work of another painter, Moses Harris, whose *Natural System of Colours* included a color circle with three primaries and three secondaries. Turner's version, which gave a dominant position to yellow (which is used heavily in Turner's painting), included two diagrams of overlapping triangles: one indicating pigment mixtures producing black, the second indicating light mixtures producing white. He wrote:

> White in prismatic order is the union or compound light, while the mixture of our material colours becomes the opposite; that is, the destruction of all, or in other words—darkness.

Turner's ideas are remarkable in that they contain the principle of additive and subtractive mixing in a qualitatively correct way, in advance of the science of his time. Furthermore, Turner's ideas are not so very far from Newton's, at least by comparison with Goethe. Turner's theory fails, however, to deal with the crucial problem of objective color specification.

Later on in the nineteenth century, the work of Helmholtz and Maxwell brought the Newtonian color theory essentially to completion, and set the stage for the development of objective systems of color specification.

As the twentieth century began, A. H. Munsell, a teacher of painting in Baltimore, was perfecting the Munsell color system that was discussed in Section 3-2. The Munsell Color Atlas became the standard reference for color specification. Since 1943, when the Munsell color system was revised and standardized against the C.I.E. system of color measurement, chips in the Atlas can be checked spectroscopically and replaced if they have faded. Thus it is Newton rather than Goethe whose ideas have come to prevail in practical questions of art.

The Munsell system of specifying Hue, Value, and Chroma (H V/C) is widely used in industry and in artists' paints, although sometimes in a disguised form. For example, Liquitex, a popular line of artists' acrylic paint, is sold in tubes labeled with "modular color specifications" of value, hue, and chroma numbers clearly related to (but not identified with) the corresponding Munsell specification.

Although the Munsell system has been widely used for paint specification, it has not proven particularly useful for predicting harmonious color combinations. Another color system devised in Germany by Wilhelm Ostwald at about the same time as the Munsell system has proven somewhat more satisfactory in this sense. In the Ostwald system discussed in Section 3-3 there is a neutral axis running vertically through the center with the "full colors" arranged on a circle; and complementary colors are placed directly across from each other (Fig. 3-4).

Starting from a "full color" point on the boundary, mixing with white moves the color through a series of *tints* toward the top of the neutral axis; mixing with black moves through a series of *shades* toward the bottom; and mixing with constant black/white ratios moves horizontally toward the neutral axis through a series of *tones*. Ostwald's aesthetic theory, which has been well received by many artists, holds that the most pleasing color combinations are produced by similar tints, shades, or tones (isotints, isoshades, isotones).

■ 11-6 COLORED GLASS

Quartz, composed of silica (SiO_2), is found in large single crystals in the earth (rock crystal) or in the form of tiny crystals as sand. When the surface of quartz is polished to avoid diffuse transmission, pure quartz is quite transparent and colorless. Large pieces of quartz have been carved into decorative objects (popularly called "crystal") since the earliest civilizations. Naturally occurring impurities can produce dramatic coloring, as in the violet amethyst, pink rose quartz, brown smoky quartz, and yellow citrine.

Quartz can be melted and then cooled to form a glass called fused quartz, but its melting temperature (around 1470°C or 2689°F) is so high and difficult to reach that it is only used in industrial and scientific applications. When quartz is mixed with a flux such as soda (Na_2O) or lime (CaO), its melting point is lowered by several hundred degrees, and the material—commonly called "glass"—becomes much easier to work. Most glass for windows and containers is made with a soda-lime-silica composition consisting of Na_2O (13 to 17% by weight), CaO (5 to 10%), and SiO_2 (70 to 73%), with the remaining weight percentage made up of proportions of alumina (Al_2O_3), magnesium oxide (MgO), iron oxides such as Fe_2O_3, and perhaps barium oxide (BaO) and other trace impurities. The exact melting point and other properties of glass depend on the quantity and identity of the flux. Lithia (Li_2O) or potash (K_2O) may be substituted for soda as a flux; and silica may be replaced by phosphorous pentoxide (PO_5) or boron oxide (B_2O_3) to produce phosphate and borate glasses. Sometimes metals such as lead or barium are also added to change the density or the refractive index of a glass.

During the seventeenth century the addition of lead oxide at concentrations of 15% or more became popular because of the enhanced refractive effect for decorative purposes, as in chandeliers. This lead-glass material has also been given the name "crystal."

In glass there seems to be no overall order to the positions of the atoms. The arrangement is said to be "amorphous." Nevertheless each atom has just the right number of neighboring atoms to ensure that most of the outermost electrons are taken up in forming chemical bonds. In order to excite one of these tightly bound electrons, relatively high energy photons must be absorbed, corresponding to the ultraviolet region of the spectrum. For this reason, glass and fused quartz are both transparent.

Glass can be colored while in the molten state by the addition of "transition element" ions as oxides (bound to oxygen). In the periodic table, the transition elements are clustered together and are distinguished physically by their atoms having a partially filled shell of electrons, called the "3d" shell, lying well within the outermost shell responsible for chemical bonding. Chromium, iron, manganese, and cobalt are examples of transition elements. Their 3d-electrons are only weakly involved in most chemical bonds and have low excitation energies. Consequently selective absorption occurs in the visible region of the spectrum.

The exact excitation energies of a transition element ion are influenced to a varying extent by the chemical environment of the ion. Therefore the color may vary with the type of host glass. For example, nickel is pale yellow in a lithia-lime-silica glass, but is relatively more absorptive in soda-lime-silica glass and so is brown. It is comparatively less absorptive at both ends of the spectrum in potash-lime-silica glass, where it provides a purple color. The color of a transition element ion is very sensitive to its valence state, that is, to how many electrons it gives to its neighbors, since that determines the electrical environment of the ion. Some important colorants, their common valence states, and their colors in soda-lime-silica glass are given in Table 11-1.

Common window glass invariably contains iron oxides as a contaminant from the sand, and iron is present in both of its valence states: some ions having a deficiency of two electrons (ferrous iron) and some having a deficiency of three (ferric iron). The combination of their weak absorption in the blue and red regions

TABLE 11-1 Colors of Transition Element Ions in Soda-Lime-Silica Glass

Element	Symbol and Valence		Color
Chromium	Cr^{+++}		green
Manganese	Mn^{++}	(manganous)	colorless
	Mn^{+++}	(manganic)	purple
Iron	Fe^{++}	(ferrous)	pale cyan
	Fe^{+++}	(ferric)	pale yellow
Cobalt	Co^{+++}		reddish blue
Nickel	Ni^{++}		brown
Copper	Cu^{+}	(cuprous)	colorless
	Cu^{++}	(cupric)	greenish blue

of the spectrum endows this glass with the well-known pale green color, best seen in a thick piece of glass. Both states of iron absorb strongly in the ultraviolet and are responsible for glass's opacity in the near-ultraviolet.

As the number of oxygen atoms that attach to a transition element ion such as iron or copper varies with temperature and the amount of available oxygen, the color of a glass may change with heating (called "striking" the color) so that a variety of colors can be created from a given set of ingredients. Many other metal oxides and nonmetallic elements can be incorporated in glass either alone or in combination to produce a wide range of colors and densities. The manufacture of glass was already well established in Egypt by the sixteenth century B.C. (Fig. 11-5) and was developed to a very high degree by the Romans who made, among other things, small glass panes for use as windows. During the Middle Ages in Europe, Romanesque architecture with its thick walls and ponderous arches gave way to the Gothic ribbed vaulting and flying buttresses. The walls of great cathedrals were relieved of their heavy load and could be opened for curtains of windows to light the interior. Techniques were developed to fabricate large windows from many small glass sections.

The term *stained glass* technically refers to glass in which color is provided by selective absorption in a particular layer, usually the surface layer. However, in popular use it refers to all colored glasses used in windows, including those which have a uniform colorant throughout, and that is the sense of the word that we adopt. Stained glass was first used to an appreciable extent during the tenth century. It became an important art form in the Middle Ages in Europe as evidenced by the glorious stained glass windows in the great cathedrals. It was first introduced during the fabrication of large windows when some pieces of colored glass were included, initially to form decorative patterns, and later to produce complete pictures for educational purposes. During the twelfth and thirteenth centuries, elaborate stained glass picture windows were made by assembling very large numbers of small pieces of glass, each piece being cut according to a pattern (a "cartoon") first drawn by the artist. The glass used in these windows came from many shops, including the great Murano glass works in Venice, and the methods of producing the colors were jealously guarded secrets, many of which were subsequently lost.

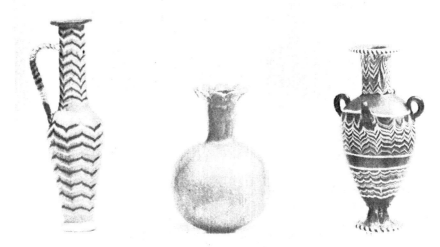

Fig. 11-5. Glass jars made in Egypt during the Eighteenth Dynasty (1500–1350 B.C.). A core of clay and dung was wrapped with coils of molten glass, perhaps with interleaved colors, and the surface was then smoothed. Once the glass cooled, the core was scraped out. (The Metropolitan Museum of Art, the Carnavon Collection purchase, gift of Edward S. Harkness, 1926.)

Historically, the transformation of stained glass windows from a minor decorative role to a centerpiece of the architecture of the great Gothic cathedrals can be traced to the influence of a single man, the monk Suger, abbot of the monastery of Saint-Denis near Paris in the twelfth century. For many centuries the abbey church had been the burial place of the kings of France. Between 1134 and 1144 Suger rebuilt the church, guided by the treatise of the abbey's patron saint, Denis (Dionysius the Areopagite). At the core of Saint Denis' theological treatise was one idea: *God is Light.*

Suger designed Saint-Denis as a monument of applied theology. The construction included a semicircular sequence of chapels that caused the church to glow with light shining through the windows, for which Suger commissioned the stained glass. These stained glass windows also displayed the iconography of the church in unprecedented detail.

The new art born at Saint-Denis, based on the equation of God with light, inspired the construction of other great cathedrals still more luminous than Saint-Denis, including those of the cathedrals of Chartres, Bourges, and Angers.

Much attention has focused on the subtle varieties of blue that were achieved by stained glass. The two principal blue colorants were copper and cobalt ions. Figure 11-6a shows the transmittance curve from a ninth century Saxon church at Jarrow, England, where the color is primarily due to copper ions. By comparison, the curve in Fig. 11-6b for a window of the thirteenth century in Canterbury Cathedral shows a much narrower band of transmission in the blue region of the spec-

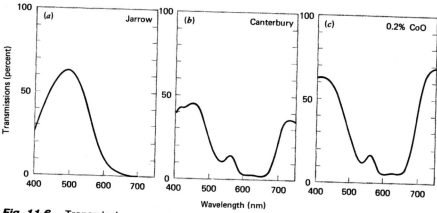

Fig. 11-6. Transmission curves for three soda-lime-silica glasses colored by (*a*) 1.8% of CuO with some iron oxide (1.43 mm thick), (*b*) 0.15% of CoO with some iron, manganese, and copper oxides (2.11 mm thick), and (*c*) 0.2% of CoO) with no other colorants (0.96 mm thick). (Courtesy of R. H. Brill, The Corning Museum of Glass.)

trum, which is primarily due to strong absorption at intermediate wavelengths by cobalt ions. However, the color is modified somewhat from that of cobalt alone, whose curve is given in Fig. 11-6c, by the presence of iron, manganese, and copper ions. A comparison of these three curves shows that cobalt "blue" actually has some red content as well as the blue, as can be seen by looking at such a window through a red filter; whereas copper "blue" has no red but includes a little green. Cobalt was prized in the early medieval period for its rich color and also, perhaps, because its source was distant Persia, as Robert Brill and Lymus Barnes have shown. Mont Saint Michel and Chartres provide additional examples of cobalt-bearing glass.

Colloidal Suspensions

The range of colors in glass that can be produced from selective absorption by ions is severely limited in that it does not include red. Virtually all of the reds and some yellows are created in a different way: by the effect of small particles of gold, copper, or silver within the glass. To create these particles, the metal is added to the molten glass as a salt (perhaps with the metal combined with chlorine), at a concentration of about 0.1% by weight. The chemical properties of the glass must be such as to separate the metal ions from the salt, and the temperature should be sufficiently high to permit the ions to migrate short distances through the glass to sites where they nucleate and form tiny particles. The presence of compounds containing tin, antimony, or arsenic serves to reduce the positive ions to neutral atoms so that they can combine in metallic form.

A suspension of particles of one substance in another substance is called a *colloid*. When the metal particles in glass are smaller than 50 nm they scatter light of short wavelengths more effectively than long wavelengths (Rayleigh scattering).

Thus while light from the extreme red end of the spectrum may pass through with little effect, light from the blue end will be scattered many times and along its tortuous route will travel a much longer distance through the glass, striking many more particles. The reflection curves for metals in Fig. 5-22 imply that when light scatters from a single particle selective absorption is relatively weak. But the process of scattering from many particles in succession leads to an accumulation of the effect and strong absorption at short wavelengths. Details of the transmittance curve for a metal colloid depend on the reflectance curve for the particular metal. Thus gold and copper, which absorb in the blue and green, produce an intense red color, from which come the names gold ruby and copper ruby. Silver, which has only weak absorption in the blue, produces a yellow color.

The brilliance of the reds in the cathedral at Chartres and in Notre Dame of Paris is an example of these colors. Scattering may arise as well from tiny air bubbles within the glass or at its surface, and the overall effect gives the impression that the glass is self-luminous.

Colloidal suspensions of other materials in glass also produce attractive colors. Tiny particles of the semiconductor compounds cadmium sulfide or cadmium selenide provide colors that are similar to those of the corresponding pigments: yellow for the sulfide and red for the selenide. A mixed compound of these elements produces a corresponding intermediate color. Nowadays "selenium ruby" glass containing cadmium selenide is commonly used for traffic signals.

The technique of assembling stained glass windows from simple colored sections was gradually replaced by other methods that allowed greater detail while requiring less channel work. First, enamels consisting of dark brown or black glass powder were used as a glaze, that is, painted on the surface and fused in an oven in order to provide dark linework. Also, the technique of flashing evolved. The glassblower first prepared a cylinder of clear glass that was then briefly dipped into a vat of colored glass to add a thin colored coat. Later, the colored surface could be ground away in places to leave some areas clear. Finally, beginning in the fourteenth century silver stains came into use. These materials (silver sulfide, oxide, or chloride) can be painted onto clear glass and then fired to give colors from pale yellow to orange.

After the fourteenth century, stained glass windows evolved toward larger panes with more and more painting. This produced windows which, while rich in detail, lacked the brilliance and saturation of the earlier simpler technique, and the art of stained glass gradually declined.

During the nineteenth century, the pre-Raphaelite movement in England rediscovered many of the early "lost" techniques of painting and other art forms. William Morris, a member of this group, reintroduced the early stained glass techniques that have enjoyed a mild renaissance during the past century.

■ 11-7 POTTERY, GLAZES, AND ENAMEL

The hardening of wet clay by heating to high temperature was known at least seven thousand years ago in the region of modern Turkey, and the resulting material is known as pottery. During this heat treatment new colors appear that arise from the presence of impurities and depend on the exact conditions of the oven (kiln). For example, with a nearly airtight kiln allowing little oxygen to enter, iron-bearing clay as found in ancient Greece generally turns black at temperatures of nearly 1000°C. This color is due to the formation of magnetite (Fe_3O_4) as the iron combines with the little oxygen that remains. But if air is allowed to circulate during the firing, the iron can combine with more oxygen to produce hematite (Fe_2O_3) and the color turns reddish-brown.

Before firing, a design can be applied with a water paint (a "slip") bearing particles of iron oxide, or perhaps manganese oxide or white clay. The artisan takes advantage of the property that different slips and the underlying clay absorb oxygen from the air at different rates, and by careful firing in alternating oxidizing and reducing atmospheres the various slips can be altered to produce different colors. Greek pottery created in this way can exhibit a fine display of reddish-browns, black, and perhaps white from a clay slip for emphasis (Color Plate 32). The production of Greek pottery in classical times was an impressive technical accomplishment.

One important use of glass, both clear and colored, is in producing a *glaze* for pottery. The glass is ground to a fine powder and mixed with a liquid binder. It is painted onto the surface of clay objects, which are then placed in the kiln. During firing, the glaze melts and fuses together to form a thin liquid layer which, on cooling, solidifies to form a coat of solid glass. By using powdered glass of various colors, a pattern can be worked into the glaze. For tableware, the presence of the glaze provides a smooth glass surface that is much more hygienic than unglazed pottery, which tends to be rather porous. However lead-bearing glass should be avoided when glazing pottery intended for storing foods or drink, because the lead can be leeched out of the glaze.

A thin layer of glass fused to a metal is called *enamel*. It provides decoration and may prevent corrosion. As in glazing, one begins with powdered glass suspended in a liquid binder, which is spread over the freshly cleaned metal surface. The object is then fired briefly to fuse the powder to the surface. As with glass, the final color of enamel is very sensitive to the firing conditions and can be dramatically affected by impurities at the metal's surface.

Porcelain is made from a white china clay (kaolin) and petuntse, a powdered feldspathic stone that when fired fuses to form a glass. It has considerable strength and can be made thin enough to be translucent. The secret of making porcelain ware remained in Chinese hands until discovered at the beginning of the eighteenth century by a young German alchemist. *Porcelain enamel* is commonly applied to iron for kitchenware, sinks, and bathtubs. It provides a hard, easily cleaned surface.

■ 11-8 PAINT

Paint consists of pigment suspended in a liquid vehicle or medium. In some cases the liquid hardens to form a protective coating as in oil paint; in others it evaporates and leaves the pigment exposed as in watercolor. The effects provided by these various media have distinct features, which we shall describe in this section.

Watercolor

A watercolor is a fine pigment mixed with an adhesive substance and water that serves as the vehicle to apply this medium to the surface. When spread on a material such as rag paper or plaster (made of sand and lime), the pigment adheres to the surface once the water has evaporated. A glue made from animal material or a gum made from plant material (e.g., gum arabic) holds the pigment quite effectively but provides no protection.

Because the pigment is applied as a thin wash, much light passes through to be reflected by the undersurface. Thus watercolors are bright and unsaturated. The texture of paper gives a matte appearance. The high transmittance of a watercolor does not allow a mistake to be overpainted. Its thin coat of pigment does not have sufficient body for light pigments to cover darker ones. This reliance on surface reflection to produce color gives the impression of less depth than is available from oil paints. Modern watercolors are available commercially in liquid form contained in a tube or in solid form. In both forms the substance contains glycerine to keep it soft.

The term *gouache* applies to a watercolor with pigments in sufficiently high concentration, perhaps mixed with white pigment, so that a thick layer is opaque. The word is commonly used to describe the appearance of opacity rather than the specific medium used to achieve it.

In the *fresco* technique, pigments mixed with water are applied to freshly prepared and still moist lime plaster (Fig. 11-7). The pigment becomes bound in the plaster as it sets. This gives some protection to the design, but the process of drying often desaturates the colors. With time a smooth skin of calcium carbonate forms at the surface, and this lends a glossy appearance. Despite the great difficulties of the fresco technique, some of the greatest artistic achievements of Western civilization—the works by Raphael and Michelangelo in the Vatican and portions of da Vinci's "Last Supper"—are frescoes.

Tempera

Egyptians since at least 2000 B.C. painted on dry plaster with pigments bound in a mixture of water and gum. Subsequently it was found that egg yolk serves admirably to bind the pigment to a surface. This is known as *tempera paint*. The yolk of hen's egg is an emulsion of a fatty substance (egg oil) together with a sticky substance (albumen) in water. Lecithin in the yolk is an emulsifier that breaks up the oil into fine particles so that they remain suspended in water. The yellow of the yolk is imperceptible when pigment is added. Its color comes from carotene that soon decomposes on exposure to air.

An attractive feature of tempera is that the mixture of pigment and egg yolk once applied to a surface is impressively adhesive. Just try to rub egg off a plate when it has dried! After six months or more, tempera dries to assume a hardness that can almost withstand a scrubbing brush.

Evidence from the Mycenaean civilization of Crete that flourished about 1500 B.C. suggests an inventive technique for using tempera: the base colors of red,

Fig. 11-7. Fresco *The Rape of Europa* by Pinturichio and Workshop (circa 1500) originally on the ceiling of the Palace of Pandolfo Petrucci in Siena, Italy: (*a*) before restoration, showing areas where the plaster surface had fallen away and (*b*) after restoration. (The Metropolitan Museum of Art, Rogers Fund, 1914.)

yellow, or blue were painted *al fresco* on the damp plaster, and the rest of the design was superimposed *al secco* in tempera or by other techniques to the dried surface. This produced a durable surface, as can be seen in the murals of the palace of Knossos (although these are largely restored).

Tempera can be applied to wood panels, as was popular during medieval times, provided that the surface is first primed with gesso, a mixture of white gypsum or chalk, and glue. It also can be applied to canvas in this way. Tempera is not a quick method of painting, because it must be applied in a thin layer and allowed a few minutes to partially dry before the next layer is worked on. Tempera holds up to detailed working and allows overpainting. The egg tempera reached a very high degree of refinement during the Italian Renaissance. With the subsequent introduction of oil painting, interest in this technique declined, but was revived in the late nineteenth century by the Pre-Raphaelites.

To provide greater protection to tempera, by the twelfth century it became customary to apply a coat of varnish. One type was the brownish-red "Vernice liquida" made of sandarac, the resin of an evergreen, dissolved in linseed oil. This provided a tough elastic layer that made the decoration resistant to the passage of moisture and harmful gases. During classical Greek times, beeswax and resin from trees had been commonly used as a protective varnish. Oil varnish made from resin from any of a large variety of trees is no longer applied to paintings because when dry it is virtually impossible to remove. Dammar varnish derived from resin of certain trees found in Malaysia and the East Indies and dissolved in turpentine is softer when dry and is commonly used on paintings today.

Oil Paint

Incorporating pigment in a transparent, protective medium was known in classical Greece, as described by Pliny. Beeswax served this purpose, being applied while the surface was kept warm to help it flow (a technique called "encaustic" painting). Each stroke had to be done rapidly, and there was little chance for correcting errors. The technique is still found in contemporary art, for example in the work of Jasper John. Since the classical period, oil vehicles have been prepared from many substances that retain their fluidity for a longer time and permit more detailed working. Linseed oil and poppyseed oil have been the most successful.

Artists know that exposing a fresh painting to sunlight hastens drying. The chemical changes that produce polymerization are accelerated by the ultraviolet component of sunlight. The oil forms a solid film over a period of time as the molecules of the oil become bonded to each other to form much longer units (polymers). Polymerization may also produce severe discoloration unless the oil has been properly prepared. It is not uncommon to see the surface acquire a yellow hue as it begins to absorb in the blue portion of the spectrum, although surface yellowing of paintings often results from changes in the protective varnish rather than the paint. Oil paints generally become less opaque and deepen in tone as the oil hardens, a serious problem that the painter must confront in using this medium. (We shall discuss this point further in the next section.)

By the fifteenth century oil paint came into prominence, initially used as a thin transparent glaze painted over a solid underpainting in egg tempera and subsequently used alone for painting on canvas. The invention of oil painting is frequently attributed to the Flemish painters, the brothers Van Eyck, but its origin remains a question of considerable controversy. Oil paint was used extensively by artists from the sixteenth century until the middle of the twentieth, when the new plastic paints were introduced.

Acrylic and Latex Paints

During the 1920s in Mexico there arose a major movement of mural painting in which artists created very large outdoor murals as public works of art. Because oil paints were poorly suited for this application, a search began for new materials among the products of industry. During and after World War II, many new plastic materials were developed, and some of these indeed proved to be suitable vehicles for paints. The most important of these are the acrylics, which were perfected around 1950.

Acrylic resins were discovered in 1901 by a German scientist, Otto Rohm. They are widely used in industry, and are the basis of the familiar clear plastic materials lucite and plexiglas. Acrylic resins can be dissolved in a solvent consisting of xylol and toluene and used directly as a medium for making paints. However, there is another preparation known as *latex,* which has proved far more convenient; here the acrylic (in the form of an acrylic ester polymer) is divided into fine droplets that are suspended in water as an "emulsion." The pigment, which is identical to those in conventional solution paints, is separately dispersed within the water. After application when the water has evaporated, the plastic particles fuse together and form a film containing the pigment. This is hard and water insoluble, and forms a porous layer that allows any trapped moisture to evaporate. Acrylics are effective for mural painting, while their superior adherence, great clarity, and ease of handling have also made them the medium of choice of many contemporary artists for studio painting.

■ 11-9 OPTICAL PRINCIPLES OF PAINT

In this section we shall explore the rather subtle question of how paints "work" optically. First, we consider white paint, turning to colored paint in the following section.

To begin, let us recall that we see a transparent object only because its refractive index differs from that of the air that surrounds it. The fraction of light reflected when light falls perpendicularly on the surface of a transparent object depends on the *difference* between the refractive indices of the object and the surrounding material. The reflected electric field is proportional to the difference $(n_1 - n_2)$, where n_1 is the refractive index of the surrounding medium and n_2 is the index of the object. We should not be concerned with a negative sign appearing if n_2 exceeds n_1: that merely expresses the fact that the direction of the electric field is reversed when reflected from a medium of greater refractive index. Our present concern is with the *intensity* of the reflected light, which is proportional to the *square* of the electric field. Therefore we deduce that the reflectance varies as $(n_1 - n_2)^2$. This is always a positive number (or zero). The exact expression for the reflectance R can be written as:

$$R = \frac{(n_1 - n_2)^2}{(n_1 + n_2)^2} \qquad (11\text{-}1)$$

Because the *square* of $(n_1 - n_2)$ appears in this formula, R is very sensitive to the difference between the values of n_1 and n_2.

A striking illustration of this principle can be observed by inserting a glass rod into a snifter half-filled with immersion oil, a transparent oil whose refractive index is the same as that of glass. The top of the glass rod which is surrounded by air

Fig. 11-8. Glass rods ($n = 1.5$) immersed in a snifter of water ($n = 1.33$) at the right and immersion oil ($n = 1.5$) at the left.

remains visible, while the bottom which is immersed in the oil seems to vanish (Fig. 11-8).

Example 11-1   Minimizing Reflection

Equation 11-1 predicts how much light is reflected at a surface for light coming straight onto it. For example, if $n_1 = 1.0$ for air and $n_2 = 1.5$ for glass or typical oils, the reflectance is:

$$R = (1 - 1.5)^2/(1 + 1.5)^2 = 0.25/6.25 = 0.04$$

This says that 4% of the incident intensity is reflected. Although this may seem small, the reflection may be objectionable with glass used to protect a painting or with lenses in a camera. Reflection can be substantially reduced by covering the material with a layer of another transparent material having a refractive index that is intermediate between the indices of the media on either side. If the layer has an index $n_3 = 1.2$, the reflectance between it and air is predicted by Eq. 11-1 to be 0.8%. And the reflectance between it and the underlying material is 1.2%. To find the total reflectance, we need first to determine the total transmittance, which for a thin layer is just the product of the transmittances of the two surfaces: $0.992 \times 0.988 = 0.980$. Therefore the total transmittance is 98%, corresponding to a total reflectance of 2%. The interposed layer has reduced the total reflectance by a factor of 2. This principle is exploited in manufacturing "nonreflecting glass" for protecting paintings and photographs. If the layer is made a quarter wavelength thick for green light, an additional reduction in the reflectance comes from destructive interference.

Next consider a ground glass plate. The surface of the plate is rough so that light is diffusely reflected from it, giving the appearance of whiteness. If a drop of immersion oil is placed on the rough surface and allowed to spread out, there will no longer be any reflection at the rough surface of the glass since it is now at the boundary between two media of equal refractive indices. Instead, because the reflection will occur at the surface of the oil, which is smooth, the ground glass plate will appear clear (Fig. 11-9)!

Now consider a white powder of ground glass. The perceptual quality of whiteness again arises from the diffuse reflection of light at the irregular boundaries between the particles and the surrounding air. If we mix the ground glass with immersion oil, the glass will disappear and the mixture will be clear. Linseed oil, whose refractive index is about 1.48, gives similar results.

How then can we make white paint? The trick is to use something like ground glass, but with a much higher refractive index so that it does not disappear when

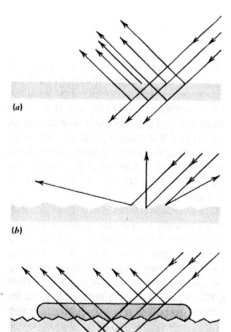

(a)

(b)

(c)

Fig. 11-9. (*a*) Transmission and specular reflection from a smooth glass plate. (*b*) Diffuse reflection from the rough surface of a ground glass plate. (*c*) Transmission and specular reflection from the surface of immersion oil covering a ground glass plate.

mixed with oil. Traditionally, zinc oxide ($n = 2.00$) and lead carbonate ($n = 2.01$), which are transparent, colorless solids, have been used for making white paint. When a pigment that has a refractive index of 2.0 is suspended in oil with $n = 1.5$, according to Eq. 11-1 about 2% of the light will be reflected each time the light enters or leaves a particle.

White lead pigment has been made synthetically since classical times by exposing sheets of lead to vapors of acetic acid and then converting the reaction product to lead carbonate. Although white lead forms a superior paint and was favored by many artists, it has the serious disadvantage of being highly toxic, and some painters have been poisoned by it. Fortunately, white lead has been largely replaced by titanium white made from titanium dioxide, which is nontoxic. The refractive index of titanium dioxide is 2.71 so that titanium white is optically superior to white lead (the reflectance at the titanium-oil interface is 7%).

In medieval times, before the technique of applying oil paints to canvas became popular, painting was done on wood. A white covering of chalk (calcium carbonate, $CaCO_3$) with a refractive index of 1.60 in a glue solution having an index of 1.35 effectively covered the wood and provided a reflective priming undercoat for paints. Whitewash is a related covering. It usually consists of hydrated lime ($Ca(OH)_2$) in water which, on exposure to air following drying, combines with atmospheric carbon dioxide to form $CaCO_3$. Similar priming materials are also used on canvas.

Opacity

Opacity describes a paint's ability to obscure an underlying surface. Related to this is the *hiding power* of a paint, which is expressed physically as the area of surface that can be concealed by a given volume of paint. A large difference in refractive index between a pigment (either white or colored) and the carrier enhances reflectance at the surfaces of each pigment particle and correspondingly enhances the paint's opacity. On the other hand, if there is only a small difference between indices, the paint will be transparent unless the pigment itself is sufficiently absorbing to attenuate light before it penetrates to the underlying surface and back. Because the refractive index of a pigment plays such an important role in its visual effect, we list some examples in Table 11-2.

In Fig. 11-10 we show a sketch of a section of white paint to illustrate its opacity. The paint consists of particles of transparent powder suspended in an oil film. If a beam of light strikes the surface of the paint layer, there will be a small reflection at the air-oil interface (about 4%). The remaining 96% of the light travels through the oil until it reaches the first pigment particle where (for the case of titanium white) 7% of the remaining light is reflected. The remainder travels through the particle until it reaches the far side where, again, 7% of the remainder is reflected. As the light travels down through the paint layer, more and more of it is reflected. In order for the paint layer to appear "solid," it is necessary that no appreciable part of the light falling on the front surface should penetrate through to the base. In the example of Fig. 11-10 with 4% reflection at the air-oil surface and six oil-particle surfaces each giving 7% reflection, about 62% of the light would reach the rear

TABLE 11-2 Refractive Index of Common Pigments and Artist's Materials[a]

Egg tempera	1.35
Polyvinyl acetates	1.46–1.47
Spirits of turpentine	1.47
Linseed oil (fresh)	1.48
Acrylics	1.48–1.52
Shellac	1.51
Cobalt blue	1.74
White lead (lead carbonate)	2.0
Red lead	2.4
Cadmium yellow	2.35–2.48
Cadmium red	2.64–2.77
Titanium dioxide	2.71
Vermilion	2.9
Galena (lead sulfide)	3.91

[a]*Source.* Adapted from T. B. Brill, *Light. Its Interaction with Art and Antiquities,* Plenum Press, New York, 1980.

surface so that it would appear transparent and cover rather poorly! It would be better to grind the pigment to a finer powder so that more reflections occur in the same distance, giving a more "solid" paint layer.

Pentimento

The optical principles we have discussed above provide one explanation for an interesting effect called *pentimento* (an artist's "repentance"). This term is used to denote the gradual reappearance over a period of time of portions of a painting

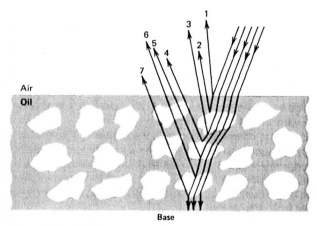

Fig. 11-10. White paint consisting of transparent particles suspended in a film of oil.

Fig. 11-11. *The Miracle of Christ Healing the Blind,* an oil painting done in the sixteenth century by El Greco, and a detail from the center displaying pentimento. (The Metropolitan Museum of Art, Gift of Mr. and Mrs. Charles Wrightsman, 1978.)

that had been subsequently painted over. It may be particularly emphasized when the overpainting consists of white or unsaturated pigments that depend more on reflection and less on absorption for their opacity. Figure 11-11 gives an example of pentimento in an oil painting by El Greco.

Pentimento appears because of a change in the refractive index of linseed oil. The refractive index of wet linseed oil is about 1.48. During the drying process the refractive index increases and may continue to increase for many years as the oil continues to gradually harden. Over a period of several centuries chemical changes within the oil cause the molecules to polymerize, and this also enhances the refractive index. Since the refractive index of the pigment particles remains constant, the result is that the difference between the two indices decreases with time and the reflectance at each oil-particle interface therefore also decreases. Thus, light will gradually penetrate further into the paint layer, and it will become increasingly translucent. An underpainting that was once completely obscured may gradually become visible.

Another effect of the chemical changes within the linseed oil is a slight yellowing. Some of the present warmth of the old paintings such as those of Rembrandt and the other Dutch masters is due to this effect, but most of this yellowing is probably due to their use of a yellow varnish that has become dirty.

A second cause of pentimento is chemical change in the pigment. The lead of lead white reacts slowly with fatty-acid components in the oil to form lead stearate or lead linoleate, which have a lower refractive index that more closely matches that of the oil. This product is commonly called "soap" by art conservators, because a traditional way to make common soap is by the same reaction between sodium in lye and fatty acids from animal fat.

■ 11-10 COLORS FROM PAINT

The optical principles we have discussed in the preceding section also apply to colored paint, but the selective absorption of colored pigments must also be taken into account to understand how they function.

In Fig. 11-12 we show schematically how light interacts with a red paint such as

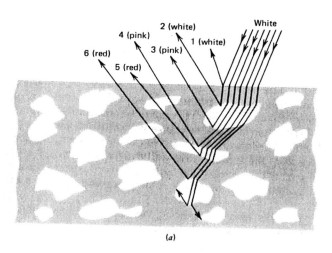

(a)

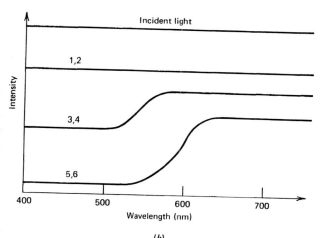

(b)

Fig. 11-12. (a) Light rays reflected from a paint have colors that depend on the total distance that each has traveled through pigment. (b) The SPDs of rays labeled 1 through 6 in the upper figure.

vermilion (mercury sulfide). This illustration and the following discussion is an extension of the brief description given in Section 2-4. Again, as in Fig. 11-10, there is a small reflection at the air-oil boundary, and additional reflections at each oil-pigment interface. But in this case, since the pigment particles absorb in the blue-green regions of the spectrum, the light reflected at each interface will have a distinct spectral power distribution. Assuming that the incident light is white, reflections 1 and 2 will be white since they involve light that has not traversed any of the pigment. Reflections 3 and 4 involve light that has traversed a single particle and will therefore have suffered some absorption in the blue-green. This reflected light will be pink—a very light pink if the pigment particle is small, or a deeper pink if the pigment particle is larger, since the amount of blue-green absorption depends on the total path length traveled by the light through the pigment particles.

Subsequent reflections will be more deeply colored since they involve light that has passed through larger numbers of particles. The appearance of the paint is determined by the additive mixture of these different reflections. If the pigment is ground to a very fine powder, then the paint will be very desaturated since the light will encounter many surfaces before traveling a given distance through pigment particles. On the average all of the reflected rays will have traveled through a very short path inside of the pigment particles. If the pigment is more coarsely ground, it will become more saturated but, as we noted in the preceding section, it may not cover as well.

The art of grinding pigments involves knowing the right amount of grinding required to produce a reasonable degree of saturation while also retaining good opacity. Prussian blue and indigo are commonly fine-grains, with sizes less than 1000 nm; whereas cobalt blue is coarse, with sizes in excess of 10,000 nm because its refractive index is so much greater that it becomes desaturated from reflection when more finely ground. Thus the coarse pigment may provide a more saturated color but less opacity. Cadmium yellow and vermilion are used in medium grain sizes of 1000 to 10,000 nm.

Orange and red pigments generally have better opacity than blue as a consequence of strong absorption at short wavelengths and strong reflection at long wavelengths. The latter results from a high refractive index that is found at the long wavelength side of a spectral region of strong absorption, as illustrated in Fig. 11-13. Within this region the refractive index of the particular material increases with increasing wavelength instead of decreasing, as is generally found. This effect, known as "anomalous dispersion," was first noticed in 1840 by William Fox Talbot, a key figure in the development of modern photography. Because the refractive index is relatively low on the short wavelength side of an absorptive region, blue paints, which absorb strongly at long wavelengths, have relatively low reflectance in the blue and correspondingly reduced opacity.

The dependence of opacity on wavelength is exploited in the study of paintings by infrared photography. Infrared radiation can penetrate some visibly opaque paints to reveal underpainting not only because the pigment may not absorb in that spectral region but also because longer wavelengths are scattered less effectively by the small particles of a pigment. This capability is sometimes important in estab-

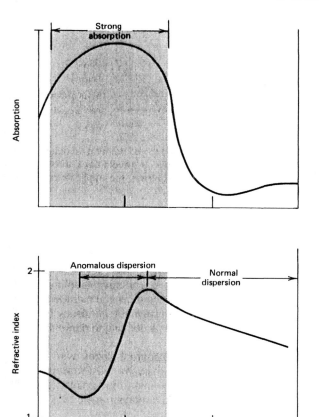

Fig. 11-13. Anomalous dispersion occurs in the spectral regions of strong absorption, producing a high refractive index at longer wavelengths.

lishing the authenticity of a work, because the underpainting may reveal a work to be a forgery if it varies from that of the purported artist. It may also reveal an important original painting that was subsequently painted over.

The early technique of oil painting that emerged in the Flemish school in the fifteenth century used thin transparent glazes of oil paint over a white ground or tempera underpainting. Although much of the light penetrates through to the ground, it again passes through the same layer of pigment particles on the way up through the paint layer, thereby becoming more saturated (Fig. 11-14). This glazing technique produces beautifully clear saturated colors, and has the advantage of being relatively insensitive to changes in the refractive index of the oil. As the oil hardens, a larger proportion of the incident light will penetrate down to the ground,

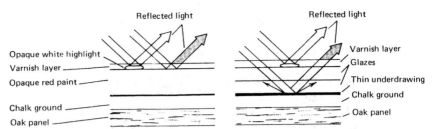

Fig. 11-14. Flemish artists like Jan van Eyck developed the oil painting technique that allows light to penetrate to the reflecting base, whereupon it travels back through the paint again, providing an illusion of depth *(right)*. By contrast, the application of one thick opaque layer *(left)* yields a rather flat appearance.

but it still emerges from the upper surface having traversed a path length of two paint layers within the pigment particles. Thus, at most, there will be a slight *increase* in saturation with time.

After the sixteenth century some painters moved away from the glazing technique in favor of impasto, or thick opaque paint layers. The seventeenth century Flemish painters learned to use very stiff paint containing a minimum amount of oil in order to achieve brilliance. Nevertheless, there is an inevitable tendency for light to penetrate more deeply into thick paint layers with time, leading to darkening and yellowing.

The sixteenth-century technique of painting thin transparent glazes over a white ground was revived in England in the nineteenth century by the Pre-Raphaelite brotherhood. In contrast to the earlier techniques, however, they painted directly onto a wet ground that was applied fresh each day in a manner analogous to and inspired by fresco painting.

Modern painting usually does not have the same appearance of depth as paintings of the seventeenth century. This is because in former times many separate layers of nearly transparent paints of different colors were applied, each layer being allowed to dry before the next one was added. The effect was to give reflections from the different layers at various depths. This many-layered glazing technique was particularly successful in the work of Titian, Rubens, and in the nineteenth century, Turner. The modern style however is to apply paint in only one or two layers, and the same degree of subtlety is usually not achieved.

■ 11-11 MIXING DYES OR PAINTS

As we discussed in Chapters 2 and 3, the perceived color of a subject can, within certain limits, be predicted if its transmittance or reflectance curves are known. In the case of colored filters or of dyes in solution, the transmittance curve of a mixture can be simply found from the transmittance curve of the individual dyes (Fig. 11-15). This is a simple application of subtractive color mixing.

When paints are mixed, it is much more difficult to predict the result since the

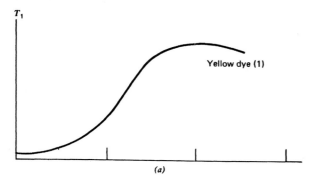

T_1

Yellow dye (1)

(a)

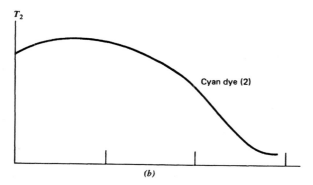

T_2

Cyan dye (2)

(b)

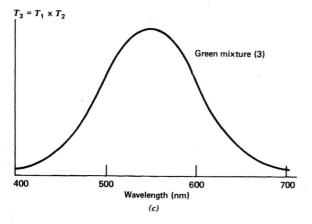

$T_3 = T_1 \times T_2$

Green mixture (3)

400   500   600   700
Wavelength (nm)
(c)

Fig. 11-15. (a) Transmittance curve for a yellow dye. (b) Transmittance curve for a cyan dye. (c) The transmittance curve for a mixture of the yellow and cyan dyes, obtained by multiplying together the values indicated by the two transmittance curves, wavelength by wavelength.

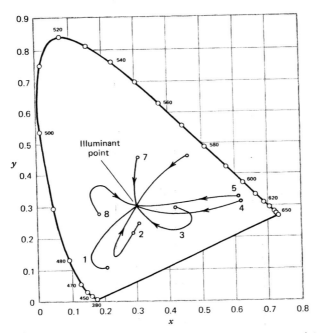

Fig. 11-16. Changes in chromaticity of some common paint pigments when mixed with increasing amounts of titanium dioxide white. The pigments are: 1, phthalocyanine blue; 2, carbazole violet; 3, quinacridone magenta; 4, BON red; 5, molybdate orange; 6, chrome yellow; 7, chromium oxide green; 8, phthalocyanine green. The paths for darker pigments (1,2,3) show a distinct *increase* in purity up to a limit as white is first added, and then become reduced again in purity with higher concentrations of white. This is because the addition of white first tends to refract and reflect light so that it travels through more colorant particles before emerging. Then for greater amounts of white the major effect is to intercept light before it can penetrate very deep and reflect it back outward. (From Johnston, "Color Theory" in *Pigments Handbook,* Vol. III, pp. 229–288, T. C. Patton Ed., Wiley-Interscience, New York, 1973.)

color of paint depends both on absorption by the pigment and on reflection by the pigment particles, as we saw in the previous section. Figures 11-16 and 11-17 show the chromaticity of mixtures of colored paint with titanium dioxide white or ivory black. With a sufficient amount of white or black the chromaticity approaches the equal power point (achromatic point). However the chromaticity need not go directly there, and the path followed by mixtures of carbazole violet and titanium oxide white is particularly surprising.

The results of mixing colored paints are sufficiently complicated so that no fully reliable theory has yet been developed. For the artist, there is no choice but to be fully familiar with the mixing properties of the paints on the palette, guided by traditional color mixing methods such as the double primary palette.

Nevertheless, approximate theories have been formulated to predict the results

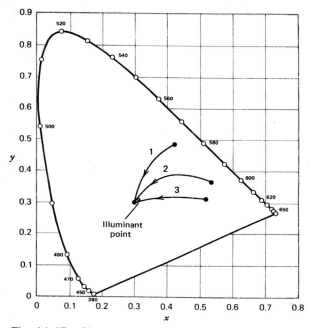

Fig. 11-17. Changes in the chromaticity of mixtures of colored pigments with increasing amounts of ivory black pigment. The pigments include: 1, zinc yellow; 2, cadmium orange; and 3, deep cadmium red. In all three cases the addition of black reduces the purity of the mixture, because the eye is unable to distinguish hue for surfaces reflecting very little light. (From R. M. Evans, *An Introduction to Color*, Wiley, New York, 1948, p. 285.)

of mixing paint that include both absorption and reflection, and have been applied with some success to artists' materials as well as to industrial materials. One widely applied theory is that of Kubelka and Munk, who worked out a mathematical description of absorption and scattering by many fine particles. This theory is described in the article by Ruth Johnston listed under Further Readings at the end of this chapter.

■ 11-12 OP ART AND MOIRÉ

In this section we shall discuss two apparently unrelated areas: optical (or "op") art and moiré patterns. At the end, we shall consider a possible connection between them.

Optical art, which enjoyed a period of great popularity during the 1960s, has been reviewed in the book by Cyril Barrett and the article by Gerald Oster in the reference list. The interested reader might consult these works for a more complete discussion.

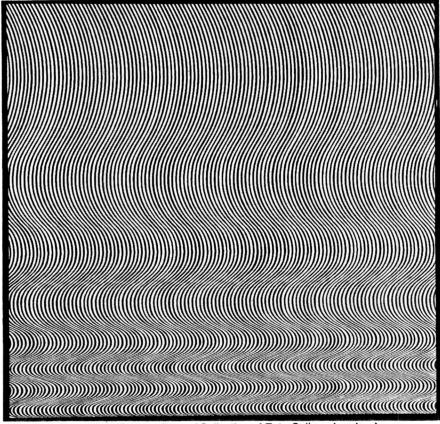

Fig. 11-18. *Fall* by Bridgett Riley. (Collection of Tate Gallery, London.)

Optical art grew out of an interest in light, color, and space that has preoccupied Western artists since the Renaissance. A distinctive feature of op art is that it relies for its effect on certain aspects of the visual process of which we are not normally aware. In an op painting, what at first confronts us is a stable, often monotonous repetition of lines, squares, and dots. But as we continue to look at this simple structure it begins to dissolve before our eyes, giving way to a sense of movement that is often dizzying. The works of Bridgett Riley, shown in Fig. 11-18, and Reginald Neal are particularly effective examples of this phenomenon.

The early roots of optical art can be found in the 1920s in the works of Marcel Duchamp and members of the Bauhaus. The real beginning of the movement occurred in the 1960s partly as a reaction to the then-prevalent Abstract Expressionism, which lacked precision in technique. The development of a new dynamic geometrical technique grew from the experiments and exercises of the Bauhaus school in Germany and of Josef Albers who had been a professor there. The name

"Op Art" was coined by a writer for *Time* magazine in 1964; in 1965 the movement obtained "official sanction" when the Museum of Modern Art in New York exhibited a show called "The Responsive Eye."

Op art has had many supporters and perhaps as many critics. Whether or not it will eventually be considered as a serious art movement or merely a technical curiosity remains to be seen. But it has certainly succeeded in making many people aware of the complexity of the processes involved in seeing. How it works is still a matter of controversy, although as Cyril Barrett states, the inability of the eye to find any focal point must play some role.

Moiré

Moiré patterns are often seen when looking through folds of a curtain or other fabric, overlapping pieces of window screening, or any similar objects. The origin of the term is in a type of silk fabric made since the fifteenth century in France, but invented long ago in China, in which two large pieces of fabric are overlapped so the ribs do not lie parallel, and the two faces are pressed together between rollers. Used in either clothing or as a wall covering, the fabric gives the impression of a "watered" surface with undulating patterns whenever the observer (or the fabric) moves. Moiré patterns have found a number of applications as methods of precision measurement to detect slight differences in alignment, and they are quite remarkable to see (Fig. 11-19).

The explanation of the moiré pattern effect together with its possible connection with optical art depends on several aspects of the visual process that were described in Chapter 10.

If a single straight line is drawn diagonally across a series of parallel straight lines, it will appear to break up into a series of short broken segments that are no longer continuous (Fig. 11-20). This effect, which is closely related to a number of important illusions studied by visual psychologists, illustrates that the points of *intersection* are perceived differently than the line segments that join them. This perceptual weighting of intersections is at the basis of the whole range of moiré effects, since the moiré pattern can readily be seen to consist of all the crossings of lines in the two screens or pieces of fabric. Since the spacing between crossings is very sensitive to the orientation of the two screens, drastic changes in the pattern are observed when one screen is rotated very slightly with respect to the other. This is why moirés can be used so effectively for purposes of measurement, and explains—at least formally—why the moiré pattern moves around so dramatically. Why the eye focuses so strongly on the points of intersection in the visual field is not yet understood, but is presumably related to the presence of specialized cells in the visual cortex that respond only when presented with the stimulus resulting from intersecting lines.

A possible connection between moiré patterns and op art has been suggested by Gerald Oster who has worked extensively with both. The explanation rests on another aspect of the visual process, which was considered in Chapter 10. This is the persistence of the visual image for a brief time (about $\frac{1}{10}$ second). Persistence is crucial for the perception of motion pictures and television as continuous rather than as a series of flickering images.

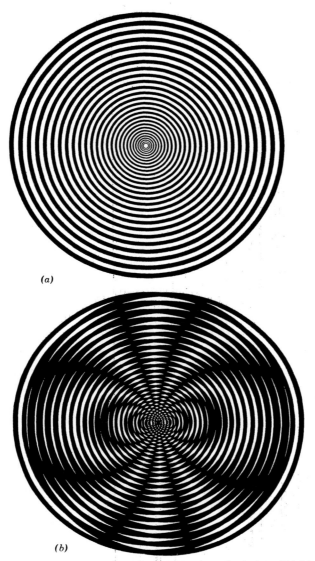

(a)

(b)

Fig. 11-19. (*a*) A single set of concentric circles; (*b*) Moiré pattern when two identical overlapping sets of concentric circles have their centers slightly displaced.

When viewing an op art painting, the eye tends to jump around, hunting for a point of focus. In fact, all "seeing" is accompanied by small jumping motions of the eye called saccades, but the motion is probably exaggerated by the complete absence of a point of focus in the picture. The visual apparatus will therefore retain a brief afterimage of the painting superimposed on the somewhat displaced imme-

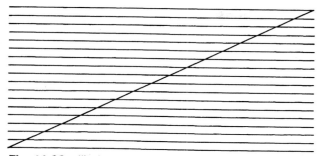

Fig. 11-20. Illusion of breaks in a straight line.

diate image. Since both are made up of mismatched regularly spaced lines or curves, they can form a transitory moiré pattern. As the eye continues to move, successive moiré patterns will be formed with little relation to each other. The kinetic aspect of optical art is thus explained as a random sequence of moiré patterns between the present image and afterimage whose motion derives from the movement of the eye. Whether the explanation is correct or not remains to be established.

SUMMARY

Many of the principles of optics and color theory underlie the functioning of materials used in the world of art. Dyes, which are the source of color in glazes, fabrics, and yarns owe their color to individual dye molecules. Pigments, which are the source of color in paints, consist of finely ground solid materials that are suspended in a liquid medium. Paints can be made by suspending pigments in water (watercolor) or egg yolk (tempera). Starting in the fifteenth century, drying oils—primarily linseed oil—which gradually harden on exposure to air by polymerization, became the dominant type of carrier. In the twentieth century acrylic and latex paints have been introduced and have gained great popularity.

The optical principles of paint involve reflection at the boundaries between pigment particle and medium, and selective absorption of light passing through each particle. The particle size plays a crucial role since if the pigment is ground too finely, the paint will be very desaturated or nearly white; if the pigment is ground too coarsely, the paint will be transparent. Transparency is also affected by the refractive index of the medium. Since the refractive index tends to increase slowly with time, there is a tendency for oil paintings to become more transparent with time, occasionally resulting in the appearance of an unwanted underpainting, an effect called pentimento. The most difficult aspect of handling dyes and paints is predicting the results of mixing. This can be carried out approximately by application of the rules of subtractive color mixing. Recent developments in the visual arts have included many other aspects of light and color outside of the traditional domain of art. One example is op art, which relies on some subtle properties of visual perception. ∎

QUESTIONS

11-1. What is the difference between a pigment and a dye?

11-2. Why do the molecules of most dyes have a structure that includes at least one carbon ring?

11-3. What side group of atoms is always found attached to the carbon ring in aniline dyes?

11-4. Of the fibers wool, flax, cotton, and silk, which are constituted of polar molecules? What is the simplest method for dyeing fibers containing polar molecules?

11-5. Traditionally, what process was used to apply the dye Turkey red to cotton and wool?

11-6. A dye having a particular color in solution may look quite different when applied to the fabric. What are some reasons for this?

11-7. What is meant by a "mordant"?

11-8. How may light from the sun promote fading of dyes?

11-9. Name six natural pigments found in the earth.

11-10. What is a "lake"?

11-11. Discuss the principal point of disagreement between Newton and Goethe concerning the nature of light.

11-12. What are the various ways that glass may be colored?

11-13. Why does the application of various slips to iron-bearing pottery produce red and black iron oxides, depending on the firing conditions?

11-14. What is the difference between watercolor and tempera painting?

11-15. Explain why oil paintings tend to yellow with time.

11-16. Latex paints are a form of acrylic paint. Why can latex paint be diluted with water if acrylic resin and water do not mix?

11-17. If a piece of glass having a refractive index of 1.6 is used to protect a painting and if the front surface is coated with a layer of transparent material having a refractive index of 1.3, what is the reflectance compared with that of the uncoated glass? (Neglect possible effects from interference.)

11-18. Explain why transparent materials may be used to make white paint.

11-19. Why have underpaintings become visible in some old oil paintings?

11-20. Explain why mixing differently colored paints together sometimes produces a color that is not predicted by the subtractive laws.

11-21. Why does op art give such a remarkable visual impression?

FURTHER READING

H. Varley, Ed., *Color* (Viking Press, New York, 1980). A beautiful book with striking illustrations for a nontechnical discussion of color, its symbolism, how it is used in design and decoration, how it is created, and how its use has evolved through the ages.

C. S. Smith, *From Art to Science* (M.I.T. Press, Cambridge, 1980). A delightful exhibition catalog showing how the urge to find expression in works of art has encouraged the development of science and technology.

R. J. Gettens and G. L. Stout, *Painting Materials* (Dover, New York, 1966). A small encyclopedia covering pigments, media, and tools used in painting.

M. Doerner, *The Materials of the Artist and Their Uses in Painting* (Harcourt, Brace, New York, 1949).

H. L. Cooke, *Painting Techniques of the Masters* (Watson-Guptill, New York, 1972).

D. V. Thompson, *The Materials and Techniques of Medieval Painting* (Dover Publications, New York, 1936). General description of the preparation and use of materials used in medieval painting.

J. Gutierrez and N. Roukes, *Painting with Acrylics* (Watson-Guptill Publications, New York, 1969). Interesting discussion of the invention, manufacture, and use of acrylics and other plastic paints.

R. Mayer, *The Artist's Handbook of Materials and Techniques* (Viking Press, New York, 1957). Largely descriptive discussion of technical aspects of paints and painting.

R. M. Johnston, "Color Theory," in *Pigments Handbook,* Vol. III, T. C. Patton, Ed. (Wiley-Interscience, New York, 1973).

T. B. Brill, *Light. Its Interaction with Art and Antiquities* (Plenum Press, New York, 1980). The ways media affect color and texture of reflected and transmitted light, as well as the way light deteriorates materials' artistic value. Topics are presented from the perspective of materials science for readers with a background of one year of college chemistry.

R. Rys and H. Zollinger, *Fundamentals of the Chemistry and Application of Dyes* (Wiley-Interscience, New York, 1972).

M. L. Joseph, *Introductory Textile Science,* Third Edition (Holt, Rinehart and Winston, New York, 1977).

C. L. Bird and W. S. Boston, *The Theory of Coloration of Textiles* (Dyers Company Publications Trust, London, 1975).

G. Duby, *The Age of the Cathedrals* (University of Chicago Press, Chicago, 1981).

R. E. Mueller, *The Science of Art* (John Day, New York, 1967). New media and materials.

Application of Science in Examination of Works of Art (Museum of Fine Arts, Boston, 1965).

    Stephen Rees Jones, "The History of the Artist's Palette in Terms of Chromaticity," p. 71.

    R. M. Johnston and R. L. Feller, "Optics of Paint Films: Glazes and Chalking." A brief account of applying the Kubelka-Munk theory for predicting the consequence of mixing oil paints of different colors, p. 86.

A. Weyl, *Coloured Glasses* (The Society of Glass Technology, Sheffield, England, 1951). Classic survey of the physical and chemical nature of glasses and the means of coloring them.

C. R. Bamford, *Color Generation and Control in Glass* (Elsevier Scientific Publishing Company, Amsterdam, 1977). An important update of Weyl's monograph.

G. O. Jones, *Glass* (Chapman and Hall, London, 1971).

D. Rhodes, *Clay and Glazes for the Potter* (Chilton, Philadelphia, 1957).

W. D. Kingery, *Introduction to Ceramics* (Wiley, New York, 1960).

C. Barrett, *An Introduction to Optical Art* (Studio Vista/Dutton Pictureback, London, 1971).

G. Oster, *The Science of Moiré Patterns* (Edmund Scientific, Barrington, New Jersey, 1969).

From Journals

D. L. MacAdam, "Color Essays," *Journal of the Optical Society of America,* Vol. 65, (1975), p. 483.

Gerald Oster, *Optical Art—Applied Optics,* Vol. 4 (1965), p. 1359.

W. H. Davenport, "Technology, Literature, and the Arts," Resource Letter TLA-2, *American Journal of Physics,* Vol. 43, (1975), p. 4. Guide to articles and books on the relationship between science, technology, and the arts.

Kent Kirby, "Art, Technology, and the Liberal Arts College," *Art Journal,* Vol. 29, No. 3, (1970), p. 330. Art technology; op, pop, and minimal art; optical theory, laser photography.

From the *Scientific American*

R. J. Charles, "The Nature of Glasses" (September, 1967), p. 126.

J. J. Gilman, "The Nature of Ceramics" (September, 1967), p. 112.

G. Oster and Y. Nishijima, "Moiré Patterns" (May, 1963), p. 54.

CHAPTER 12

HOLOGRAPHY

An exciting development during the past 15 years provides a means for seeing a three-dimensional image from optical information recorded on a two-dimensional sheet of photographic film. The technique called *holography* was invented by Dennis Gabor in 1947 but did not become useful until the advent of the laser in the 1960s. It has applications in many areas of science, engineering, and the visual arts. The name derives from two Greek words: *holo* for "entire" and *graph* for "picture." The developed film or hologram contains more information than an ordinary photograph; it indicates not only the pattern of illumination over the negative but also the relative directions from which the light was incident. To explain this sophisticated technique we shall apply many concepts introduced earlier concerning the wave nature of light, photography, and optics.

12-1 THE HOLOGRAM

Let us first consider a subject being illuminated by light from a laser (Fig. 12-1). A diverging lens placed in the beam ensures that the subject is uniformly illuminated. Because the light is coherent the "crests" representing the electric field of the light form a regular, diverging pattern at any instant. This is also true of light reflected by the subject, but here the pattern is more complicated because light waves from different positions on the subject overlap. Someone looking at the subject will intercept some of this light. It is obvious but important to emphasize that if we imagine a "window" placed as shown in the figure, all of the light entering the observer's eyes must first pass through this window. All of the information about the scene must thus be contained in the properties of the light at the

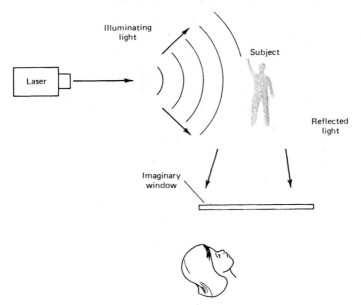

Fig. 12-1. Overhead view of a subject illuminated by laser light, showing the pattern of crests of the illuminating and reflected light waves.

window. What properties are these? The *illuminance* and *phase* at each point across the window. The uneven distribution of light intensity from the subject produces more illumination at some places on the window than others. And because wave crests arrive at some places earlier than others, the phase at any instant varies from one location to the next. If you place a piece of photographic film on the window it will record only the pattern of illuminance. In normal photography a lens is placed before the film to focus an image on it. All phase information is lost, and the resulting picture is two dimensional.

Phase and Intensity

If we could put a film on the window and somehow record *both* illuminance and phase, all of the original three-dimensional information would be retained. This can be done by simultaneously exposing the film both to the reflected light *and* to light from a second, coherent reference beam as schematically illustrated in Fig. 12-2. An actual arrangement of a mirror and lenses is shown in Fig. 12-3. The pattern of constructive and destructive interference between the two sets of waves records the variation in phase across the film for the reflected light. Because the illuminance of the light at places of constructive interference depends on the amplitude of the light at that place, this amplitude information is also retained when the film is developed. In effect, the amplitude is recorded in terms of the illuminance in regions of constructive interference, and phase variation is indicated by how closely spaced these regions are. The developed film is called a *hologram*.

The hologram exhibits nothing recognizable, because an image is not formed on the film. Perhaps a few swirls are evident (Fig. 12-4a), but these large-scale features are actually caused by light that is diffracted around tiny dust particles on one of the mirrors or lenses. The information of interest is registered as a seemingly chaotic pattern of fine light and dark regions arrayed across the film.

To retrieve the image the hologram is illuminated by coherent light (Fig. 12-5). The pat-

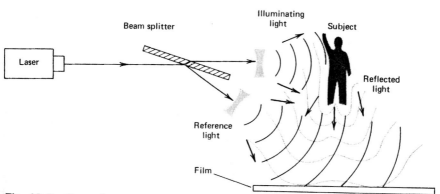

Fig. 12-2. Recording a hologram on a film by illuminating it with coherent light reflected from the subject and a reference beam. A "beam splitter" (or "half-silvered mirror") transmits half of the light and reflects the other half, thereby permitting the same laser to illuminate the subject and also provide the reference beam. A diverging lens in the reference beam gives even illumination across the film.

Fig. 12-3. Setup for recording a hologram. Light from the laser at the back comes along the left side of the table, is reflected through 90° and illuminates the subject at the right. The film holder located at the front is also illuminated by the reference beam which is reflected from the beam splitter immediately in front of the laser. (Courtesy of Newport Research Corporation.)

tern of closely spaced silver grains in the hologram diffracts the light as it passes through, thus causing the spreading waves to overlap and interfere as the light emerges from the far side. The emerging wave patterns mimic those which had originally been reflected from the subject and intercepted by the film during recording. Consequently, on looking through the hologram as though it were a window, we can see an image when these waves enter our eye. The image appears in back of the film and is a virtual image. The spreading waves from the hologram have all the features of the original light reflected by the subject, so the virtual image is "three-dimensional." That is, the image is seen stereoscopically with the views seen by the left and right eyes differing slightly. Also, by moving your head horizontally or vertically, the phenomenon of parallax is experienced whereby nearby subjects appear to be displaced more than distant ones. Features of the scene that may be blocked from view by a nearby object when we look through the hologram in a particular direction may be revealed by moving to a new vantage point and "looking around" the obstruction.

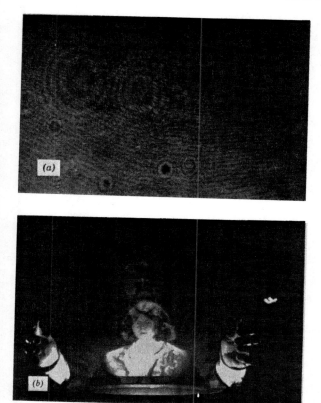

Fig. 12-4. (a) A hologram containing both amplitude and phase information of coherent light reflected from a subject has no discernable pattern; the various concentric rings are caused by light diffracted by dust on the lenses of the optical system; (b) virtual image produced by a hologram held over a source of light; the technique for producing this hologram is described in Section 12-5. ("The Kiss" 120° integral hologram by Lloyd Cross, 1975. Photo by Daniel Quat. collection of the Museum of Holography, New York City)

12-2 HOLOGRAM OF A POINT SOURCE

The underlying principles of holography can be most easily appreciated by recognizing that the subject of which the hologram is being made consists of many small independent points that scatter incident laser light toward the film. Figure 12-6a illustrates the spherical waves spreading out from such a point and meeting the reference beam (a "plane" wave with parallel crests) at the film. In this case the reference beam was directed onto the back of the film to provide the simplest example of an interference pattern. The film will be strongly exposed wherever the spherical and plane waves arrive in phase (crests coin-

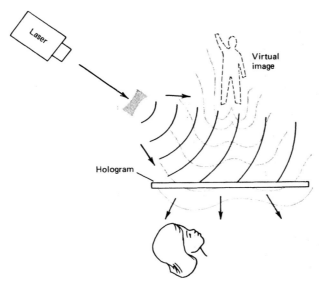

Fig. 12-5. When a hologram is illuminated from the same direction as the original reference beam, a virtual image is formed at the same location as the original subject.

ciding), and will be weakly exposed where they arrive out of phase (crest of the spherical wave coinciding with a trough of the plane wave). The resulting interference pattern recorded on the developed film is called the hologram of the point subject and is shown in Fig. 12-6*b*. It consists of a series of transparent rings separated by opaque rings, all centered directly in front of the point. The transparent rings have particular sizes because they denote where light from the point traveled just the right distance to arrive at the film with the same phase at each ring. This feature of the pattern is crucial for reconstructing an image from the hologram.

Reconstructing the Real Image

To form a real image from the hologram we can illuminate the film with the same reference beam. Since this beam is a plane wave the light transmitted through each of the narrow transparent rings has the same phase. Therefore the portions of the diffracted waves that overlap in back of the film, at the position where the point subject was originally placed, also have the same phase and interfere constructively. If a white paper were placed there, a bright spot would appear.

If the direction of travel of the plane wave in Fig. 12-6*a* used in producing the original hologram had been tilted, the rings in the film would be elliptical rather than circular and the reconstructed image would appear off to one side rather than directly in front of the film.

The reason we gave this explanation of how to make the holographic image of a point subject is simply that *any* subject can be imagined as consisting of individual reflecting points, placed one beside the other over the surface of the subject. When recording the hologram of the subject, each point produces its own set of concentric rings, the pattern

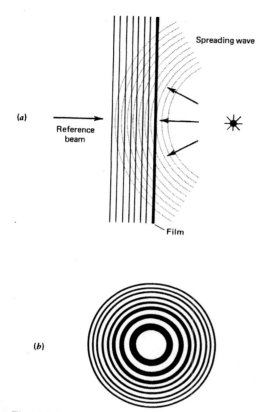

Spreading wave

(a)

Reference
beam

Film

(b)

Fig. 12-6. (a) Side view of interference between spherical waves reflected from a point and plane waves of the reference beam meeting at the plane of the film (b) Front view of the pattern on the developed film showing the many fine alternating transparent and opaque rings.

from one point being superimposed on the others. To reconstruct the real image, the hologram is illuminated with a coherent beam of light, and each of the patterns then produces its corresponding spot behind the film. This reconstructs the original pattern of light reflected by various areas of the subject.

Virtual Image

A hologram produces both a real image and a virtual image. The reason is that although the hologram records both phase and intensity information across the film, this does not specify whether the reference and reflected light passed through the film in the forward or reverse direction. As a result, when illuminating the hologram with coherent light both possibilities are realized: the virtual image is produced by light diffracted by the film that sets up a *spreading* wave pattern; and a real image is formed by diffraction that sets up

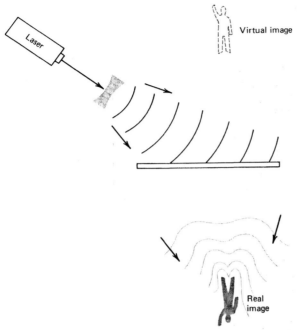

Fig. 12-7. A virtual image in addition to the real image is created when a hologram is illuminated by coherent light. The converging wave pattern forms a real image at the indicated position, and a diverging pattern (illustrated in Fig. 12-5) forms a virtual image behind the hologram.

a *converging* pattern. Figure 12-7 shows the formation of the real image. Each point of the hologram that during recording had received light from a particular point of the subject now sends light to converge onto a corresponding point at the real image. A photographic film could record this image if placed at its position.

Because no lens is used to form an image on the film when recording a hologram, information about the light from any point of the subject is spread across the entire hologram. In other words each region of the hologram contains information about the entire scene. As a remarkable consequence, were the hologram to be cut into several pieces, viewing any one of them could reproduce the entire image! The effect of cutting away portions of a hologram is not to remove subjects from view—as would happen in a photograph—but to reduce the resolution, or amount of detail. Edges become blurred and sharp features are lost.

12-3   VOLUME HOLOGRAM

The recent surge of interest in holography as an art form in large measure is due to the introduction of a technique that permits holograms to be viewed under incoherent white light. The innovation is credited to the Russian scientist Yuri Denisyuk who in 1962 exploited an idea originally advanced in 1891 by Gabriel Lippmann for a method of color photography. In the holographic application of Denisyuk's idea the hologram is formed within a thick, high resolution emulsion. The directions of reference and reflected beams are arranged so that interference patterns occur not only across the film but also many times *within* the emulsion (Fig. 12-8a). An emulsion perhaps 10 μm thick then contains more than 20 periodically spaced layers exposed by high intensity light at successive locations of constructive interference. When developed, the silver grains in each layer reflect a small percentage of any incident light and transmit the rest. If white light illuminates the hologram, the layers of silver grains give rise to iridescence (Section 4-5) whereby the component of the light at the original laser wavelength is selectively reflected (Fig. 12-8b). Because of the diffraction and overlap of waves reflected from various locations across the hologram, diverging patterns of light similar to those originally produced by the subject enter the eyes of an observer, and a three-dimensional virtual image can be seen. Unfortunately, just as in iridescence, various wavelengths interfere constructively in differing directions. Therefore a volume hologram recorded with one coherent source cannot reproduce a full color virtual image when viewed under white light, because the images for various components of the spectrum are not in register. In fact, a hologram recorded with a red helium-neon laser appears green with white illumination, because the thickness of the film shrinks during development. Furthermore, the light is diffracted over such a broad span of directions that the image from any one vantage point is dim.

Three Colors

Nevertheless a full-color image can be obtained by a different technique. The hologram is recorded by illuminating the subject successively with coherent sources of the three additive primary colors. This forms three sets of interference patterns within the emulsion.

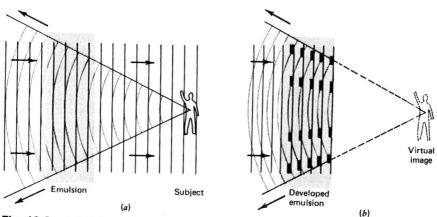

Fig. 12-8. (a) Volume hologram is formed when a film is exposed at equally spaced depths by constructive interference between the illuminating light (which first passes through the film before striking the subject) and the reflected light. (b) After development when the film is illuminated with white light, the component of light at the wavelength of the original laser is reflected by the periodic layers and reforms the original spreading wave pattern.

With white light illumination, constructive interference of light reflected by the silver layers will occur simultaneously in one direction only at wavelengths that correspond to these primary colors. The simultaneous iridescence in the same direction of these three colors in various proportions can reproduce a wide gamut of colors.

Volume holographs can also be viewed in a transmission mode whereby the light source is behind the hologram instead of in front. To avoid the loss of intensity caused by the reflecting silver layers, the developed emulsion is chemically treated to bleach away the silver grains and replace them by a transparent material having a lower refractive index than the gelatin. The resulting uneven texture of the emulsion diffracts light as it travels through. The interference between spreading, overlapping waves again mimics the pattern of light originally reflected by the subject when the hologram was recorded.

12-4 RAINBOW HOLOGRAM

Much brighter images can be enjoyed with white illumination by using another type of transmission hologram that has proven popular in recent years. Introduced in 1969 by Stephen Benton, the technique restricts the spreading light at each wavelength to a narrow range of angles. As an observer moves his head up and down to view the hologram from various angles, the same image is seen but in differing bright colors corresponding to the spectral sequence. Hence the name "rainbow" hologram. To achieve this brightness without having images from various wavelengths overlap, parallax in the vertical direction is sacrificed when the hologram is recorded.

In creating a rainbow hologram a "master" hologram is first prepared in the conventional way. The master is then illuminated by coherent light to form a second hologram by

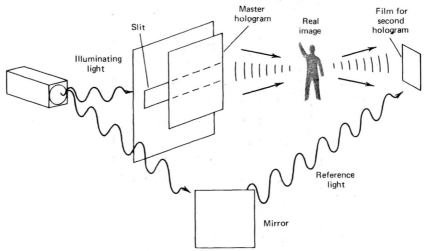

Fig. 12-9. Arrangement for recording a "rainbow" hologram on film from a master hologram. Only a narrow horizontal section of the master is illuminated, and the real image is formed from light diffracted from this section. This light spreads out on leaving the position of the real image and then exposes the film.

exposing a piece of film to light that comes from the master's *real* image. However during the exposure most of the master is blocked off and light is allowed to pass through only a narrow horizontal slit (Fig. 12-9). The second hologram therefore records only the information from the portion of the master illuminated by the slit. Because each point on a hologram contains information from only that viewpoint, information concerning horizontal parallax is retained because of the large range of viewpoints along the length of the slit. But vertical parallax is lost because the slit is so narrow. Now when the film is developed and this second hologram is illuminated with coherent illumination, both a virtual image and real image are formed as before. The virtual image will be dim except when viewed from a narrow range of directions. These correspond to the position of the real image of the slit. Light is concentrated there, with all of the information contained in the portion of the master hologram that had been illuminated. Even when illuminated by a white light (Fig. 12-10), a real image of the slit is formed at the same location by components of the spectrum with nearly the same wavelength as the orginal laser's. Other wavelengths form real images of the slit at higher and lower positions. Thus, when we look through the hologram from a given height, we see a bright virtual image of the original subject rendered in the color corresponding to the wavelength of light concentrated at that height. By looking from successively higher viewpoints, we intercept light of differing wavelengths and see the virtual image in a sequence of spectral colors. No parallax in this direction is perceived. This is not a major disadvantage, because ordinarily when standing upright we use horizontal parallax to judge depth, which is retained in the image.

12-5 COMPOSITE HOLOGRAMS

A fascinating type of hologram can be exhibited on a transparent cylindrical form that rotates so that it presents different portions of the surface as it revolves. Illuminating the inside with white light creates a virtual image that displays action much like a movie, but it also retains horizontal parallax. However, moving your head up and down reveals that it lacks parallax in the vertical direction. The composite hologram actually consists of

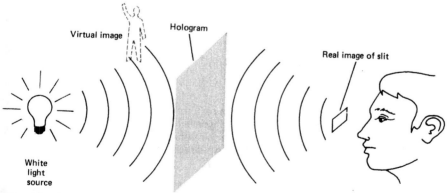

Fig. 12-10. A rainbow hologram illuminated by white light produces a bright virtual image of the subject when viewed from directions where the real image of the slit is formed. Each component of the spectrum forms the real image at a different height.

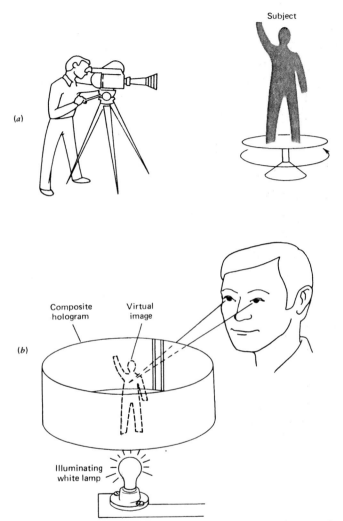

Fig. 12-11. (*a*) Recording a movie as a performer is rotated completely around one time. (*b*) After conversion into a 360° composite hologram, a different virtual image is seen by each eye when looking through the composite hologram, corresponding to frames of the movie recorded from these two directions.

many extremely narrow holograms arranged vertically next to each other around the circumference, each of which represents the scene in one frame of a movie, as we shall now explain.

How the movie is made depends on the subject matter. In one effective presentation the camera is mounted so that it rotates once about a performer as the sequence of his actions is recorded (or the performer could be rotated instead, as in Fig. 12-11a). Each

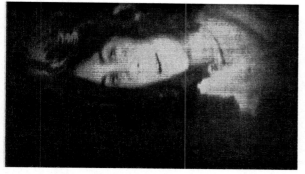

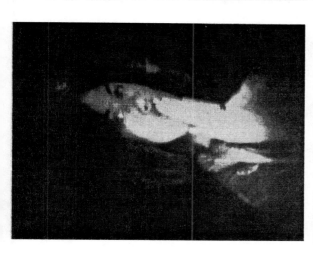

Fig. 12-12. Looking through a composite hologram gives a different image for each position of view, three of which are shown here for the *Kiss II*. (These are closeups of the hologram displayed in fig. 12-4*b*)

frame of the developed film with its two-dimensional scene is then recorded as a portion of a hologram. For recording on the emulsion, a cylindrical lens vertically enlarges the holographic pattern to a height of perhaps 20 cm, whereas a narrow vertical slit restricts the width of the hologram to about 1 mm. Then the movie is advanced one frame and the emulsion is shifted by 1 mm for recording the next hologram. This continues until all the frames are represented by contiguous holograms on a long strip of emulsion.

The developed composite hologram is then bent into an arc corresponding to the angle of rotation that was included when the movie was shot (Fig. 12-11b). With illumination by coherent light each individual hologram presents a virtual image without parallax, faithful to the original two-dimensional frame. But an observer looking at the image behind the composite has his eyes directed through *different* holograms, so the light entering the eyes offers scenes from different viewpoints as in a stereoscopic viewer. Moving the head sideways shifts the views to new holograms, with the result that parallax is experienced. Although the parallax shifts occur in steps, so many frames are recorded close together that the steps should not be perceptible. The action during filming should be sufficiently slow that frames simultaneously viewed by the two eyes are virtually identical aside from parallax.

For viewing under white light the movie can be converted into a rainbow hologram by the Benton technique and displayed in the way just described (Figs. 12-11b and 12-12). Nothing further is lost, since no vertical parallax is available in the original movie.

This brief description of holographic techniques reveals some of the ingenuity that has led to this exciting form of visual art. We can anticipate that even more remarkable effects will be possible as its practitioners advance in sophistication and exploit the advantages of other optical effects.

SUMMARY

A hologram is a piece of film which, on being properly illuminated, presents the viewer with a three-dimensional reconstruction of the scene recorded on it. A hologram, unlike a photograph, bears no resemblance to the original scene.

Holography, which became practical about 15 years ago with the invention of lasers, depends on the properties of coherent light to record both amplitude and phase of the light reaching the film.

The first holograms were both recorded and reconstructed with monochromatic laser light. Subsequently, holograms were formed by successive exposure to three primary colors and reconstructed with white light, resulting in full color holographic reconstruction.

Holograms recorded with a cylindrical lens can form a series of time lapse images. When viewed, the composite hologram, which is mounted on a rotating cylinder, gives the impression of a three-dimensional movie.

QUESTIONS

12-1. Explain why monochromatic light and not white light is used to record a hologram.

12-2. What property of the light reflected by the subject is recorded in a hologram that is not recorded in a photograph?

12-3. What is the purpose of the reference beam when recording a hologram?

12-4. Explain why a lens is not needed to focus an image on the film when recording a hologram.

12-5. Why must vibration be avoided when making a hologram?

12-6. If you destroy half of a photographic print you see only half of the picture. But if you destroy half of a hologram you still can see the entire picture. Why?

12-7. When viewing a rainbow hologram under white light, why are different colors seen from different directions of view?

12-8. What type of hologram is recorded by shining light through the film to illuminate the subject? How may it be made so that a full color image may be seen?

12-9. When viewing a volume hologram, is the image real or virtual?

12-10. Explain why holograms recorded with a neon (red) laser appear green when illuminated by white light.

12-11. What is parallax?

12-12. What special precautions must be taken when recording a composite hologram for the image to appear three-dimensional?

12-13. When viewing a rainbow hologram, if you tip your head to the side so that one eye is higher than the other and receives slightly different color, is the image three dimensional?

12-14. A composite hologram that presents images of a moving subject is commonly thought to provide a three-dimensional view, but is it really three-dimensional? Explain.

FURTHER READING

G. Saxby, *Holograms. How to Make and Display Them* (Focal Press, New York, 1980). A lucid, practical discussion of various types of holograms, how to arrange equipment to record them, find the image, and display the image.

H. M. Smith, *Principles of Holography,* Second Edition (Wiley-Interscience, New York, 1975). A clear introduction to the principles and methods of producing holograms.

M. Francon, *Holography* (Academic Press, New York, 1974). A broad introduction to the topic at an elementary level.

R. J. Collier, C. B. Burckhardt, and L. H. Lin, *Optical Holography,* (Academic Press, New York, 1971). A complete treatment at the advanced level, including the properties and processing of emulsions to form composite and full color holograms.

H. A. Klein, *Holography* (Lippincott, Philadelphia, Pennsylvania, 1970).

M. Lehman, *Holography: Theory and Practice* (Focal Press, London, 1970). Intended primarily for researchers.

From the *Scientific American*

E. N. Leith and J. Upatnieks, "Photography by Laser" (June, 1965), p. 24.
K. S. Pennington, "Advances in Holography" (February, 1968), p. 40.
E. N. Leith, "White Light Holograms" (October, 1976), p. 80.
From the "Amateur Scientist" column, "Light and Its Uses, Making and Using Lasers, Holograms, Interferometers and Instruments of Dispersion," (Freeman Press, San Francisco, California, 1980).

▫ CHAPTER 13 ▫

TELEVISION

The popular medium of television (TV), which serves as an important means of communication and entertainment, is entering its third stage of technological development. The first stage was the invention and commercial development of monochrome TV. The second was the introduction of color TV, which in some countries was made compatible with the pre-existing monochrome system. And the third stage, a logical extension of the first two, is the concept of high-definition television (HDTV). HDTV is expected to provide a resolution rivaling that provided by 35-mm movie film.

13-1 PICTURE QUALITY

To appreciate the term "high-definition" we must consider what is meant by "picture quality." The TV image on the screen of a receiver is formed by a scanning method. A narrow beam of electrons produced in the back of the TV tube is accelerated forward and strikes a small region of a thin layer of phosphorescent material coating the inside face of the screen. The energy given to the material excites the phosphor, which in turn gives up much of this excess energy by emitting light. If the electron beam is deflected horizontally by a set of electromagnets attached to the tube, the spot of light moves across the screen. In a monochrome set this spot appears essentially white. In practice the movement of the spot is so fast that it cannot be perceived, and only a bright horizontal line is seen. To present an image on the screen this sequence is repeated at successively lower positions, with the beam moving from the viewer's left to right to make each successive line. When returning quickly from the end of one line to the beginning of the next, the intensity of the beam is reduced to zero ("blanked") so that this flyback is not seen. If the entire sequence of lines from the top to the bottom of the screen is presented rapidly and is repeated at a rapid rate, the persistence of both the phosphor and the human eye does not permit the movement of the spot to be detected. One will simply see a steady pattern consisting of a series of closely spaced horizontal white lines filling the screen.

   In addition, if the intensity of the electron beam is varied by a separate electronic circuit within the tube as the beam traverses the screen, the intensity of light is correspondingly varied: some regions appear brighter than others because the electron beam is more intense as it strikes these regions. When the intensity is varied in a consistent way as the beam traces one line after another, a picture can be formed (Fig. 13-1).

Fig. 13-1. Depiction of a television image formed by horizontal lines whose intensity varies from one place to another across the screen.

Linear Resolution

From experience, we know that when viewing a TV image from a distance we cannot see the individual lines, and the features in the image appear to have sharp edges. This is considered an image of good quality. But, if we stand close to the screen, the individual horizontal lines are evident, with thin black lines separating them where the phosphor was not excited, and the picture quality is poor. The vertical resolution, or the smallest feature that can be portrayed, is limited by the spacing of the lines. This limitation of picture quality is particularly evident when the image is projected onto a large screen and we view it from nearby.

Other factors being equal, a picture formed by more closely spaced lines is said to have greater linear resolution. In North America and Japan the TV system is that chosen by the National Television Systems Committee (NTSC) and the broadcast signal is based on 525 lines that cover the complete screen. This number is the same for all TV screens regardless of size. Thus for a 10 cm (4 in.) screen the lines are separated by only about 0.2 mm, while for a 50 cm (20 in.) screen, they are separated by 0.8 mm. Most of continental Europe, the United Kingdom, and Brazil use the "Phase Alternation Line" (PAL) system with 625 lines. France and the Soviet Union chose a different 625 line system known as Séquentiel Couleur Avec Mémoire (SECAM). In the United States, the set of 525 lines needed for a complete presentation is called a *frame*.

The present quality of images on color TV sets is comparable to that provided by movie films that are popularly used by amateurs. The projected image of a film is limited by its graininess (see Section 9-5), and if you look carefully at the image you can always see a degree of blurring. Today's color TV images lie between what can be achieved with super-8-mm and 16-mm size films. The greater sharpness provided by a commercial 35-mm film is the present target of the TV industry. This would require about 1100 lines on a TV screen at normal viewing distance, or about double the present linear resolution. It is hoped that perhaps 1500 lines may be achieved in the more distant future.

Pixels

Brightness and contrast also determine image quality. The maximum achievable brightness must be great enough to compete with the background lighting in the room and with the portion of this light reflected from the face of the screen. By the term "contrast" we really mean two different things. One use of the word is in an overall sense: the total range of brightness that can be spanned from extreme black to the brightest white. If black shadows of the original scene appear as light gray on the image, the overall contrast is low and the image loses dramatic appeal.

Contrast also refers to "local contrast": as a line is traced by the electron beam, the quality of contrast is measured by the maximum change in brightness that can be achieved within a very short horizontal distance. This determines the sharpness of an edge. A helpful way to think about this concept of local contrast is to imagine the TV screen as divided into rectangular divisions, or picture elements. The height of each element represents the thickness of the line traced by the electron beam, and the horizontal distance between the centers of neighboring elements designates the shortest distance over which the brightness can be adjusted from its maximum to minimum value. To represent the image in terms of these individual elements, each element, or *pixel,* is considered to have uniform brightness across its area (Fig. 13-2). The pattern formed by the

Fig. 13-2. A picture represented by pixels. Image quality improves when seen at a distance. (Leon D. Harmon, from *Scientific American,* November 1973)

variation in the brightness across the array of pixels comprises the image. Thus a pixel represents the smallest area of the image that can be resolved. If we are close to the image the individual pixels can be seen; if far away the image appears to have a continuous variation in brightness across its surface, and the quality of the image is improved.

There is little point in providing better resolution in one direction across the image without providing it in the other direction; thus each pixel should have the same height as width. A 35-mm film image is made up of about one million pixels. To achieve that resolution on a TV screen with 1000 lines would require dividing each line into 1000 pixels. The total number of pixels that is equivalent to the entire image provided by the NTSC system in North America is only about 200,000 (about 500 per line).

To provide better resolution not only must the number of lines be increased but also the horizontal size of each pixel must be reduced. The TV system must be capable of varying the intensity of the electron beam more rapidly than in present systems. This can be done in principle, but there are important limitations to be considered. These will be taken up in the next section.

13-2 MONOCHROME BROADCAST SYSTEM

The principles of television broadcasting are reasonably straightforward. Separate video (picture) and audio (sound) signals are sent by conducting cables to a transmitter, where each of the two signals is encoded on a separate high frequency wave (a ''carrier'') as

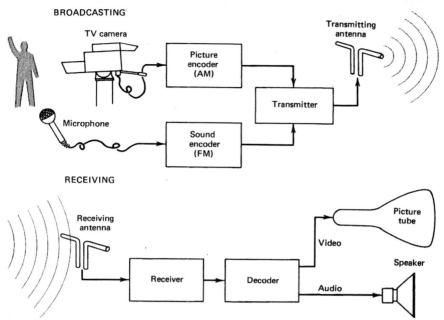

Fig. 13-3. Schematic of the television broadcasting system.

illustrated in Fig. 13-3. The frequencies of the carriers are nearly equal, and the two carriers with their encoded information are added together, amplified, and sent to a transmitting antenna. The antenna radiates this complex signal as an electromagnetic wave, which spreads out as it travels away from the antenna. At each home an antenna intercepts a small portion of the power of this radiation, which provides a corresponding voltage on the antenna that is conducted by wires to the TV set. There the signal is amplified, the video and audio signals are separated and decoded, and these two signals are sent to the picture tube and loudspeaker.

Each receiving set has two electronic circuits for deflecting the electron beam horizontally and vertically, and they are controlled by the broadcast signal, which also varies the intensity of the electron beam. For practical reasons in North America it was desired that the rate at which the beam is moved vertically be synchronized with the frequency of the power line. That is, the beam would be steadily moved downward while quickly tracing one horizontal line after another during a total period of nearly $\frac{1}{60}$ second and then would be abruptly returned to the top of the screen to begin the next vertical scan. However, for a reason we shall shortly make apparent, it was undesirable to use such a method to trace a full 525 line frame in only $\frac{1}{60}$ second. It was also unacceptable to trace the top half of the frame in $\frac{1}{60}$ second and then the bottom half in the subsequent $\frac{1}{60}$ second. That procedure could also be synchronized with the power line frequency, but the resulting 30 Hz rate for presenting frames would not exceed the flicker fusion frequency for the eye, so that the screen would appear to flicker unpleasantly. A clever solution for this dilemma is known as the *interlaced scan.*

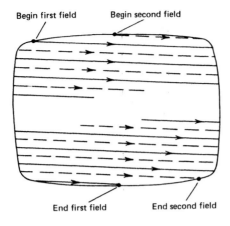

Begin first field   Begin second field

End first field   End second field

Fig. 13-4. Interlaced scan on a TV screen, with the solid lines showing the horizontal scans comprising the first field and the dashed lines the second field.

To produce the interlaced scan (Fig. 13-4) the beam starts from the top corner and scans every other line of the image during the first $\frac{1}{60}$ second. This displays 262 $\frac{1}{2}$ lines, which is called a *field*. Then the beam immediately returns to the center of the top line and continues to finish the remaining half line and the subsequent 262 lines of the second field. Thus two fields comprise one frame. The field presentation rate is 60 Hz, and flicker is avoided, even though the frame presentation rate is 30 Hz. This latter rate is still slightly faster than the presentation rate for a commercial movie, which is projected at 24 different frames per second. We should keep in mind, however, that since each frame of a movie is shown twice, the rate at which an image is flashed onto the screen is actually 48 Hz. For both TV and movies, our visual persistence ensures that there is no significant flicker, and moving subjects are perceived as moving continuously and not in a series of jerks.

A problem that caused considerable trouble initially was how to show standard 24 frame per second movies on 30 frame per second television. This was solved by designing a machine that projects one frame of film during two $\frac{1}{60}$ second fields, and the next frame during three fields. Thus two frames of the film are covered in $\frac{5}{60} = \frac{1}{12}$ second so that 24 frames are seen each second as required.

A TV set acts as a slave to the camera that is recording the image. The lens of the camera focuses a real image of the subject on a light sensitive surface whose electrical properties at each point vary with the level of illuminance. In some types of cameras, a corresponding pattern of electrical charge is distributed over the surface. Within the camera there is a scanning arrangement so that this pattern can be detected, with the scan performed in the interlaced fashion described above at 30 frames per second. Part of the broadcast signal serves to synchronize the TV sets of the viewing audience with the scanning rate of the camera. This is done by devoting a few of the lines of each frame to synchronizing information, and that portion of the signal is not shown on the screen. Only about 500 lines actually present the video image.

Video Bandwidth

With 525 lines being scanned in $\frac{1}{30}$ second, the rate or frequency at which individual lines are scanned must be $525 \times 30 = 15.75$ kHz. The corresponding time to scan a given line completely is about 64 microseconds (μsec). If we want about 525 pixels to fit into each line, the electronic circuits must be capable of varying the intensity of the beam from its maximum to its minimum value within $\frac{1}{525}$ of 64 μsec, or about 0.12 μsec. This short interval represents half a cycle of the highest frequency signal needed to control the brightness as the line is scanned. In other words, it is the time for this signal to go from its maximum value to its minimum. A full cycle would take twice this interval, or 0.24 μ sec, and the corresponding frequency is about 4 MHz. We thus conclude that a relatively large band of frequencies must be available to control the brightness of the image as each line is scanned: the span from very low frequencies (representing large regions of uniform brightness on the screen) up to 4 MHz. Thus the frequency bandwidth for the video signal is said to be 4 MHz. If the frame presentation rate were 60 Hz instead of 30 Hz, the required bandwidth would be twice as great.

This bandwidth is remarkably broad. By comparison the bandwidth of an electrical signal representing the audio portion of a broadcast need be only 20 kHz. That is because the range of frequencies of sound waves to which our ears are sensitive extends only up to 20 kHz. Most individuals cannot even hear tones that high in frequency, especially in the presence of normal background noise. The broadcast of an FM station, for example, includes only the audio bandwidth up to 15 kHz. AM radio and telephone lines are even more restrictive, with the upper limit being about 5 kHz. It should now be apparent that transmitting a TV image of acceptable quality demands significantly greater electronic capabilities than transmitting audio signals.

Broadcasting Bandwidth

We shall not go into the details of how the video and audio portions of a TV signal are broadcast. It will suffice to note that each broadcasting station is assigned a band of frequencies within the electromagnetic spectrum for transmitting its program. In North America each station is assigned a frequency range with a total bandwidth of 6 MHz. Assignments are made so that any two stations close enough geographically to have their signals received by the same TV set have their broadcast frequencies at least 6 MHz apart to prevent interference.

In the *very high frequency* portion of the spectrum (the VHF portion), TV broadcasts are found in the regions from 54 to 88 MHz and from 174 to 216 MHz. Channels labeled 2 through 6 are found in the lower of these regions (a band between 72 and 76 MHz is reserved for other purposes), while channels 7 through 13 are in the upper region. Between the two regions (88 to 108 MHz) lies the bandwidth assigned to FM stations. Another region reserved for TV broadcasts is in the *ultra high frequency* portion of the spectrum (the UHF portion): the span from 470 to 890 MHz for channels 14 through 83.

Within the 6 MHz bandwidth assigned to each station must fit both the video and audio information. The video signal from the camera (Fig. 13-5a) extends from essentially zero frequency to 4 MHz, and this must be transferred to the band of frequencies assigned to the station. This is easily done by using the video signal to "modulate" the amplitude of another electrical signal of appropriate high frequency, which has the "sinusoidal" shape displayed in Fig. 13-5b. By "modulate" we mean that the amplitude of this sinusoidal signal will be increased in proportion to the video signal when the video signal is positive, and the amplitude will be decreased when the video signal is negative. This is depicted in Fig. 13-5c. The sinusoidal signal is called the "carrier" for the broadcast, because after being modulated it carries the video information according to the variation of its amplitude from one moment to the next. The frequency of the carrier must be chosen so that all of the information fits within the allocated broadcasting bandwidth for the station. This type

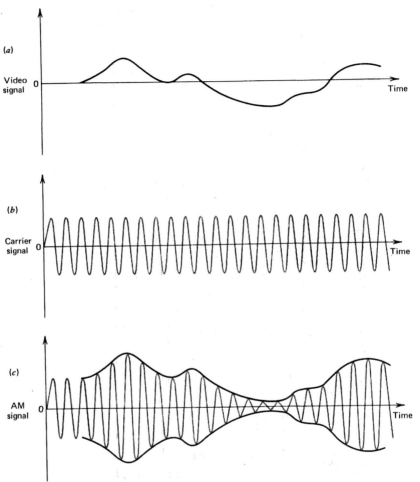

Fig. 13-5. Preparing the video portion of a TV broadcast by the amplitude modulation technique: (*a*) video signal with its picture information, (*b*) carrier, (*c*) amplitude-modulated carrier.

of broadcasting is known as *"amplitude modulation"* or AM. It is also popularly used by radio stations for broadcasting purely audio programs, in which case the carrier frequencies are much lower. The dial used to tune a station on an AM radio receiver indicates the frequency of the station's carrier.

Figure 13-6 shows how the power in a TV broadcast is spread across the spectrum. We may think of this diagram as the SPD for a TV broadcast. The only difference between this and our curves showing the SPD for light in previous chapters is the convention used in broadcasting that the horizontal axis indicates frequency instead of wavelength. The

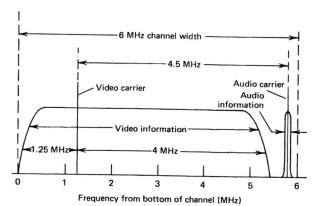

Fig. 13-6. Location of video and audio information wtihin the 6 MHz bandwidth allocated to each television station in the United States. The actual SPDs of the video and audio portions of the band vary from one moment to the next according to the program material.

SPD for an ordinary AM signal extends upward and downward in frequency from the carrier by an amount matching the highest frequency used to modulate the carrier. For TV this would be a spread of 4 MHz upward and 4 MHz downward. However the information in the region of the spectrum below the carrier frequency duplicates the information in the region above; thus most of the former is eliminated with no loss in picture quality. For technical reasons a small portion of this lower region is retained. Thus, as shown in Fig. 13-6, the video portion of the TV broadcast extends from the bottom end of the allocated band, lying 1.25 MHz below the carrier, to about 4 MHz above the carrier.

The audio signal of a TV broadcast is treated separately. It is encoded on its own carrier by the process of *frequency modulation* (FM). As Fig. 13-7 shows, in this technique the frequency of the audio carrier is increased in proportion to the strength of the audio signal when the latter is positive, and the frequency is decreased when the audio signal is negative. This FM encoded information is more immune than an AM broadcast to atmospheric electrical disturbances such as lightning, which is heard as "static," and is used for audio programs of higher quality. The FM technique is not needed for the video portion of the broadcast because our eyes' persistence enables us to block out the noise on the screen.

The audio carrier is placed 4.5 MHz above the video carrier in the broadcast, so it lies precisely 0.25 MHz below the upper end of the station's bandwidth. Since an FM radio station is allocated a bandwidth of 200 kHz for its program, there is plenty of bandwidth available in the TV band for transmitting sound of equivalently high quality. The reason that the audio quality of the TV broadcasts is so poor is simply the inferior electrical circuits in the transmitting and receiving sets and the cheap speakers that are usually installed in the sets. High quality sound is possible with TV, however, and some manufacturers are beginning to exploit this possibility.

When the signal with the SPD illustrated in Fig. 13-6 is received by a TV set, the video signal is extracted from the video carrier, and the audio signal from the audio carrier. A few lines of the 4 MHz video signal are used to synchronize the vertical and horizontal scanning circuits that control the deflection of the electron beam in the TV tube, and the remaining signal modulates the intensity of the beam as it is swept.

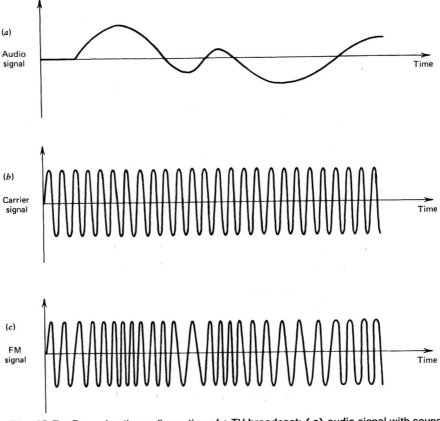

Fig. 13-7. Preparing the audio portion of a TV broadcast: (*a*) audio signal with sound information, (*b*) carrier, (*c*) frequency-modulated carrier (the increase and decrease in frequency is exaggerated to emphasize the effect).

High Definition TV

To improve the resolution of a video image and attain HDTV equivalent to one million pixels, both the number of lines per frame and the video bandwidth must be increased. An increased bandwidth for each broadcasting station necessarily means that fewer channels could be allowed to broadcast, since the available portion of the electromagnetic spectrum is limited. This trade-off between image quality and bandwidth is what led the North American countries to adopt an image that is inferior to that in Europe, because by so doing more stations could be accommodated. Presently a bandwidth of 6 MHz is required for the NTSC system and from 7 to 8 MHz for PAL and SECAM systems. An HDTV system with 1125 line resolution would require a video bandwidth of about 27 MHz. There are ways to "compress" the video signal electrically so it could be fit into a narrower bandwidth; special circuits would then be added to the TV set to "expand" the signal before viewing. In addition, broadcasting via satellite and by closed cable systems offer other possibilities for HDTV in the relatively near future. Manufacturers are presently developing systems to replace or supplement the present ones by HDTV.

Concurrent with the desire to improve resolution is a growing interest in changing the aspect ratio—the ratio of width to height of the viewing screen. For the present systems it is 4:3 (that is, 4/3 times wider than high). However viewers tend to prefer a wider screen, with an "aspect ratio" of 5:3, or even 2:1 as found in cinemas. The popularity of the TV medium is raising many questions of this type. As more complex electronic circuits become available at lower cost, there is little doubt that we shall see major changes in the methods of broadcasting that will improve both the image and the sound.

13-3   COLOR BROADCASTING SYSTEM

For a color broadcast the only needed change from the monochrome system is to include three simultaneous signals with the picture information in order to express the amounts of three primary colors required for reproducing the desired color. The other aspects of the television broadcast system can remain essentially the same. The conventional NTSC color system in the United States is based on a set of blue, green, and red primaries. The color TV camera is actually "three cameras in one" (Fig. 13-8). The image of a scene is directed through a set of "dichroic mirrors" that separate the short, intermediate, and long wavelength components and direct them toward three camera tubes to yield blue, green, and red signals. The camera signals, when adjusted by electronic circuits called gamma correctors to correspond to the relative sensitivity of the human eye, are labeled by the symbols B, G, and R.

Upon receiving the broadcast signal, a color TV set extracts the B, G, and R signals encoded in the video signal and feeds them to the picture tube, which controls the intensities of three separate electron beams. Figure 13-9 shows how the beams are directed toward the respective phosphor dots that pattern the face of the tube. A mask ensures that each beam strikes only its proper phosphor as the three beams simultaneously make a horizontal scan (Fig. 13-10). The phosphors producing blue, green, and red colors are

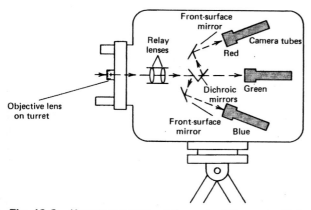

Fig. 13-8.   Key components within a color TV camera. Dichroic mirrors reflect blue light to one camera tube, red light to another, and allow the remaining green light to pass through to a third.

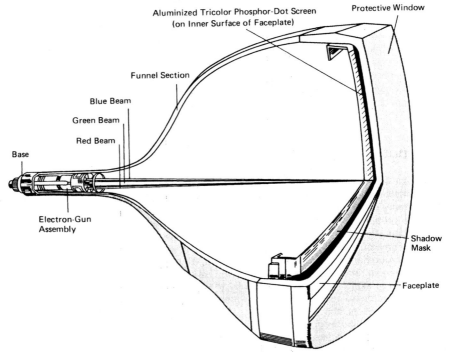

Fig. 13-9. Arrangement within a modern color TV tube. Electrons boiled off from a small metallic plate are accelerated forward in three separate beams that are focused so as to strike the proper color phosphor dot on the inner surface of the faceplate of the tube. (Courtesy of Radio Corporation of America.)

spaced sufficiently close together on the face of the picture tube that by partitive color mixing the eye detects only the net effect predicted by additive color mixing rules. Color Plate 4 shows a color TV picture at successively greater magnifications. In the last picture the array of blue, green, and red dots that form the picture is clearly visible. Different manufacturers apply the dots in various patterns to obtain the best performance with the type of control they use to guide the electron beams, but the principles of additive color mixing always govern the production of the color. On a typical screen there are more than 200,000 triads of phosphor dots.

The gamut of colors that can be produced by B, G, and R signals was shown in Fig. 3-7. Typical phosphors produce nearly saturated colors; this is shown in the chromaticity diagram by how closely B, G, and R lie to the boundary of the diagram. In comparison, the gamut obtained in four-color printing is much more restricted.

Luminance and Chrominance

According to the preceding discussion, what happens in the TV studio and TV tube of a receiver is a straightforward process of separating a scene into its B, G, and R primaries and displaying B, G, and R signals by additive mixing. However, what actually happens to these signals in preparing them for broadcast is the result of engineering compromises

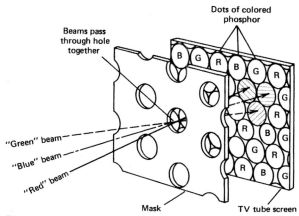

Dots of colored
phosphor

Beams pass
through hole
together

B G R
B G R
G
R
R B G
G R
R B G

"Green" beam
"Blue" beam
"Red" beam

Mask

TV tube screen

Fig. 13-10. The three electron beams in a color television tube are kept on their courses by passing through a hole in a mask, thereby ensuring that they strike the appropriate phosphor dots of the correct colors on the face of the tube.

that significantly influence the quality of the color presentation. In the United States the compromises were forced by two restrictions: the bandwidth of each TV channel had to remain at 6 MHz, and the format of the color broadcast had to be compatible with the existing monochrome system. The latter means that a color program must be viewable in monochrome on a monochrome set. Compatibility was achieved by a method based on the notion that the psychological attributes of color are hue, saturation, and brightness and that brightness is what is portrayed in a monochrome image. Thus a signal known as the *luminance signal* is formed by adding B, G, and R signals together in proper proportions as indicated by the relative sensitivity of the eye to short, intermediate, and long wavelengths (Section 6-1). The luminance signal is encoded by amplitude modulation of the carrier and is broadcast in the same way as for a monochrome program. Consequently it can be received and displayed by monochrome TV sets.

This leaves two remaining signals to be dealt with. The two are known collectively as the *chrominance signal,* since they describe the additional color information relating to hue and saturation. It is an interesting problem as to how they could be added to the luminance signal so as not to interfere with it and still be detectable by a color set. In addition, the bandwidth restriction is of central importance. A method suggested by P. Mertz and F. Gray in 1934 provided the solution. Of concern to us is the resulting specification that the highest frequency of one of the two chrominance signals (whose symbol is Q) must not exceed 1.5 MHz and the highest frequency of the other (whose symbol is I) must not exceed 0.5 MHz for the technique to work satisfactorily. These allowed bandwidths for Q and I are illustrated in Fig. 13-11 for comparison with the much broader bandwidth allocated to the luminance signal (Y).

Fine Color Detail

The reason these relatively narrow bandwidths are acceptable for the two components of the chrominance signal is that colorimetry experiments have shown that fine detail in color is not usually noticed by the average observer. It has also been found that the eye has better resolution for certain colors, particularly orange and cyan whose combinations

(a)
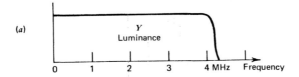

(b)
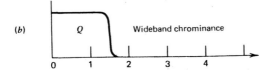

(c)
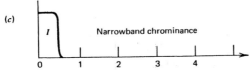

Fig. 13-11. Range of video signal frequencies that can be broadcast to describe (*a*) the variation of luminance across the TV screen as a line is scanned and (*b,c*) the variation in chrominance as described by the *Q* and *I* signals.

produce a wide range of skin tones. This resolution is shown by experiments in which an observer compares a number of small patches of different colors with a patch of similar size and luminance which is gray. As the observer moves away or the patches are made smaller, a point is reached where the blue patch becomes indistinguishable from gray. At greater distance this happens for yellow also, and at still greater distance orange and cyan also become confused with gray. Eventually all perception of hue is lost, but yet the patches may remain resolved to the eye as a result of their contrast in luminance with the background.

These results suggest that the color TV system would best match the resolution of the eye if the wideband chrominance component *Q* is dedicated to a combination of B, G, and R whose chromaticity can match the region of the chromaticity diagram along a line running from orange to cyan; and the narrow band chrominance component *I* would supply the additional information that together with *Y* and *Q* is needed to specify B, G, and R signals in the image. The TV set has electronic circuits that on receiving the broadcast add and subtract *Y*, *Q*, and *I* signals according to specified formulas, and the three resulting signals control the intensity of the electron beams striking the blue, green, and red phosphors.

Figure 13-12 shows the range of chromaticities matched by the full span of values of the *Q* signal and a different range matched by the span of values of the *I* signal. A curious feature of Fig. 13-12 is the fact that the *Q* and *I* lines are not at right angles. The *Y, Q,* and *I* signals would provide the largest gamut of colors if the *I* signal is defined so that the chromaticities it produces for *Q* = 0 and all possible values of *Y* fall on a line that is perpendicular to the *Q* line. In fact, if the chromaticity diagram is made into a three-dimensional color space in which the third axis perpendicular to both *x* and *y* represents *Y*, then the *I* and *Q* lines in this three-dimensional space are indeed perpendicular, and the greatest range of color can be reproduced.

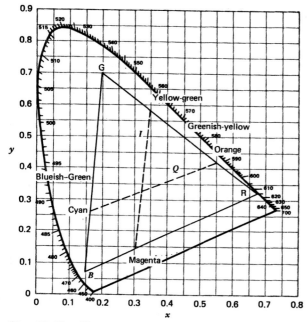

Fig. 13-12. Chromaticity diagram illustrating colors described by all possible values of the signals Q and I separately. These chromaticities fall within the gamut of the original set of R, G, and B primaries.

This section has described how the electrical signals representing the B, G, and R content of the scene being broadcast are combined by circuits that add and subtract the signals in various proportions to yield three independent signals: a luminance signal Y with the equivalent monochrome information determining brightness; a wideband chrominance signal Q that yields colors ranging from orange through white to cyan on the chromaticity diagram; and a narrowband chrominance signal I that combined with Y and Q produces the full gamut of the original B, G, and R primaries. The color TV set extracts the Y, Q, and I signals from the video portion of the broadcast bandwidth, adds and subtracts them in proper proportion to yield the original B, G, and R signals, and applies them to control the intensity of the three respective electron beams in the TV tube. For a typical screen size of about 50 cm, fine details involving features smaller than 1 mm are represented only by the luminance signal; details in the range of a few millimeters have color if their chromaticities fall near the Q-line on the chromaticity diagram; and larger scale features in the scene are represented in full color. The loss of color detail becomes easily apparent in a scene involving a large crowd, as during a sporting event.

SUMMARY

The image on a television screen is formed by a scanning electron beam that excites phosphorescent materials coated on the inside of the screen. In monochrome sets there is a single white phosphor and a single electron beam; in color sets, three beams excite an array of dots of three phosphors that are additive primaries. The scanning beam sweeps out the image of 525 lines in $\frac{1}{30}$ second, using two interlaced $\frac{1}{60}$ second scans so that the persistence of vision gives the impression of a continuous image as successive frames are presented. The image is considered to be composed of individual elements called pixels whose size determines the quality of the picture, just as grain size limits the resolution of a photograph. Thus the ultimate quality of the picture is limited by the bandwidth of the electromagnetic spectrum that is devoted to broadcasting its signal. The video signal is broadcast by an amplitude modulation method because the method is relatively efficient and simple, but the audio signal is broadcast by the more complicated method of frequency modulation to provide acceptable sound quality. In a color broadcast the signals specifying the intensity of the blue, green, and red primaries at each moment as the picture is scanned are encoded as the luminance signal and two chrominance signals. Only the luminance signal is decoded by a monochrome set.

QUESTIONS

13-1. What is a pixel? How many pixels comprise the image on the screen of a television set using the NTSC system?

13-2. Why can a TV antenna also serve to receive FM radio signals?

13-3. In a television presentation, how many fields are presented each second? How many frames?

13-4. How is an interlaced scan presented?

13-5. What is the bandwidth set aside for each TV channel? What bandwidth does the video portion occupy?

13-6. Explain the difference between a frequency modulated signal and an amplitude modulated signal.

13-7. What produces the color on the tube of a television receiver?

13-8. Compare the maximum saturation that may be achieved with color TV with four-color printing.

13-9. The "luminance" component of a color video signal represents what aspect of the scene?

13-10. The "chrominance" component of a color broadcast contains two separate signals. The level of each of the two signals designates a particular location on the C.I.E. chromaticity diagram. What colors correspond to the range of chromaticities indicated by the chrominance signal called Q and what is spanned by I?

13-11. Why is a wider bandwidth devoted to the chrominance Q signal than the chrominance I signal?

FURTHER READING

R. W. G. Hunt, *The Reproduction of Colour,* Third Edition (Fountain Press, London, 1975). Authoritative explanation of color reproduction techniques in television, printing, and photography.

There are many texts on the technical aspects of television. The two books listed below explain the details of the electrical circuits at the level of understanding needed for commercial or amateur broadcasting licenses.

G. J. Angerbauer, *Electronics for Modern Communications* (Prentice Hall, Englewood Cliffs, New Jersey, 1974).

R. L. Shrader, *Electronic Communication,* Second Edition (McGraw-Hill, New York, 1967).

From the *Scientific American*

N. Smith, "Color Television" (December, 1950), p. 13.

CHAPTER 14

LIGHT AND COLOR IN THE ATMOSPHERE

Blue skies, red sunsets, rainbows, twinkling stars, halos, brightly colored flowers and insects and innumerable other manifestations of light and color in the natural world have been a source of wonder since earliest history. Although many discussions of the origin of some of these effects exist in the early literature of many cultures, it was not until the seventeenth century that there began to emerge a systematic interpretation of these phenomena.

Most of the colors that we see in terrestrial objects—green leaves, flowers, etc.—result from selective absorption of sunlight in particular spectral regions, controlled by the pigments present in these objects, as we have discussed extensively in Chapters 2 and 5. Some iridescent colors seen, for example, in hummingbirds, ducks and beetles, depend on interference effects, which we considered in Chapter 4. A few particularly saturated flowers such as azaleas contain pigments that can fluoresce and emit more light in particular spectral regions than that present in the incident sunlight, as discussed in Chapter 5.

Colors observed in the atmosphere are somewhat more difficult to explain. For this reason we have waited until this chapter to discuss these effects since the reader has now had a broad exposure to the optical principles involved. Our discussion will be limited to a small selection of natural phenomena. For a much broader coverage we recommend the delightful book by M. Minnaert, *The Nature of Light and Colour in the Open Air* (see Further Reading).

■ **14-1 REFRACTION IN THE AIR**

When riding along an asphalt road on a hot day, we often see far ahead on the road's surface a shimmering that looks like a puddle of water, but as we approach the water disappears. Similarly in the desert the Bedouin may see what appears to be a blue oasis in the distance (Fig. 14-1), but it is something that he never reaches. These are examples of the same type of *mirage.* They are phenomena

Fig. 14-1. Mirage in the Mojave desert of California. (Richard Weymouth Brooks/ Photo Researchers.)

caused by refraction in the air. As explained in Chapter 5, refraction is the bending of the path of light at a boundary between media of differing refractive indices. But refraction also occurs within a single medium if its refractive index varies from one place to another. This nonuniformity is quite common in air that is heated unevenly. As a region of air is heated it expands. The resulting decrease in density leaves fewer molecules to influence the passage of light, and this decreases the refractive index. In this way we understand why a warmer region of air has a lower refractive index than a cooler one. This unevenness causes light to be refracted as it passes between regions of differing temperature.

Inferior Mirage

Shimmering on a road's surface and the blue oasis in the desert are examples of *inferior mirage*. Air coming into contact with a hot surface is warmed more than the air just above. Consequently the refractive index is lowest next to the surface and increases with increasing height. This characteristic can extend several hundred meters above the ground. Imagine that this region near the ground can be divided into distinct layers of increasing refractive index, denoted by n_1, n_2, n_3, etc., as shown in Fig. 14-2a. A light ray traveling downward from the sky will be refracted at each boundary between successive layers as illustrated, with the amount of bending determined by Snell's law (Eq. 5-2). In reality, of course, there are no distinct layers. Instead there is a continuous increase of density and refractive index with height. We can imagine this by making the layers more numerous and thinner as shown in Fig. 14-2b, until ultimately the variation becomes continuous, as in Fig. 14-2c. Then we conclude that there will be a continuous upward bending of light rays in the air over a heated surface.

Now we can understand the origin of the phenomenon of inferior mirage. In Fig. 14-3 as the rider looks ahead toward the hot road surface, he will receive light rays from the sky that follow paths such as the one labeled *A*, originating at point *M*, which are bent because of the uneven heating of the air. Because of the mind's process of visual projection, he will perceive at point *M'* the image of the subject from which the ray originates. By looking at the ground he sees the sky! Looking at the surface much closer ahead (toward *N*), the illusion is not seen, because of the limited amount of bending that is possible. This mirage is particularly strong on asphalt roads because the black surface absorbs light more efficiently than concrete and hence becomes hotter. Similarly, in the desert mirage (Fig. 14-1), the oasis of blue water is really the blue sky. The term *inferior mirage* is used to describe these effects because the image (the sky) is perceived at a position *beneath* the true position of the subject.

Images in an inferior mirage may appear stretched and distorted in the vertical direction according to how the air temperature varies with height. A portion of an image may even appear inverted. Figure 14-4a depicts the paths taken by several representative rays reflected from a telephone pole. First we notice that the ray *B* coming horizontally toward the observer takes a straight path because it travels through a layer of uniform refractive index. But ray *B'*, which left the same point *N* of the pole traveling downward, is refracted upward and also can be seen. So

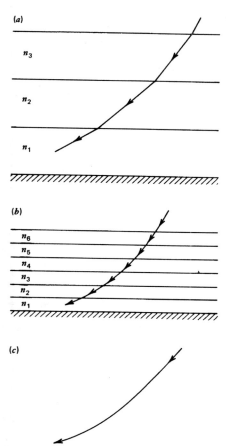

(a)

n_3

n_2

n_1

(b)

n_6
n_5
n_4
n_3
n_2
n_1

(c)

Fig. 14-2. Refraction of a light ray traveling downward into layers of successively warmer (and less dense) layers of air.

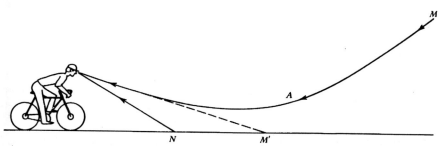

Fig. 14-3. The inferior mirage gives the illusion that the ray A originating at M seems to come from M'.

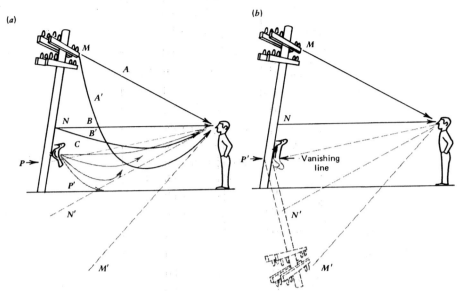

Fig. 14-4. (*a*) Hotter, less dense air near the ground causes rays to be refracted upward, so that a distorted upright image and an inverted image—the inferior mirage—are formed. (*b*) The inferior image meets the upright image at the vanishing line.

the same portion of the pole would be seen when the eyes are directed toward point *N* as downward toward point *N'*. More remarkably, ray *A'* leaving from a point *M* higher on the pole will be refracted by a correspondingly greater amount as it enters a layer of very hot air near the ground and appears to come from *M'*. The effect is to provide an inverted image, with some vertical distortion. Visual projection gives the impression that the top of the pole lies below the ground!

There is one height on the pole, labeled *P,* from which only one ray enters the eye. Rays originally leaving toward lower angles are refracted sharply upward and never reach the observer. This height marks the *vanishing line* below which no portion of the subject can be seen, because none of their rays reach the observer. For telephone poles seen further away, the vanishing line is higher, and very distant poles may not be seen at all!

Superior Mirage

A related but much less frequent type of mirage occurs over a cold surface such as the sea. If the air is generally warmer than the water, the air density and thus the refractive index will be greatest close to the surface where the cooling effect of the water is greatest. The refractive index in this case decreases with height. By repeating the analysis of the preceding section, we find that rays bend downward rather than upward. The result is called the *superior mirage,* and it can

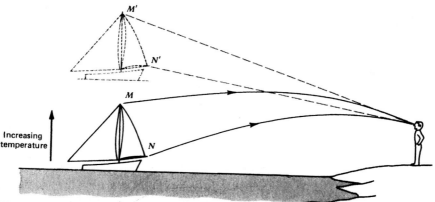

Fig. 14-5. When temperature increases rapidly with height, the paths of light rays traveling from a point on the sailboat to the viewer are bent downward, and an upright superior mirage appears.

appear in either of two forms. It is most commonly an upright image, as illustrated in Fig. 14-5. Because a superior mirage appears suspended in air the phenomenon is also called *looming*. When the effect is pronounced, ships below the geometrical horizon can be seen. The image may be vertically stretched or compressed, depending on the way in which temperature varies with altitude. Sometimes the sun itself appears in superior mirages just at sunset, when viewed over the sea. As the sun sets, a segment near the top may appear to become detached from the rest of the sun's disc and hover over the water after the rest has disappeared.

Certain conditions of the atmosphere produce combinations of superior and inferior mirages. Figure 14-6*a* shows a situation where air temperature decreases with height except within an "inversion layer" where it increases with height. Therefore light rays are refracted upward everywhere but within the inversion layer where they refract downward. Thus it is possible for rays from the top of the mast of the boat to take three different paths and arrive at the observer. Figure 14-6*b* shows the three corresponding images: an upright inferior mirage, an upside down superior mirage, and an upright superior mirage. The latter appears only if the temperature in the inversion layer increases more rapidly with height at greater heights, so that the topmost ray traces a more sharply curved path than the others.

Castles in the Air and Monsters in the Lake

Occasionally, superior mirages of the opposite shore can be seen over the surface of a cold lake. The mirage is usually broken into a series of uneven roughly rectangular shapes by the local variations in temperature that may refract light horizontally as well as vertically. The mirage is seen hanging in the air in the form of buildings or castles. This was often seen by mariners sailing the strait between Sicily and the Italian mainland, where hot air from the land flows over the cool

(a)

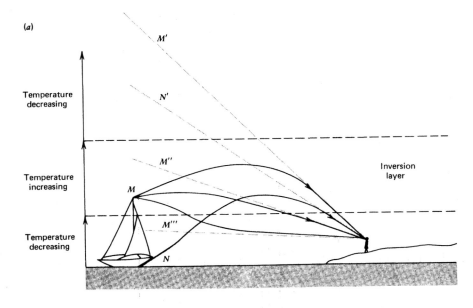

(b)

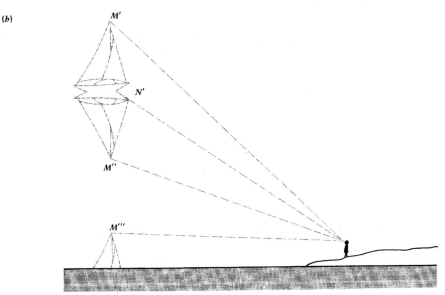

Fig. 14-6. (a) Rays curving downward in an inversion layer where temperature increases with height and curving upward where it decreases with height; (b) the resulting two superior and one inferior mirages.

Fig. 14-7. Fata Morgan over Arctic ice. (Robert Greenler)

water. As the vagaries of wind and water change the pattern of temperature, the columns change shape and shift around. This effect is known as "castles in the air," or "Fata Morgana" in Italy, named after the fairy Morgan le Fey who according to Italian tradition lived in a palace beneath the sea. Similar effects can be seen over ice (Fig. 14-7).

The sighting of "lake monsters" in Scotland and elsewhere may very well be a superior mirage. Steep-shored lakes such as Loch Ness encourage the accumulation of cold air just over the surface. When temperature does not increase uni-

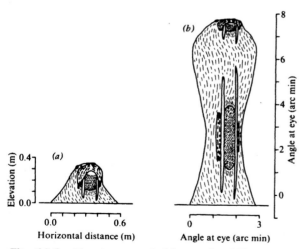

Fig. 14-8. Monsters reported by medieval Norse seafarers may well have been a superior mirage that distorted the appearance of a walrus: (*a*) subject; (*b*) image. (By W. H. Lehn and I. Schroeder with permission of *Nature.*)

formly with height, a single point on an object such as a stick protruding at an angle from the water can be the source for several rays that enter the observer's eye at different vertical angles. This smearing of the image can produce fanciful shapes (Fig. 14-8).

Lingering Sunset

There is a phenomenon related to the superior mirage that has nothing to do with air temperature but depends primarily on gravity. The blanket of air held close to the earth's surface is comparatively thin, and within it the density of air generally decreases with increasing altitude. Consequently, as light from the sun penetrates the atmosphere it is refracted downward. This can be quite pronounced when the sun lies near the horizon so that its light travels through an appreciable distance within the atmosphere. Seeing the image of the sun, our backward projection gives us the impression that the sun is higher in the sky than it actually is! As illustrated in Fig. 14-9, when we see the sun touch the visible horizon it actually lies below the actual, or geometrical, horizon.

Twinkling Stars

The expansion or contraction of air when it is locally heated or cooled gives rise to vertical movement and winds. A lighted candle is a good example, as we are reminded when holding our hand above the flame and feeling the hot, rising air. The marked change in refractive index between regions of hot and cold air produces intriguing optical effects. One is the everchanging image of a subject seen through the air just over the flame. The image appears to shimmer, dancing back

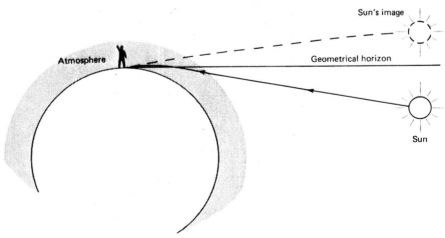

Fig. 14-9. Refraction in the atmosphere makes the sun appear higher than it actually is.

and forth as its shape is distorted in fanciful ways according to the pattern of hot air movement. This lenslike effect of air is similar to what produces the mirages, but the light paths may be curved in both vertical and horizontal directions.

A similar but more subtle effect occurs when we look upward through the air at a star. Air above us is in continual motion, bringing across our line of sight a shifting pattern of warmer and cooler regions. Refraction bends the star's rays ever so slightly, and this movement of its image on the retina we see as "twinkling." The scientific term is *scintillation*. It is particularly conspicuous for stars near the horizon. Dispersion may also create subtle changes in color. This movement of the image degrades the sharpness of a photograph when made through a telescope over a period of seconds or minutes. For this reason astronomers prefer to place their observatories on high mountains to have less air between them and the stars. (A major improvement will soon take place with the establishment of a space telescope.) For subjects having larger images, such as a planet's, the size is sufficiently large so that the slight movement of the image is not perceptible to the eye. Planets appear quite steady, and the nearby ones are easily distinguished from stars in this way. Twinkling can also be seen for streetlights and other small sources of light at a distance.

Another more subtle effect also contributes to the twinkling of a star. There is an actual fluctuation in the brightness of the star's image on the retina. This effect is related to the pattern of shadow bands seen on the bottom of an outdoor swimming pool in bright sunlight. These complicated patterns move about as the changing pattern of waves refracts light so that it is concentrated in differing regions. A similar phenomenon is produced by the atmosphere because of fluctuations in its air density.

■ 14-2 REFRACTION BY WATER DROPLETS AND ICE

The Rainbows

The spectacular multicolored sweep of the rainbow has been a source of endless curiosity since the beginning of history. Most often before or after a rainstorm one bow is visible, and it is called the *primary bow*. Occasionally two bows appear, one outside the other (Color Plate 33), and the larger one is called the *secondary bow*. The bright inner primary bow is violet on the inside, with hue changing continuously through the spectrum to red on the outside. The secondary bow is red on the inside and violet on the outside. Occasionally, there may be extra partial bows inside of the primary bows, known as *supernumerary bows*. The region between the primary and secondary bows is darker than the rest of the sky and is called *Alexander's dark band* after the Greek philosopher Alexander of Aphrodesia who first described it around 200 A.D. All of these features are visible in Color Plate 33, except the supernumerary bows.

The positions of the primary and secondary bows were first measured in 1266 by Roger Bacon. The primary and secondary bows are each readily seen to be

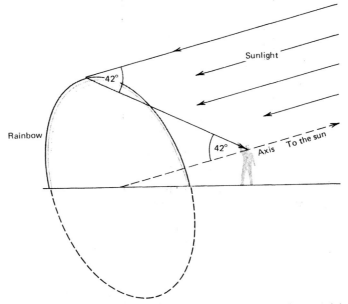

Fig. 14-10. Rays reaching the observer from the primary rainbow make an angle of 42° with respect to the incident sunlight, and arriving at the observer's eye make the same angle with respect to the axis between the observer and the sun.

part of a circle whose center lies on the line or axis directed from the sun through the observer, as depicted in Fig. 14-10. Rays reaching the observer from the primary bow make an angle of approximately 42° with this axis. Similarly, rays from the secondary bow make an angle with the axis of 51°. When the sun is just at the horizon, an observer standing in an open field with the sun behind him can see half of the rainbow's full circle. If the sun is higher in the sky, less of the arc is visible; when the sun's elevation is greater than 42°, the primary rainbow can no longer be seen.

To understand the phenomenon of the rainbow we must first explain how a bright arc of light is formed and then explain the separation of its colors. Aristotle proposed that the rainbow is caused by the reflection of sunlight from clouds, which is close to the truth except that the bow can be seen even without clouds present. In 1304 Theodoric of Freiberg suggested that the origin of the rainbow is in the reflection of light rays inside of the tiny drops of water present in the air after a rainstorm. The drops are spaced too far apart to be seen as a cloud but can remain airborne if smaller than about 4 mm in diameter. Those larger than about 0.01 mm contribute to the rainbow, while smaller ones produce mainly diffraction. Theodoric was able to demonstrate the effects with a spherical flask filled with water, which behaved like a giant raindrop. Some 300 years later, his results were rediscovered by Descartes who carried through the mathematical analysis that is the basis of our understanding of the rainbow.

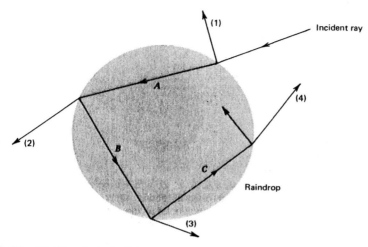

Fig. 14-11. A single light ray meeting a raindrop produces a series of rays by refraction and reflection.

The analysis begins with the ray optics of a single raindrop, assumed to be a perfect sphere. We consider a single ray of sunlight that is incident on the drop (Fig. 14-11). When the ray first strikes the drop, part of it will be reflected (1) and the rest refracted into the drop (A) at an angle determined by Snell's law. When this refracted ray reaches the far surface of the drop, part of it will be refracted out through the surface (2) and the rest reflected back inside (B). When ray B reaches the surface, part of it will be refracted through the surface (3) and the rest again will be reflected (C). Thus, a series of rays (1, 2, 3, etc.) of rapidly decreasing intensity emerges from the drop.

Next, consider just rays of type 3. The direction at which the ray leaves the drop—relative to the direction of the incident ray—can be calculated by the methods of ray optics discussed in Chapter 8. As shown in Fig. 14-12a, the direction is easily seen to depend on a single factor: the distance between the center of the drop and the original path of the incident ray. For the ray A in Fig. 14-12a whose original path is headed straight along the axis toward the center of the drop, the reflected ray A' will head straight back along the incident ray's original direction. Ray B, which is slightly off axis, is reflected as B' at a small angle b with the axis. Ray C, which is farther off axis, is reflected as C' at the larger angle c. For rays increasingly farther off axis, the angle of deviation continues to increase. But at a particular distance off axis (roughly shown by ray D) the angle of deviation reaches a maximum value, indicated by d. For rays that arrive yet farther off axis such as E, the reflected ray emerges with an angular deviation e, which is *less* than d.

Although rays of class 3 emerge from the raindrop at all angles between 0° and d, careful tracing of the rays shows that there is a "bunching" of many of them close to the maximum angle d. Thus, there will be a particularly strong reflection of incident sunlight at an angle d to the incident direction. The angle d turns out to

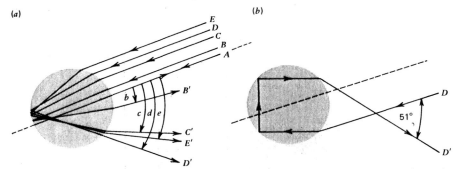

Fig. 14-12. (*a*) A group of parallel rays (*A, B, C, D,* and *E*) incident on a drop of water at differing distances from the axis through its center. Only rays that exit after having been reflected once within the drop are shown. Of all the rays, *D'* emerges at the greatest angle from the axis (42°) and together with its neighbors is responsible for the primary rainbow; (*b*) Incident rays that are reflected twice within the drop emerge with a minimum angle of 51° (ray *D'*) and form the secondary bow.

be 42°. There is also a bunching around 0°, for rays reflected back directly toward the sun, and these are responsible for some of the other phenomena to be discussed later.

The origin of the primary bow is just this bunching of type 3 reflected rays at an angle of approximately 42°. Referring to Fig. 14-10, imagine that the air is filled with droplets. The sun is so far away that its rays are parallel as they come into this region. Those droplets which are located so that sunlight reflected from them at an angle of 42° can reach the observer will be seen by their reflected light and will thus be perceived as forming a bright arc in the sky. Rays that do not participate in the bunching (*A, B, C, E* in Fig. 14-12*a*) will all be reflected at angles less than 42°, and so we must look *lower* in the sky to see them. They will give rise to the generally higher level of luminance in the region of sky inside of the primary bow, which can be clearly seen in Color Plate 33. Since no rays of this type reflect at angles greater than 42°, the outside of the bow should be darker than the inside as it is seen to be.

A similar analysis of rays of class 4 in Fig. 14-11 shows them to be the origin of the *secondary bow*. Again there is a bunching of reflected rays, at an angle of 51° (Fig. 14-12*b*). This time, however, the bunching occurs at a minimum rather than a maximum angle to the axis. Those rays not included in the bunching will be reflected at angles greater than the bow-angle of 51°; therefore the atmosphere will be brighter on the outside and darker on the inside. Since there are no rays of either type 3 or 4 with angles between 42° and 51°, the region between the primary and secondary bows should be darker than the rest of the sky. This is the origin of Alexander's dark band.

Descartes' analysis of the optics of raindrops was able to explain the primary and secondary bows and Alexander's dark band. It remained for Newton to show that dispersion—the small change in refractive index of the water with wavelength—would cause the maximum angle of deviation of 42° (and the minimum angle of 51°) to be slightly different for each wavelength so that the bow is spread out into a continuous sequence of colored bows forming a spectrum. Since short wavelengths are refracted more than long, the inside of the primary bow is violet and outside is red, whereas for the secondary bow the order of colors is reversed.

This Descartes–Newton optics of rainbows successfully explained all the major features of this remarkable natural phenomenon. But some of the details, including supernumerary bows lying inside the primary bow and the distribution of intensity along a rainbow, were not explained until quite recently.

The rainbow is one of a large group of related natural phenomena that result from refraction and dispersion of light by airborne particles. Other examples caused by water droplets are fog bows and dew bows. Ice crystals give rise to a similar effect called the halo, as we shall explain shortly.

Heiligenschein

Dew on the grass early in the morning produces a curious optical effect if you know where to look for it. Strictly speaking this is not an atmospheric phenomenon, but it does illustrate the principles explained in the preceding section. If you stand with your back to the sun and look near the edge of the shadow of your head, the ground appears somewhat brighter than elsewhere (Fig. 14-13). This effect is given a German name, *heiligenschein* (pronounced hī′lĕ·gĕn·shīn), meaning halo. In fact, its origin is quite different from that of the atmospheric halo. Heiligenschein is

Fig. 14-13. The heiligenschein seen by the photographer around the shadow of his head on a dew-covered lawn. (Robert Greenler.)

due to reflection from the front and back surfaces of the droplets of dew. Because these droplets are nearly spherical, the rays from the sun that head straight toward the center of a droplet are reflected straight back (ray *A* in Fig. 14-12). Similarly, rays that are slightly to the side (ray *B* in Fig. 14-12) are reflected nearly straight back. Thus there is a bunching of rays so that a considerable proportion of the light is reflected back within a few degrees of the direction of the sun. These rays provide the heiligenschein around your head's shadow. Light that strikes a droplet farther from the axis through its center is reflected and refracted into a wide range of angles and just provides a more uniform, weaker intensity in these directions. If you look at the shadow of someone else you see this latter light, and there is no heiligenschein because your eye is not intercepting the backward reflected light from that region of the grass. Although having your personal halo may give illusions of grandeur, unfortunately no one else can see it!

Halos

Anyone who looks carefully at the sky has occasionally seen a circular luminous ring around the sun or moon when a thin, high cloud intervenes. This is called the *halo*. Tradition says that it is a forerunner of rainy weather. Because the halo is caused by refraction from ice crystals, its connection with impending bad weather can be explained simply. The warm air associated with an oncoming, rainy region will rise over the preceding cooler air because its lower density gives it buoyancy. As it pushes upward, it reaches altitudes in excess of 10 kilometers where the temperature is sufficiently low to freeze the moisture it contains and form clouds of ice crystals. When such a cloud in front of the sun or moon is sufficiently thin to transmit appreciable light, a halo forms in the cloud (Fig. 14-14). The most common halo is merely white, but others are brightly colored, with red on the inside and blue on the outside. The inner edge whose radius subtends an angle of 22° is rather sharp, but the outer edge is diffuse, and the sky has somewhat greater luminance outside the halo.

The shape of ice crystals is responsible for these features of a halo. Although ice crystals may form in a variety of shapes depending on how they begin to grow and the nature of air currents, only a few shapes predominate. All share a common feature: a hexagonal (six-sided) cross section. Figure 14-15*a* shows one type whose length greatly exceeds its width and is known as a *pencil crystal* because of its resemblance to the common wood pencil. Another type has the opposite extreme shape where the length is much shorter than its width, and it is called a *plate crystal*. The 22° halo comes from light refracted from pencil crystals. If a crystal should be oriented so that a ray of sunlight passes through alternate sides of the hexagon, as shown in the first illustration in Fig. 14-15*b*, its path is refracted once on entering the crystal and again on leaving. It is a simple matter of using Snell's law to show that because these two sides make an angle of 60°, the least possible total deflection is 22°, for ice whose refractive index is 1.31. This minimum deflection is found when light enters and leaves at the same angle to the surfaces. Other incident angles, either smaller or larger, as illustrated in the bottom

Fig. 14-14. Common halo surrounding the sun, formed by ice crystals in a high cirrus cloud. (Copyright Alistair Fraser.)

two illustrations of Fig. 14-15*b*, result in greater deflection. The consequence of this minimum deflection is simply that when crystals are randomly oriented as in a cloud there is a bunching of rays at the angle of 22°. In a cloud, pencil crystals whose lengths exceed approximately 20 micrometers are too heavy to be supported by the random buffeting of air molecules. In still air they fall slowly and tend to become oriented with their long axes all horizontal, and this helps to intensify the effect.

Figure 14-16 shows how the 22° deflection angle produces a luminous circle of the same angular radius when we view an appropriate cloud in front of the sun. The region enclosed by the halo appears dark because light cannot be deflected by smaller angles. The region outside has greater luminance because of the haphazard orientations of the ice crystals and larger deflection angles that are possible. Since dispersion causes long wavelength light to be refracted less than short, the inner edge of the halo is red. The colors are less saturated toward the outer edge because some of the red through green portion of the spectrum is also refracted by these greater angles.

Occasionally a much larger 46° halo may be seen as well. This is caused when light passes through one hexagonal-shaped end and one of the adjacent sides, which meet at a 90° angle. It is generally much dimmer because light must enter or leave the small end, and thus the crystal does not intercept much light. Also the

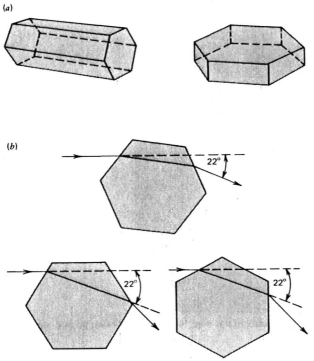

Fig. 14-15. (*a*) Common types of ice crystals in high clouds: the pencil and the plate. (*b*) Crystals with differing orientations refract light by varying amounts but never less than 22°.

incident angle must be large and consequently much more intensity is lost by reflection at the surface. When atmospheric conditions are special, making ice crystals better aligned, other phenomena may also appear. In very still air, as occasionally found in Arctic regions, one of the flat side faces of pencil crystals tends to remain horizontal. *Perry arcs,* described by W. E. Perry during his voyage in 1919–1920 in search of a northwest passage, are caused by rays that are only 3° from the horizontal. Other fascinating effects—with the names of sun dogs, parahelic arc, circumscribed halo, and circumzenith arc—are described in the readings given at the end of this chapter.

■ 14-3 DIFFRACTION BY AIRBORNE PARTICLES

The preceding discussion has shown how the rainbow and halo can be understood in terms of the concepts of ray optics. Some other phenomena, however, are due to the wave aspects of light. Such is the case when light is affected by water drop-

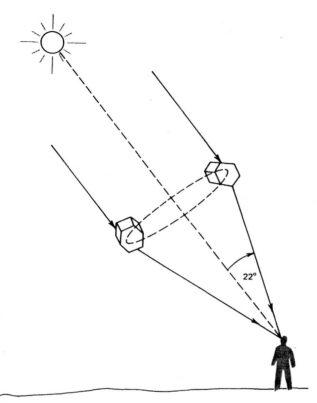

Fig. 14-16. The 22° halo is caused by refraction of sunlight or moonlight from ice crystals of pencil shape oriented in approximately the directions shown.

lets whose size is somewhat larger than the wavelength of light. More specifically, we shall limit our attention to droplets whose diameters range from about 5000 nm to 20,000 nm. Then the phenomenon of *diffraction* (described in Sections 4-6 and 4-7) is important, and we can no longer deal with the situation in terms of ray optics.

Corona

When moonlight or sunlight is incident on a droplet of this size, the light diffracted away from the original direction produces a pattern that is identical to that produced by a circular hole of the same size in an opaque screen. (This remarkable equivalence is known in optics as Babinet's principle.) In Section 4-6 we saw that the diffraction pattern produced by a small circular hole consists of a series of bright rings (Fig. 4-14); the angular sizes (in radians) of the rings are given in Table 4-2. The first bright ring away from the forward direction has an angular size $\alpha =$

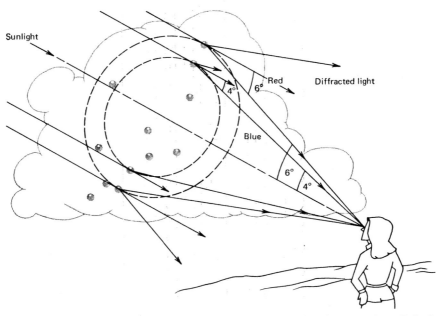

Fig. 14-17. Formation of a corona when droplets in a cloud diffract sunlight. Only the paths taken by red and blue diffracted light for the innermost ring are shown.

$1.64\lambda / d$. If the diameter of the droplet is 0.01 mm 10,000 nm), then blue light with $\lambda = 400$ nm is preferentially diffracted by an angle of $\alpha = 0.066$ radian or about 4°. And red light with $\lambda = 700$ nm is preferentially diffracted by $\alpha = 0.11$ radian or about 6.5°. Thus the first bright ring will be spread out to form a spectrum with blue on the inside and red on the outside, ranging in angular size from 4° to 6.5° (Fig. 14-17). The spectral sequence from blue through red repeats in each successive ring. The size of the rings is reduced if the cloud has larger droplets and is increased for smaller droplets.

These sequences of colored diffraction rings can be seen when a thin cloud passes in front of the moon or the sun. This phenomenon is called the *corona* (literally, crown). The most common form of corona is seen before the moon as a comparatively bright disk of white light which becomes blueish at the edge of the moon (the counterpart by Babinet's principle to Airy's disk when light passes through a small hole). The tenuous edge of the disk is reddish, because long wavelengths are diffracted by greater angles than short wavelengths, and around the disk may appear the successive colored rings that were just explained. Sun coronas are much more intense, but in order to observe them the bright disc of the sun must be blocked out to avoid dazzling one's eyes. This can be done with one's hand or the edge of a building or other obstruction (Color Plate 34).

The quality of the corona depends primarily on the nature of the cloud. If the droplets are of different sizes, the overlapping of different patterns tends to wash out the individual rings and only a faint inner ring with desaturated colors is seen. If the droplet size is nearly uniform, the corona is well developed with saturated colors, and several rings may be visible. Typically, the size of the corona is from 1° to 5°. The most beautifully developed coronas occur in thin cirro-cumulus or cirrus clouds, the latter containing tiny ice crystals that can also produce this diffraction effect.

Finally, we note that the colored diffraction rings called the corona, which originate in the earth's atmosphere, should not be confused with the solar corona, which is the outer layer of the sun's atmosphere that can be seen during a solar eclipse.

Glory

The *glory* is another effect of diffraction by atmospheric water droplets and is caused by the backward diffraction of light. Until the advent of the airplane it was necessary to climb to a mountaintop to see it. When standing with back toward the setting or rising sun, the observer should look toward his shadow projected onto a nearby cloud (Fig. 14-18). Under favorable conditions, when the cloud contains droplets of a fairly uniform size, a series of colored rings surrounds the shadow, and this is called the glory. Anyone of course can see another person's shadow on the cloud, but only that person (or someone standing very close) can see another person's glory. An airplane trip provides a much more likely opportunity to see the glory (Color Plate 35). Look for the shadow of the airplane when it appears on a cloud with a fairly smooth top and you may be fortunate enough to see the striking diffraction colors.

Red
Green
Blue

Fig. 14-18. The glory is created by backward diffraction of sunlight from droplets in a cloud.

■ 14-4   SCATTERING BY AIR AND AIRBORNE PARTICLES

If the thin layer of air blanketing the earth were truly uniform in density and did not contain tiny particles of solid or liquid matter, we would see quite a different world. Of course there would be no clouds, but less obvious is the fact that the sky would appear black and sunsets would be devoid of color. These colorful effects result from *light scattering* in the air. In Section 4-7 we discussed how light is scattered from tiny particles. This may be important in the atmosphere, because bits of matter of a few micrometers or less in size can remain airborne for many days or even months before eventually settling out or being removed by rain. However, even in the absence of such particles the gases of the air would scatter light. How this happens is subtle but fascinating.

The molecules that make up the earth's atmosphere are mainly nitrogen and oxygen. The electrons of these molecules, and of the many others which are present in smaller concentrations, absorb principally in the ultraviolet spectral region and do not absorb appreciable visible light. However, when a molecule is illuminated by light that is not absorbed, its electrons can still be driven to oscillate slightly at the frequency of the incident electromagnetic light wave, as we explained in Section 5-2. The oscillating electrons then become a weak source of radiation, reemitting light at the frequency of the driving light wave. Most of the incident light goes right past, but a small fraction is scattered.

The fraction of the incident light that the molecule scatters depends on the frequency (or wavelength) of the light. Short wavelengths are scattered more effectively than long, just as for scattering by small particles. A calculation by Lord Rayleigh shows that the scattering efficiency is proportional to $1/\lambda^4$. This result, known as the *Rayleigh inverse fourth power law,* shows that blue light ($\lambda = 400$ nm) is scattered about 10 times more efficiently than red light ($\lambda = 700$ nm).

Blue Sky

It is just this Rayleigh scattering that produces the blue color of the sky. Sunlight at all visible wavelengths penetrates the earth's atmosphere. When we look up at the sky, we observe sunlight that has been scattered by the molecules of the air, and because of the $1/\lambda^4$ law the blue light dominates. Note that an observer outside of the earth's atmosphere (an astronaut in orbit) looking upward sees black.

Although molecular scattering is the fundamental reason for the blue sky, how it comes about is more subtle than just described. The distance between molecules in the earth's atmosphere is typically only about 50 nm, which is much less than the wavelength of light. If the air density were completely uniform, the light scattered from different molecules would tend to interfere destructively, drastically reducing the scattered intensity. However, since the gas molecules are in constant motion, over small regions the density of the air fluctuates in a random way because at any moment there happen to be more molecules in some places than others. Stronger scattering from those regions with momentarily higher density largely eliminates the effects of destructive interference.

In liquids, where the molecular separation is typically 1 nm (much smaller than in gases), density fluctuations are much less pronounced and are correspondingly less effective in overcoming the effects of destructive interference. The scattered intensity per molecule is therefore much smaller in liquids than in gases. In solids where molecular motion is severely limited the scattering is extremely weak.

Haze

This same process of Rayleigh scattering, but from small airborne particles, also accounts for the blueish color of a puff of smoke from a sawmill or cigarette and the blue haze seen when viewing distant mountains in clear weather. As shown in Fig. 14-19 a distant mountain is seen through a long path of air, all of which scatters light, which is seen superimposed on the mountain. The further the mountain the bluer is its color (Color Plate 36), an effect known to painters as *aerial perspective*. It is frequently employed in painting to provide a sense of distance: the greater the distance, the bluer the appearance of subjects on which it is superimposed.

As long as the particles in air are small (with diameters less than the wavelength of light) they will scatter in accordance with Rayleigh's $1/\lambda^4$ law and add to the blue molecular scattering of air molecules. As particle size increases, however, the dependence of scattering efficiency on wavelength gradually weakens. For particles whose size is comparable to or somewhat larger than the wavelength of light, diffraction becomes more important than Rayleigh scattering. When the particle size is much greater, reflection and refraction describe the process of scattering. Figure 14-20 shows how the pattern of scattered light varies with the size of the particle. For large particles, there is little difference between the refraction and reflection efficiency for short and long wavelengths, so the scattered light looks essentially white, the same color as the incident light (Fig. 14-21). This whiteness

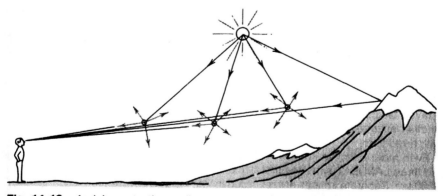

Fig. 14-19. Aerial perspective: a distant mountain is seen through the blue light scattered by air along the line of sight.

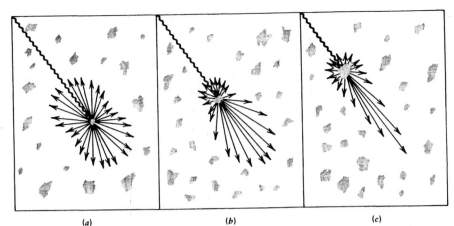

(a) (b) (c)

Fig. 14-20. (*a*) When the incident light is not polarized, small particles scatter light into every direction, but the lowest intensity is in directions at right angles to the beam; (*b*) particles comparable in size to, or somewhat larger than the wavelength of light tend to scatter light preferentially in directions that make only a small angle to the beam; (*c*) large particles scatter light by very small angles from the beam, so that the direction is only slightly changed.

is familiar in mist and fog that are caused by water droplets suspended in the air. It is also seen in heavily polluted air.

In the summer, the blue sky appears most saturated after a heavy rainstorm when the dust has been "washed" from the air leaving only air molecules to scatter. In dry weather, the amount of dust in the air gradually increases leading to an increase in the brightness of the sky. But since the dust particles are often large, their scattering will add white to the blue molecular scattering and produce a marked desaturation of the blue. This notion that particles suspended in the air produce the brightness of the daytime sky was originally formulated in the early eleventh century by Al Hazen of Basra, who carried out most of his scientific work in Cairo.

The "haze" filter used over the lens of a camera permits greater clarity for distant views. It is transparent through most of the visible but suppresses transmission in the far blue and ultraviolet to reduce the effect of scattered light. Polarizing filters may also be useful for reducing scattered light in certain directions if the sun is in a favorable position. The polarization aspects of skylight will be discussed below. Another alternative for reducing the effect of scattered light is use of infrared film with a red filter.

Even more effective in penetrating haze are some of the images recorded by the three LANDSAT satellites that circle the earth in polar orbits at an altitude of 1000 km. Separate images are recorded in four ranges of wavelength: 500 to 600, 600 to 700, 700 to 800, and 800 to 1000 nm. The first is particularly useful for indicating features of sediment-laden water; the second for cultural features such

Fig. 14-21. Light scattered by large airborne particles is white, nearly the color of sunlight. The greater amount of air in front of more distant mountains makes them more difficult to see in comparison with the greater amount of scattered light.

as villages, roads, and cities; the third for the distribution of vegetation; and the last is most effective in penetrating atmospheric haze and delineating landform.

Red Sunsets

The Rayleigh scattering of blue sunlight by the atmosphere also accounts for the redness of the sun at sunset. As we can see in Fig. 14-22, the distance that sunlight travels through the earth's atmosphere is shortest at noon and is longest at sunset when the sun is at the horizon. Since blue light is scattered by the air much more efficiently than red light, sunlight reaching the observer through a long atmospheric path will be deficient in blue light, and relatively rich in the long wavelength components that are less strongly scattered. The resulting red appearance of the setting sun becomes even more pronounced when the atmospheric scattering is enhanced by the presence of small dust or smoke particles. This is most remarkable following volcanic eruptions. The 1883 eruption of Krakatoa in Indonesia produced spectacular sunsets around the world for more than a year.

Several of the effects we have just discussed can be visualized in a demonstration that we can set up. Light from a slide projector passes through a glass water tank and is focused on a distant screen. The water in the tank contains a small amount of hydrochloric acid mixed with photographic "hypo" (sodium thiosulphate, $Na_2S_2O_3$). Initially, there is only very weak scattering by the water and the inevitable suspended dust particles. After a few minutes, a chemical reaction between the acid and the hypo begins to release free sulfur, which forms tiny grow-

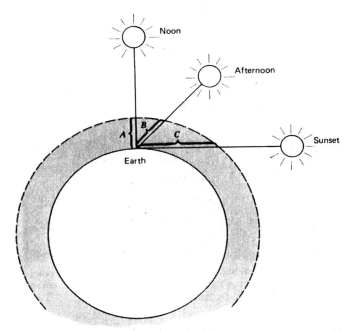

Fig. 14-22. The distance that sunlight travels through the earth's atmosphere to reach an observer at sea level is shortest at noon (*A*), increases throughout the afternoon (*B*), and is longest at sunset (*C*).

ing sulfur particles in the water. Since these particles are initially very small, they scatter according to the $1/\lambda^4$ law and the light scattered out of the beam as it crosses the tank is distinctly blue. As the size of the particles continues to increase, they pass out of the range of the $1/\lambda^4$ law and the scattered light becomes nearly white, resembling a mist or fog. While the sulfur particles grow and the scattering becomes more and more intense, the transmitted beam becomes increasingly deficient in blue light. Watching the circle of light on the screen gradually redden, the viewer can see first hand what happens to sunlight as the sun sets.

Green Flash

Just as the top of the sun is about to disappear below our visual horizon, a green flash is occasionally seen from the rim. It lasts only a second or two, and few people are sufficiently attentive to notice it. The green flash may also be seen at sunrise, but it is harder to judge when and where to look. The green flash is made possible by dispersion in the refractive index of the atmosphere. This causes short wavelengths of light to be refracted more than long wavelengths.

We can understand the origin of the flash by thinking of what we see of the sun as comprised of three images: one for the red portion of the spectrum, one for the

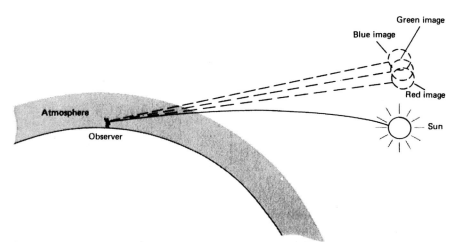

Fig. 14-23. Blue, green, and red images of the sun at sunset are displaced upward by differing angles owing to dispersive refraction in the atmosphere.

green, and one for the blue. So, according to our discussion of the lingering sunset in Section 14-1, dispersion causes the blue image to lie higher in the sky than the green and red images (Fig. 14-23). At sunset we do not see the blue image because most of that light is scattered away. Furthermore, much of the yellow portion of the spectrum is absorbed by water vapor, whose effect is appreciable only because of the very long distance the light is traveling through the atmosphere. This nicely separates the green image from the red image, so that as the sun sets and the bright disk disappears it is possible to see the weaker rim of the green image as a saturated color, and this is the green flash. It is best seen when the variation of air temperature with altitude is appropriate to magnify the angular size of the green image so that it is large enough to be distinguished by the eye.

Polarized Skylight

Figure 14-24 is a "snapshot" of a tiny dust particle in the air when light is incident on it. The electric field of the electromagnetic wave (whose strength at each position is indicated by the length of the heavy arrows) forces the electrons in the particle to oscillate back and forth so that they are displaced from their normal atomic orbits at the same frequency as the light wave. The acceleration of these electrons in turn launches another electromagnetic wave that we call the "scattered light." This light is scattered outward in nearly all directions. The only exception is the direction of acceleration of the electrons, that is, the direction of the electric field of the incident light (indicated by the arrows). This one restriction on the direction of scattered light follows from the fact that the strength of the electric field of the wave scattered in any direction is proportional to the length of the electric field arrow seen from that direction. Thus the intensity of the scattered light is

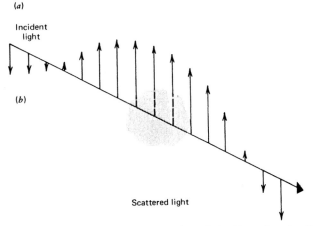

(a)

Incident
light

(b)

Scattered light

Fig. 14-24. The electric field of sunlight (here for simplicity shown polarized vertically) forces electrons bound to atoms in an airborne particle to oscillate in the direction
parallel to the field.

greatest along directions *perpendicular* to the arrow (in all horizontal directions)
and is zero in the two directions along the arrow. Light scattered from air molecules
has the same directional property.

Skylight is polarized for this reason (Fig. 14-25). The polarization direction of
light that is scattered from a particular region of the sky must be perpendicular to
the directions of travel of both the incident and scattered light (light is a transverse
wave!). From the figure it is clear that these conditions require the polarization to
be perpendicular to a plane that passes through the location of the observer, the
location in the sky being observed, and the position of the sun.

For example, if the observer faces south in the morning with the sun off to the
left and views the horizon, the scattered sunlight is vertically polarized. Figure 14-
26 shows that much more vertically polarized light than horizontally polarized light
comes from the sky under these conditions. If our observer looks directly overhead
during the early morning, the light that is scattered down is polarized in the north-
south direction. On a clear day, take polarizing sunglasses outside and confirm that
this is so! If you look in directions closer toward the sun the percentage of polarization decreases. That of course is because sunlight itself is unpolarized and so
there is as much scattered light polarized in the horizontal direction as the vertical
direction. Figure 14-27 summarizes the state of polarization of light coming from
various positions in the sky at sunrise.

There is evidence that the Vikings took navigational advantage of polarized skylight on overcast days in their voyages out of sight of land between Iceland, Green-
land, and Newfoundland. The Danish archeologist Thorkild Ramskon has interpreted the "sunstones" mentioned in medieval sagas as birefringent crystals such
as cordierite, which can be found as pebbles on the Norwegian coast. These could

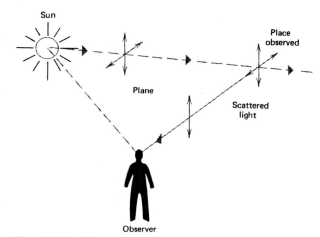

Fig. 14-25. The component of sunlight polarized toward an observer cannot give rise to scattered light in his direction. Only the other component perpendicular to this component can do it; therefore the observed scattered light is polarized perpendicular to the plane that passes through the observer, the place in the sky being observed, and the sun.

Fig. 14-26. Polarization of skylight from the southern sky in the morning is demonstrated by these photographs taken through a polarizing filter that was oriented to pass only vertically polarized light for the left view and horizontally polarized light for the right one. (These views also demonstrate that light reflected by the metal corners of the World Trade Center in New York City is not polarized as a result of reflection, unlike the case of reflection from an insulating material such as glass.)

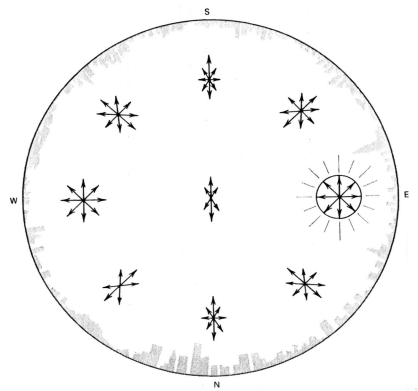

Fig. 14-27. View of the overhead sky at sunrise, as seen through a "fisheye" lens. The nature of polarization of skylight on a clear day is indicated for representative directions.

be used as polarizers to detect polarized light. Of course the Vikings lacked the modern notions about polarization, but they couldn't fail to notice that the sky when viewed through one of these transparent cordierite crystals was darker in some regions than others, and the patterns varied with the orientation of the crystal and sun. In the biological world vertebrates such as man lack effective polarizing structures in the eye, but some invertebrates such as the honeybee and desert ant use polarized skylight as a primary navigational reference.

Invertebrates detect the direction of polarization of skylight through birefringent properties of the eye. We would expect an advantage in sensitivity to scattered light compared with unpolarized light on a hazy day if their photodetectors respond preferentially to short wavelength radiation. Indeed, both the honeybee and desert ant have maximum sensitivity in the ultraviolet, with little response to wavelengths greater than about 410 nm.

It should be clear why a polarizing filter can be useful in outdoor photography.

When properly oriented, it removes scattered light and permits more distant viewing when the sun is to the side. It also darkens the sky to provide more contrast with clouds and buildings, an effect of considerable interest for architectural photography.

■ 14-5 LIGHTNING

A flash of lightning and its resounding thunder are one of nature's most impressive displays. The chain of events that leads to this climax begins inside a cloud where water vapor condenses to form water droplets. As this occurs, energy is released, which warms the region of surrounding air. This air expands and rises very quickly if the concentration of water vapor is sufficiently great. In some clouds the vertical speeds of updrafts can reach 100 km/hr. This creates friction against particles of ice and water that form the cloud. As a result, electrons are stripped away from their parent atoms. The details of how this occurs are yet to be explained, but the end result usually produces a net concentration of electrons at the bottom of the cloud and positively charged ions at the top.

When a localized concentration of electrons develops, the strong electrical force ionizes the nearby air molecules and creates a conducting path. The repulsive

Fig. 14-28. Air currents displace the conducting path of lightning during exposure and show the multiple flashes of successive strokes. (John Hendry,Jr. / Photo Researchers.)

force between the electrons pushes them through this path, and they advance as the path steps its way in a series of jerks until making contact with the earth (or another cloud). This path is called the *stepped leader,* and it contains a high concentration of electrons that were pushed downward from the cloud. Once the stepped leader touches ground, these electrons spill from its bottom and spread out over the earth. Electrons from successively higher regions of the leader follow behind, and this procession of high electrical current intensely excites the air molecules. Correspondingly, the light emitted by the excited molecules first appears at the bottom where the high current began to flow into the earth, and the source of light quickly proceeds upward. This so-called *return stroke* is what we see. The sequence of forming a leader and then the return stroke repeats as many as ten times or more within a fraction of a second (Fig. 14-28). The individual strokes can be recorded by a camera if it is moved during exposure. With each stroke the sudden heating of the air causes it to expand dramatically, and this pressure wave is heard as thunder.

Because clouds are often charged in this manner and lie closer to each other than to the ground, it is no surprise that there are many more lightning strokes from one cloud to another than from clouds to the ground. It is estimated that strokes wholly within the air occur five times more frequently than those between a cloud and the earth. The flash of lightning comes from incandescent nitrogen and oxygen atoms and ions, constituents of the air that have been altered from their normal molecular form by the impact of high-speed electrons. Because the peak temperature in the flash reaches 30,000 K, the color has a distinctly bluish hue. There is also a strong emission line in the red end of the spectrum contributed by hydrogen atoms. These are not normally found in air but result from the dissociation of water during the return stroke.

SUMMARY

Most optical effects in the atmosphere are due to reflection, refraction, diffraction, or Rayleigh scattering. Mirages arise from refraction, when the air temperature and corresponding density and refractive index vary markedly with altitude. This is most pronounced when the earth's surface is much warmer (or cooler) than the overlying air. Then the atmosphere serves as a lens to produce a virtual image below (or above) the subject. In addition the uplift of air from some warmer regions and descent into adjacent regions may break up such an image into columns as in the Fata Morgana. The scintillation of starlight is also due in large measure to refraction by uneven variations in air density associated with warm and cold currents of air overhead. Rainbows arise from a combination of refraction and reflection of sunlight by airborne droplets of water. The primary bow is formed by rays that have been reflected once within the droplet, and the secondary bow by rays that have been reflected twice. The spectral colors are spread out by dispersion as rays are refracted on entering and leaving the droplets. The bows are bright because there is a comparatively large range of distances from the center of a droplet where rays can enter and be refracted outward in a nearly common direction This effect of "bunching" of the emerging rays in a particular direction is also responsible for the 22° halo when a cirrus cloud covers the sun or moon but, in this case, the refraction is caused by ice crystals with a hexagonal cross section. Rainbows and halos are examples of light being scattered by large particles where the effect can be described by ray optics. Scattering by particles that are smaller than roughly 100 times the wavelength of light, but more than one-tenth of a wavelength, is described as diffraction. When a cloud has droplets of fairly uniform size, light of a given wavelength is diffracted in the same direction by all the droplets, and the corresponding color is seen on that portion of the cloud. Since the direction varies with wavelength, a sequence of spectral colors is seen at larger angles, and the spectrum may be repeated several times. This effect forms the corona, which may be seen encircling the sun when a cloud passes in front of it, and the glory, which may appear around the observer's shadow on a cloud when facing away from the sun. But if the droplets of a cloud are of uneven size, only white light is seen as a consequence of the overlapping diffraction patterns from individual droplets. Particles much smaller than the wavelength of light preferentially scatter blue light. Thus this Rayleigh scattering by air molecules and by tiny airborne particles accounts for the blue of the sky. Rayleigh scattering also explains why sunsets are red, because the blue and green components of sunlight have been scattered away from the line of sight. Rayleigh scattering also accounts for the fact that skylight from certain directions is polarized. ■

QUESTIONS

14-1. What determines whether a mirage will be inferior or superior?

14-2. Explain whether mirages are real or virtual images.

14-3. Most superior mirages are seen upright. Under what conditions will an inverted image be seen?

14-4. The refractive index of air depends solely on its density. This is also true of water. Why are mirages not commonly seen by swimmers underwater?

14-5. When you look at the sunset, is the sun's actual position above or below its image? Why?

14-6. If sunlight on entering different locations on a raindrop is refracted by differing amounts, why is it that with a rainbow we see an arc in the sky corresponding to essentially only one value for the refraction angle?

14-7. What is the origin of Alexander's dark band seen between the primary and secondary rainbows?

14-8. Why is a rainbow seen with our back to the sun whereas a halo appears when we face the sun?

14-9. What different conditions are responsible for a halo, corona, and the blue of the sky?

14-10. How is it possible for pure air to scatter light just as very tiny airborne particles do?

14-11. Why does the distant horizon on some days appear blue and on others white?

14-12. Sometimes the setting sun appears redder at the bottom than at the top. Why is this?

14-13. On a clear day, under what conditions is the skylight from directly overhead polarized? In which direction is it polarized?

14-14. With the sun rising in the east, when you view the north sky near the horizon, in what direction is the light polarized? If you view the sun directly, is the light polarized? Why?

14-15. What excites the air molecules that give us the flash of lightning?

FURTHER READING

R. Greenler, *Rainbows, Halos, and Glories* (Cambridge University Press, 1980). A superb introduction to atmospheric effects, covering much more than suggested by the title. Beautifully illustrated.

M. Minnaert, *The Nature of Light and Colour in the Open Air* (Dover, New York, 1954). The standard reference for optical effects in the atmosphere.

K. Heuer, *Rainbows, Halos, and Other Wonders* (Dodd, Mead, and Company, New York, 1978).

E. J. McCartney, *Optics of the Atmosphere* (Wiley-Interscience, New York, 1976).

R. A. R. Tricker, *Introduction to Meteorological Optics* (Sidgwick, London, 1960). Standard reference work with a strong reliance on mathematics.

L. E. Salanave, *Lightning and Its Spectrum: An Atlas of Photographs* (University of Arizona Press, 1980). Dramatic photographs and a lucid explanation of the rich variety of effects found in lightning.

W. H. Lehn and I. Schroeder, *The Norse Merman as an Optical Phenomenon* (*Nature*, Vol. 289, 1981), p. 362.

From the *Scientific American*

A. B. Fraser and W. H. Mach, "Mirages" (January, 1976), p. 102.

H. M. Nussensveig, "The Theory of the Rainbow" (April, 1977), p. 116.

D. K. Lynch, "Atmospheric Halos" (April, 1978), p. 144.

H. C. Bryant and J. Jarmie, "The Glory" (July, 1974), p. 60.

V. K. La Mer and M. Kerker, "Light Scattered by Small Particles" (February, 1953), p. 69.

T. H. Waterman, "Polarized Light and Animal Navigation" (July, 1955), p. 88.

R. Wehner, "Polarized-Light Navigation by Insects" (July, 1976), p. 106.

D. J. K. O'Connell, "The Green Flash" (January, 1960), p. 112.

A. A. Few, "Thunder" (July, 1975), p. 80.

CHAPTER 15

A REFLECTION ON LIGHT AND COLOR

In the beginning of this book we suggested that many diverse phenomena involving light and color could be understood on the basis of a few unifying principles, and that the logical procedures we would develop in explaining these principles and in analyzing particular examples would also equip the reader to interpret many other optical phenomena. Now, having reached our conclusion, we should briefly summarize and evaluate the main elements of the approach that has been developed in the preceding chapters.

We have seen that light is a form of electromagnetic energy and is characterized by a spectral power distribution (SPD) that can be measured by appropriate instruments. The SPD is determined by the process by which light was created at the source. The SPD is modified when reflected from an object in accordance with the object's reflectance curve, and the SPD of the light finally reaching the eye determines our visual perception. This predictive capability for perception, based on the SPD of light, is the crux of colorimetry.

With modern instruments, the chromaticity of a color can be automatically calculated from the measured SPD. The chromaticity is represented by its position on a map, the C.I.E. Chromaticity Diagram, as an economical way to visualize the relationship between different colors. The procedures of colorimetry also permit us to predict the color produced when two or more colors are added together. Just as colorimetry is the science of measuring color, photometry is the science of measuring light. For instance, the luminance of a surface determines whether it appears brighter than another, independent of the hue and saturation of its color. By applying the procedures of photometry and colorimetry, we can quite precisely describe how objects affect the perceptual quality of light that they reflect or transmit.

The laws of ray optics that govern reflection and refraction, and those of wave optics that govern diffraction, interference, and scattering, combine with fundamental color theory to explain the optics of the media of the visual arts, photography, holography and light and color phenomena observed in the atmosphere.

Selective absorption is the most common process by which materials affect color. Absorption occurs at only certain wavelengths because the corresponding energy of the photons of these wavelengths is what is just needed to excite an electron in the material to one of its quantum states of higher energy. Even if absorption is not strong in the visible portion of the electromagnetic spectrum but instead is strong in the ultraviolet, the strong interaction with light of nearby wavelengths causes dispersion; thus, refracted light may be colored, as seen in the spectrum formed by a prism or the "flash" of a diamond.

It is perhaps surprising that the structure of an object often plays a more important role than selective absorption in determining the color of light it reflects. This is true for interference colors when light reflects from the two surfaces of a thin film such as an oil slick, or the multiple surfaces of veins on a butterfly wing. It is also true for diffraction colors as displayed by the sun's glory and corona and it is perhaps most dramatically presented by the image from a hologram illuminated by white light. Scattering by small airborne particles, and indeed by air itself, is another example of coloration without significant absorption. The preferential scattering of blue light as sunlight travels through the atmosphere provides a blue sky for people below and a red sunset for those far to the east.

This book has stressed the physical effects that shape the SPD of light. However, many important effects are not explained solely by the nature of the physical stimulus. Perception may be strongly influenced by the nature of our visual system itself. For instance, by considering some elements of visual physiology, we have also summarized explanations of such subtle effects as induced contrast, color constancy, and afterimages.

Many interesting and important aspects of light and color have been touched on only briefly or else omitted entirely from our discussion. We omitted, for example, the topics of color preference and psychological effects of perception, as well as the esthetics of color (what is an effective painting?), which are questions properly in the realms of psychology and art criticism. Similarly we have entirely ignored the creative aspects of photography and holography; for example, when discussing focusing we dealt only with how to achieve a technically "correct" focus. In practice, however, the esthetics of a photograph often involve intentional blurring of the image. Among the important topics we treated only peripherally in the visual arts are the effect on perception of surface texture and the practical difficulties of predicting the result of mixing paints. Because both reflection and absorption depend on the size of pigment particles, predicting the resulting color is readily recognized as ultimately dependent on the skill of the experienced artist, for whom the laws of colorimetry can only provide a qualitative guide.

The various chapters of our text have been designed for rather different purposes. Some are intended to be practical: ray optics and photometry are quantitative topics that can be used by the director of theatrical lighting to determine the necessary lamps for effective stage lighting, and the designer of light displays to choose an appropriate projection lens. Similarly we devoted considerable attention to the properties of monochrome and color films, since an appreciation of the capabilities of these media is essential for the serious photographer. In these and related chapters we have provided a considerable range of practical material that can be put to use by the reader with particular applications in mind.

We hope that this book has stimulated the reader's curiosity to look carefully at new visual effects and seek explanations for them. The physical principles of most of them have been explained here. For instance the myriad round spots of light often seen beneath a tree on a summer day are in reality images of the sun, formed by the tiny openings between the leaves that serve as natural "pinhole" lenses. Again, the pattern of bright lines seen moving about on the bottom of a swimming pool is due to the curvature of the water's surface, which focuses the illumination on the bottom when the water's depth corresponds to the focal length of the waves. Finally, halos commonly may be seen surrounding streetlights, and colorful diffraction patterns appear through window screens, umbrellas, or curtains, or even our own eyelashes when looking toward a light with eyes nearly closed. These are but a few additional examples. The reader who has found the ideas presented here interesting will find endless examples of light and color phenomena in the natural world as well as in the world of the visual arts. ∎

2A: DENSITY OF FILTERS

The notion of optical *density* provides a useful characterization of the light absorptive property of a filter or of a particular region of a photographic negative. The greater the density, the less light is transmitted. A precise definition of density D can be given in terms of the transmittance T as follows:

$$D = \log_{10} \frac{1}{T} \tag{2A-1}$$

For filters where transmittance depends upon wavelength, this relation defines a different density appropriate for each wavelength. More commonly, the notion of density is used with neutral density filters, for example, the black and white photographic negative. Here the same density applies to all wavelengths. The value of T in the above formula should be expressed in decimal form. From the properties of logarithms, a transmittance of 10% (or $T = 0.1$) corresponds to a density of $D = 1$; 1% or $T = 0.01$ corresponds to $D = 2$; 0.1% or $T = 0.001$ corresponds to $D = 3$, and so on. Large reductions in T produce small increases in D. More precisely, reducing T successively by the same *multiplicative* factor causes D to increase successfully by equal *additive* amounts. In the preceding example each reduction of T by a factor of 10 caused D to additively increase by one (Table 2A-1). If the density of a filter is known, it follows from Eq. 2A-1 that the transmittance is:

$$T = \frac{1}{10^D} \tag{2A-2}$$

Because the net transmittance of filters in cascade is the product of the individual transmittances (Bouguer's law), the logarithmic relation in the above equation tells us that the net density of filters in cascade is the sum of the individual densities. For two filters with transmittances of T_1 and T_2 we have for the net density:

$$\begin{aligned}
D_{net} &= \log_{10} \frac{1}{T_{net}} \\
&= \log_{10} \frac{1}{T_1 T_2} \\
&= \log_{10} \frac{1}{T_1} + \log_{10} \frac{1}{T_2} \\
&= D_1 + D_2
\end{aligned} \tag{2A-3}$$

TABLE 2A-1 Filter Transmittance and Density[a]

Transmittance (%)	Density	Transmittance (%)	Density
100	0.0	10	1.0
79	0.1	7.9	1.1
63	0.2	6.3	1.2
50	0.3	5.0	1.3
40	0.4	4.0	1.4
32	0.5	3.2	1.5
25	0.6	2.5	1.6
20	0.7	2.0	1.7
16	0.8	1.6	1.8
13	0.9	1.3	1.9

[a]*Note.* Each factor of 10 reduction in transmittance adds 1.0 to the density.

Thus the densities of neutral density filters are additive. For three filters with densities of $D_1 = 1.2$, $D_2 = 0.5$, and $D_3 = 0.2$, the total density when placed together is $D_1 + D_2 + D_3 = 1.9$. It follows that the density of a filter is proportional to its thickness (Bouguer's law). Also the density of a filter composed of a dye is proportional to the concentration of the dye (Beer's law).

The density of dark regions in a photographic negative can be measured with a *densitometer*. This instrument measures the intensity of the transmitted light for a known incident intensity and computes either the transmittance or density.

QUESTIONS

2A-1 If a display light is too bright and only 10% of the light is needed, what should be the density of the filter placed in front of the lamp?

2A-2 What is the transmittance of a neutral density filter having a density of 2? Suppose that this filter and another having a density of 1 are both inserted in a beam of white light. What is the total density of the two used together?

3A: TRICHROMATIC THEORY

PREREQUISITES: Chapters 3 and 6

In this section we shall explain the mathematical foundation that underlies the formulation of tristimulus values and the C.I.E. chromaticity diagram.

We begin by recalling from Chapter 6 that the relation between the amount of *power* coming from a surface, which is a physical quantity called the radiant flux (measured in watts), and the amount of *light,* which is a psychophysical quantity called the luminous flux (measured in lumens), is fixed by the SPD of the radiant flux. This relation depends on how the visual sensitivity of the standard observer varies with wavelength, and it was determined experimentally by brightness matching experiments with monochromatic lights. The shape of the relative sensitivity curve V_λ of the standard observer (Table 6-1 and Fig. 6-1) had already been adopted by the C.I.E. before it completed its considerations on colorimetry, and it is an essential component of the present color measurement theory.

The maximum spectral luminous efficacy (K_m = 683 lumens per watt at 555 nm) defines the unit of luminous flux called the lumen. The combination $K_m V_\lambda$ indicates the luminous flux corresponding to one watt of radiant flux at any other wavelength λ. If the actual radiant flux is ϕ_e watts at a particular wavelength λ, then $K_m V_\lambda \phi_e$ is the luminous flux at that wavelength. The total luminous flux coming from a surface is calculated by adding together these individual contributions from all wavelengths. Two surfaces reflecting the same luminous flux into the eye from a given perceived area of surface appear equally bright, even if their SPDs differ. That is, they appear equally bright if the *luminance* of each surface is the same.

Color Matching Functions

The color matching experiments discussed in Section 3-5 and illustrated in Fig. 3-8 were performed using monochromatic blue, green, and red primaries (at 435.8, 546.1, and 700.00 nm) to achieve color matches to monochromatic test samples. The results of these experiments are summarized by the color matching functions $\bar{b}_\lambda$, $\bar{g}_\lambda$, $\bar{r}_\lambda$ listed in Table 3-1 and plotted in Fig. 3-9. The relative intensities of the primaries **B, G, R** used in determining the values of the color matching functions were selected so that for white light the tristimulus values B, G, R are numerically equal.

An alternative choice of relative intensities is more useful in working out the mathematics of trichromatic theory. In this second choice, the relative intensities of **B, G, R** are chosen so that the *luminances* they produce on the viewing screen are equal. Then the screen illuminated with one unit of **B** has the same visual brightness as when illuminated with one unit of **G** or **R**. White will no longer be matched by the same number of units of **B** and **G** and **R**; that is, the tristimulus values for white are no longer equal. There are also new color matching functions $\bar{b}_\lambda$, $\bar{g}_\lambda$, $\bar{r}_\lambda$ describing the amounts of these new primaries **B, G, R** needed to match monochromatic test samples M_λ of the same radiance at each wavelength λ:

$$M_\lambda \blacktriangleleft \bar{b}_\lambda \mathbf{B} + \bar{g}_\lambda \mathbf{G} + \bar{r}_\lambda \mathbf{R} \qquad (3A\text{-}1)$$

This expression represents a full-color match—for brightness as well as chromaticity. If the monochromatic test samples M_λ are arranged for convenience to have the same radiance amounting to L_e = (1/683) watt/steradian-meter^2, then by Eq. 6-1 the luminance of each sample is:

$$L_v = K_m V_\lambda L_e = (683 V_\lambda \text{ lm/W}) \times (1/683 \text{ W/sr-m}^2) \qquad (3A\text{-}2)$$
$$= V_\lambda \text{ lm/sr-m}^2$$

This value for the luminance of the test sample equals the luminance provided by the primaries when there is a full-color match. For convenience in computation, we suppose the primary lights to be constructed so that one unit of **B** or **G** or **R** provides a luminance of 1 lm/sr-m². Then the matching condition described by Eq. 3A-1 says that $\bar{b}_\lambda$, $\bar{g}_\lambda$, and $\bar{r}_\lambda$ are the numerical values of the luminance provided by each of the primaries **B, G,** and **R** to achieve a brightness match; and the total of these amounts equals the numerical value V_λ of the luminance provided by the monochromatic test sample:

$$V_\lambda = \bar{b}_\lambda + \bar{g}_\lambda + \bar{r}_\lambda$$

(3A-3)

The reader can verify that this equality holds for the values of the color matching functions in Table 3A-1 at each wavelength. That is, at each wavelength the sum of the color matching functions equals the relative sensitivity of the C.I.E. Standard Observer for photopic vision. There is nothing profound about this equality—the luminances of the primary colors and the radiance of the monochromatic test samples were chosen to achieve it.

Color Matching Relations and Grassmann's Laws

Equation 3-1 which specifies a full-color match between a sample **S** and appropriate amounts of the primaries

$$\mathbf{S} \; \blacktriangleleft \; B\,\mathbf{B} + G\,\mathbf{G} + R\,\mathbf{R}$$

(3A-4)

can now be understood as follows: **B, G,** and **R** represent one unit of each of the primary lights, and this unit corresponds to a luminance of 1 lm/sr-m². The tristimulus values B, G, and R represent the number of lumens from each primary required to achieve a full-color match to the sample **S**.

To avoid carrying out an actual color matching experiment, we can calculate the tristimulus values if the SPD of the radiance of the sample is known. To calculate B, for each wavelength the color matching function $\bar{b}_\lambda$ must be multiplied by the ratio of the actual radiance of the sample **S** at that wavelength to the radiance of the original monochromatic test sample **M** (1/683 W/sr-m²), and the contributions from all wavelengths are added together. We let P_{400} designate the radiance in the SPD of **S** found within the range of wavelengths from 395 to 405 nm, P_{410} the radiance from 405 to 415 nm, etc. Thus the tristimulus value B is given by the formula:

$$B = (683 P_{400})\bar{b}_{400} + (683 P_{410})\bar{b}_{410} + \text{etc.}$$

(3A-5a)

Similar expressions specify G and R:

$$G = (683 P_{400})\bar{g}_{400} + (683 P_{410})\bar{g}_{410} + \text{etc.}$$

(3A-5b)

$$R = (683 P_{400})\bar{r}_{400} + (683 P_{410})\bar{r}_{410} + \text{etc.}$$

(3A-5c)

Equations 3A-5 have the important property of being *linear* in that the net perceptual result is equal to the arithmetic sum of the individual contributions from each interval of the SPD, and each of these contributions is linearly proportional to the radiance P_λ in the interval. Additivity and linearity are consequences of a stronger form of trichromatic colorimetry that follows from rules described and developed by many workers, culminating in the publications by Herman von Helmholtz. The rules were attributed by Helmholtz to another psychophysicist Grassmann and are known as "Grassmann's laws of additive color mixture." It follows from these laws that for a considerable range of observing con-

TABLE 3A-1 Alternative Average Color-Matching Functions $\bar{b}_\lambda$, $\bar{g}_\lambda$, and $\bar{r}_\lambda$ of Observers with Normal Color Vision Viewing a Two Degree Visual Field.[a]

λ (nm)	$\bar{r}_\lambda$	$\bar{g}_\lambda$	$\bar{b}_\lambda$
380	0·0000	0·0001	0·0001
390	−0·0001	0·0002	0·0002
400	−0·0003	0·0006	0·0007
410	−0·0008	0·0019	0·0022
420	−0·0021	0·0050	0·0069
430	−0·0022	0·0055	0·0149
440	−0·0026	0·0068	0·0188
450	−0·0121	0·0311	0·0190
460	−0·0261	0·0682	0·0179
470	−0·0393	0·1165	0·0138
480	−0·0494	0·1797	0·0087
490	−0·0581	0·2612	0·0050
500	−0·0717	0·3919	0·0029
510	−0·0890	0·5904	0·0016
520	−0·0926	0·8019	0·0007
530	−0·0710	0·9327	0·0003
540	−0·0315	0·9854	0·0001
550	+0·0228	0·9722	0·0000
560	0·0906	0·9045	−0·0001
570	0·1677	0·7844	−0·0001
580	0·2453	0·6248	0·0000
590	0·3093	0·4478	0.0000
600	0·3443	0·2867	0·0000
610	0·3397	0·1633	0·0000
620	0·2971	0·0839	0·0000
630	0·2268	0·0382	0·0000
640	0·1597	0·0153	0·0000
650	0·1017	0·0053	0·0000
660	0·0593	0·0017	0·0000
670	0·0315	0·0005	0·0000
680	0·0169	0·0001	0·0000
690	0·0082	0·0000	0·0000
700	0·0041	0·0000	0·0000
710	0·0021	0·0000	0·0000
720	0·0010	0·0000	0·0000
730	0·0005	0·0000	0·0000
740	0·0902	0·0000	0·0000
750	0·0001	0·0000	0·0000
760	0·0001	0·0000	0·0000
770	0·0000	0·0000	0·0000
Totals	1·8911	8·6812	0·1136

[a]The primaries are monochromatic **B** (at 435.8 nm), **G** (at 546.1 nm), and **R** (at 700.0 nm). This set is chosen to give $\bar{b}_\lambda + \bar{g}_\lambda + \bar{r}_\lambda = V_\lambda$.

Source. From P. J. Bouma, *Physical Aspects of Color*, Second Edition (St. Martin's Press, New York, 1971).

ditions: (1) a match between two colors continues to hold if the radiant flux from each is increased or reduced in magnitude by the same constant factor at each wavelength; and (2) if colors **C** and **D** match and colors **E** and **F** match, then the additive mixtures (**C** + **E**) and (**D** + **F**) also match. Grassmann's laws permit the rules of algebra to be used in color matching equations.

The C.I.E. Coordinates

The 1931 C.I.E. recommendation for a new set of primaries to replace the **B, G, R** set was designed to deal with the following issues:

1. The tristimulus values in the **B, G, R** system are negative over part of the wavelength range, making computation clumsy.
2. There is no single quantity in the **B, G, R** system that corresponds to the luminance of the sample being characterized.

Because of the linearity of Eqs. 3A-5 any new set of primaries can be defined mathematically as linear combinations of the **B, G, R** primaries and the combinations chosen then allow the new tristimulus values to be computed from the original values of $\bar{b}_\lambda$, $\bar{g}_\lambda$, and $\bar{r}_\lambda$. The C.I.E. chose as new primaries three "nonphysical colors" denoted by **X, Y,** and **Z**. They are "nonphysical" in the sense that none of them can be constructed from lamps and filters. Each has only a mathematical definition. The definition is expressed by a full-color matching condition to each of the **B, G, R** primaries:

$$
\begin{aligned}
\mathbf{B} &\quad 18.801\ \mathbf{X} + \mathbf{Y} + \quad 93.066\ \mathbf{Z} \\
\mathbf{G} &\quad 0.38159\ \mathbf{X} + \mathbf{Y} + 0.012307\ \mathbf{Z} \\
\mathbf{R} &\quad 2.7689\ \mathbf{X} + \mathbf{Y} + \quad 0
\end{aligned}
\qquad (3A\text{-}6)
$$

If a sample **S** is matched by B units of **B,** G units of **G,** and R units of **R,** then the amounts of **X, Y,** and **Z** shown in Eq. 3A-6 must be increased accordingly to achieve a match with **X, Y, Z** primaries

$$
\begin{aligned}
\mathbf{S} &\quad (18.801\ B + 0.38159\ G + 2.7689\ R)\ \mathbf{X} + \\
&\quad (\ 1.0 \quad B + 1.0 \quad\quad G + 1.0 \quad\quad R)\ \mathbf{Y} + \\
&\quad (\ 93.066\ B + 0.012307\ G + 0 \quad\quad)\ \mathbf{Z}
\end{aligned}
\qquad (3A\text{-}7)
$$

From these equations, the expressions in parentheses give the values for the tristimulus values X, Y, Z in the **X, Y, Z** system for the sample **S** in terms of the tristimulus values B, G, R in the **B, G, R** system:

$$
\begin{aligned}
X &= 18.801\ B + 3.38159\ G + 2.7689\ R \\
Y &= \quad\quad B + \quad\quad G + \quad\quad R \\
Z &= 93.066\ B + 0.012307\ G
\end{aligned}
\qquad (3A\text{-}8)
$$

Equations 3A-8 allow the tristimulus values X, Y, Z to be computed directly from B, G, R. In particular if **S** is the monochromatic source $\mathbf{M}_\lambda$ having a radiance of $\frac{1}{683}$ W/sr-m² then the corresponding values of B, G, R are just $\bar{b}_\lambda$, $\bar{g}_\lambda$, $\bar{r}_\lambda$, and Eqs. 3A-8 specify the C.I.E. color matching functions $\bar{x}_\lambda$, $\bar{y}_\lambda$, $\bar{z}_\lambda$. The values of these functions are listed in Table 3-2.

In practice the color matching functions $\bar{x}_\lambda$, $\bar{y}_\lambda$, $\bar{z}_\lambda$ can be used directly to compute the tristimulus values X, Y, Z without further reference to the **B, G, R** system. By analogy with Eqs. 3A-5 we have

$$X = (683P_{400})\bar{x}_{400} + (683P_{410})\bar{x}_{410} + \text{etc.} \qquad (3A\text{-}9a)$$
$$Y = (683P_{400})\bar{y}_{400} + (683P_{410})\bar{y}_{410} + \text{etc.} \qquad (3A\text{-}9b)$$
$$Z = (683P_{400})\bar{z}_{400} + (683P_{410})\bar{z}_{410} + \text{etc.} \qquad (3A\text{-}9c)$$

Equations 3A-9 are the fundamental working equations of colorimetry. They may be used to calculate the tristimulus values X, Y, Z for any color whose SPD, denoted here by P_λ, is known.

As a consequence of the way the definitions of **X, Y, Z** were chosen (Eqs. 3A-6), the value of the tristimulus value Y in Eq. 3A-8 is just $Y = B + G + R$. Therefore the color matching functions obey the relation:

$$\bar{y}_\lambda = \bar{b}_\lambda + \bar{g}_\lambda + \bar{r}_\lambda \qquad (3A\text{-}10)$$

But Eq. 3A-3 shows that the left-hand side is equal to V_λ, so the color matching function $\bar{y}_\lambda$ is identical to the relative sensitivity of the standard observer. Thus the tristimulus value Y calculated according to Eq. 3A-9b gives the luminance of the sample **S** directly, which resolves issue #2 of concern to the C.I.E. Furthermore, as can be seen in Table 3-2, the choice of the **X, Y, Z** primaries resulted in color matching functions that are entirely positive, thereby resolving issue #1. An added feature of the **X, Y, Z** system can be seen by noting that the sum of values of the color matching function $\bar{x}_\lambda$ in the first column of Table 3-2 is the same as the sum of the values of $\bar{y}_\lambda$ in the second column and the values of $\bar{z}_\lambda$ in the third column. Therefore, an "equal power stimulus" (whose SPD is flat across the spectrum) is described by equal tristimulus values $X = Y = Z$. This color is white.

Color Addition

It is useful to think about additive mixtures of color in terms of where the chromaticities of the colors lie on the C.I.E. chromaticity diagram (Fig. 3A-1). Recalling that $x = X/(X + Y + Z)$ and $y = Y/(X + Y + Z)$ from Eq. 3-7, our point of reference is the "equal power point," which lies at $x = \frac{1}{3}$ and $y = \frac{1}{3}$. All colors are represented by chromaticities that lie within the boundary of monochromatic colors, specified by the chromaticity coordinates $x = \bar{x}_\lambda/(\bar{x}_\lambda + \bar{y}_\lambda + \bar{z}_\lambda)$ and $y = \bar{y}_\lambda/(\bar{x}_\lambda + \bar{y}_\lambda + \bar{z}_\lambda)$ at each wavelength, and the line of purples that runs from one end of the spectrum to the other.

Suppose that two colors **S₁** and **S₂** have tristimulus values expressed by the full-color matching relations:

$$\mathbf{S_1} \blacktriangleleft X_1\mathbf{X} + Y_1\mathbf{Y} + Z_1\mathbf{Z} \qquad (3A\text{-}11)$$
$$\mathbf{S_2} \blacktriangleleft X_2\mathbf{X} + Y_2\mathbf{Y} + Z_2\mathbf{Z}$$

If **S₁** and **S₂** are mixed additively to give $\mathbf{S_3} = \mathbf{S_1} + \mathbf{S_2}$, then

$$\mathbf{S_3} \blacktriangleleft \mathbf{S_1} + \mathbf{S_2} \qquad (3A\text{-}12)$$
$$\blacktriangleleft (X_1 + X_2)\mathbf{X} + (Y_1 + Y_2)\mathbf{Y} + (Z_1 + Z_2)\mathbf{Z}$$

The corresponding chromaticity coordinates are:

$$x_3 = \frac{X_1 + X_2}{X_1 + Y_1 + Z_1 + X_2 + Y_2 + Z_2}$$

$$= \frac{x_1(X_1 + Y_1 + Z_1)}{(X_1 + Y_1 + Z_1 + X_2 + Y_2 + Z_2)} + \frac{x_2(X_2 + Y_2 + Z_2)}{(X_1 + Y_1 + Z_1 + X_2 + Y_2 + Z_2)}$$ (3A-13)

$$y_3 = \frac{Y_1 + Y_2}{X_1 + Y_1 + Z_1 + X_2 + Y_2 + Z_2}$$

$$= \frac{y_1(X_1 + Y_1 + Z_1)}{(X_1 + Y_1 + Z_1 + X_2 + Y_2 + Z_2)} + \frac{y_2(X_2 + Y_2 + Z_2)}{(X_1 + Y_1 + Z_1 + X_2 + Y_2 + Z_2)}$$

To simplify such cumbersome expressions we introduce a new quantity Y' defined as $Y' = Y/y = (X + Y + Z)$. Then Eqs. 3A-13 can be rewritten as:

$$x_3 = x_1 \left(\frac{Y_1'}{Y_1' + Y_2'} \right) + x_2 \left(\frac{Y_2'}{Y_1' + Y_2'} \right)$$ (3A-14)

$$y_3 = y_1 \left(\frac{Y_1'}{Y_1' + Y_2'} \right) + y_2 \left(\frac{Y_2'}{Y_1' + Y_2'} \right)$$

These relations predict the new chromaticity coordinates when two colors are additively mixed. With reference to Fig. 3A-1 they imply the following: If two sources S_1 and S_2 whose individual chromaticity coordinates are (x_1, y_1) and (x_2, y_2) are mixed additively, the resulting color S_3 will have coordinates (x_3, y_3) that specify a point lying on the straight line connecting S_1 and S_2. The distances d_1 and d_2 of this point from S_1 and S_2 are in the ratio $d_1/d_2 = Y_2'/Y_1'$. This is sometimes called the center of gravity rule: if we interpret Y_1' and

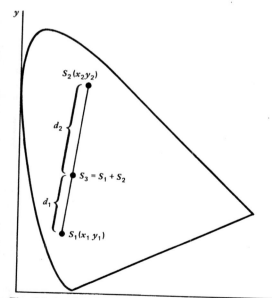

Fig. 3A-1. Additive color mixing of S_1 and S_2 in the C.I.E. system.

Y_2' as the "weights" of two objects located at the coordinates (x_1, y_1) and (x_2, y_2), then the "center of gravity" of the pair is at (x_3, y_3).

This result is slightly more complicated than the approximate result given in Eq. 3-8 and is a minor disadvantage of the C.I.E. system. A similar calculation in the **B, G, R** system where $\bar{b}_\lambda + \bar{g}_\lambda + \bar{r}_\lambda = V_\lambda$ results in the same straight-line rule in any plane in the space of **B, G,** and **R** for additive mixing, but with the simpler conclusion that the distances are in the ratio $d_1/d_2 = L_2/L_1$, where L_1 and L_2 represent the luminances of samples **S₁** and **S₂**.

Finally, a practical remark is in order. Usually the SPD of a color is not given in terms of its radiance, but just indicates radiant flux, or perhaps only "relative power" with no units expressed. Then Eqs. 3A-9 can still be used to calculate tristimulus values with the given values of the SPD substituting for P_{400}, P_{410}, etc. However the calculated value for Y will not represent the luminance of the sample. Nevertheless, the chromaticity coordinates calculated by Eq. 3-7 remain independent of the choice of units. Only the shape of the SPD and not the physical units used to describe it determine where the chromaticity of a color lies on the C.I.E. diagram.

QUESTION

3A-1 If the color with a chromaticity given by $x = 0.4$, $y = 0.55$ is added to another with twice the luminance (twice the value of Y) having $x = 0.1$, $y = 0.7$, what are the coordinates (x, y) of the resulting color?

4A: ANGLES MEASURED IN RADIANS

We are accustomed to measuring angles in *degrees*, as when we say that a right angle is 90° and a full circle contains 360°. The unit of measure of the degree is used for historical reasons. A more useful unit for many purposes is the *radian*. Figure 4A-1 shows how to determine the value of the angle θ between the two straight lines in radians. Imagine a circle drawn with its center at the intersection of the lines. Because the ratio of the included arc length of the circle (say l) to the radius (r) is independent of our choice of the radius r, we can use this ratio to specify the value of the angle:

$$\theta = \frac{\text{arc length}}{\text{radius}} \qquad (4A-1)$$

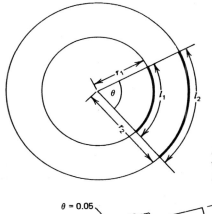

Fig. 4A-1. The angle θ in radian measure is the same whether calculated as l_1/r_1 or as l_2/r_2.

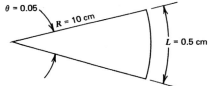

(a)

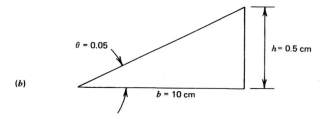

(b)

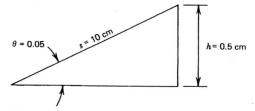

(c)

Fig. 4A-2. Examples of the use of angles measured in radians to determine certain lengths: (a) the arc length of a circle when the radius R is known; (b) the short side h of a triangle when the angle θ is small and the base b is known; (c) the short side h when the angle θ is small and the hypotenuse s is known.

TABLE 4A-1 Values for the Sine and
Tangent of a Small Angle Are Nearly Equal
to the Value of the Angle Itself

θ (Degrees)	θ (Radians)	Sin θ	Tan θ
1°	0.018	0.018	0.018
5°	0.087	0.086	0.087
10°	0.175	0.174	0.176
15°	0.262	0.259	0.268
20°	0.349	0.342	0.364
25°	0.436	0.423	0.466

As an example of a familiar angle expressed in radians, we calculate the total angle contained in a full circle. Here the arc length for any circle of radius r is $2\pi r$; therefore $\theta = 2\pi r/r = 2\pi$. This answer clearly shows that the calculated angle of 2π is independent of what radius we choose for the size of our imagined circle. From the result that 2π radians = 360°, we deduce that 1 radian is about 57.3°.

The particular utility of radian measure becomes apparent when we want to calculate a length that is determined by the size of an angle. Suppose we want to determine the arc length L depicted in Fig. 4A-2a when we know the angle θ and distance to the arc R. The answer comes immediately from Eq. 4A-1 and is $L = R\theta$. This simple relation is *not* true if the value of the angle is given in degrees.

As another example given in Fig. 4A-2b, suppose we know how far away a subject is (b = 10 cm) and the visual angle it intercepts (θ = 0.05 radian). Then we can deduce the height h of this subject, and it is given exactly as $h = b \tan \theta$. When θ is small (say, less than 0.4 radian) it is a very good approximation, as illustrated in Table 4A-1, to say that tan θ is essentially equal to θ. Therefore the height of the subject can be approximated as $h = b\theta$. This so-called "small-angle approximation" is very useful to remember. In essence it says that when θ is small the two long sides of a right triangle are very nearly equal, and the short side opposite θ is about the same length as an arc of a circle.

A similar example is shown in Fig. 4A-2c where the hypoteneuse of a right triangle is known (s = 10 cm) and the angle between it and the second long side is θ = 0.05 radian. Then the opposite side $h = s \sin \theta$ can be approximated by $h = s\theta$. The fact that both $b\theta$ and $s\theta$ are very nearly equal to h for a small angle θ is true only when θ is measured in radians.

4B: TWO-SLIT INTERFERENCE

PREREQUISITES: Chapter 4 and Advanced Topic 4A

A simple mathematical derivation predicts the size of the interference pattern when monochromatic light passing through two closely spaced slits illuminates a screen (Fig.

4B-1). Several simplifying approximations will let us avoid unnecessary details. If the screen is far from the slits (L is usually one or more meters in length) and the slits are close together (S is a fraction of a millimeter), then the paths taken by the beam from the upper slit and the beam from the lower one are very nearly parallel. The angle denoted by the Greek letter alpha (α) is then essentially the same whether measured to the upper or lower path. This angle designates various positions on the screen, and we shall predict for which values of α the beams interfere constructively and for which values destructively. For constructive interference the difference in path lengths q must be an integral multiple of the wavelength λ. This means $q = N\lambda$, where N stands for any integer (N = 1, 2, or 3, etc.). When $q = N\lambda$, the lower beam goes through exactly N complete cycles more than the upper beam during its travel along the longer path to reach the screen; thus the two beams arrive in phase.

What does the condition $q = N\lambda$ tell us about the distance a between the center of the screen right behind the slits and where the Nth bright fringe is seen? If we restrict our interest to positions near the center of the screen, the angle α is small. The path of either beam together with the length a and the length L (at a right angle to a) form a right triangle. Therefore by trigonometry we can relate α to the other two lengths: $\tan \alpha = a/L$. Because

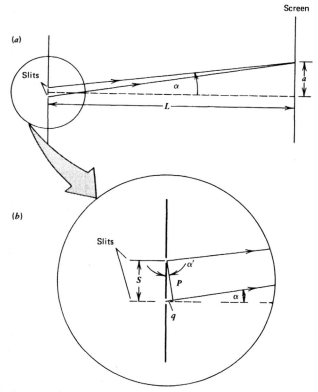

Fig. 4B-1. Relevant angles and dimensions for describing the interference pattern when light passes through two slits and illuminates a screen. The slit spacing S is much smaller than the distance L between the slits and screen.

α is small, this exact expression can be replaced by an approximate one that for practical purposes is just as useful:

$$\alpha = a/L \qquad (4B\text{-}1)$$

where α is measured in radians. This approximation is accurate to within a couple of percent if α is smaller than about 0.2 radian, or about $10°$ (Table 4A-1).

Now we turn to a more detailed view of the slits in order to relate the path difference q to α. As illustrated in Fig. 4B-1b, when α is small the upper beam and lower beam leave the slits heading along nearly parallel paths. The distance q is therefore marked off by a line P from the upper slit that meets the lower path at right angles. These two lines together with the line of length S denoting the distance between the slits form a right triangle. Thus by trigonometry two of the sides can be related to the angle α': sin α' = q/S. Again when α is small, so is α', and this expression can be approximated by

$$\alpha' = q/S \qquad (4B\text{-}2)$$

Now Eqs. 4B-1 and 4B-2 can be related by noticing that the two lines of lengths P and S that include the angle α', are respectively perpendicular to the beam path and the line of length L that include the angle α. Therefore the two angles are equal: $\alpha' = \alpha$. The right-hand sides of Eqs. 4B-1 and 4B-2 can therefore be set equal to each other:

$$\frac{a}{L} = \frac{q}{S} \qquad (4B\text{-}3)$$

When this is multiplied on left and right by L, and we insert the condition for constructive interference ($q = N\lambda$), we obtain our desired result:

$$a = N\lambda \left(\frac{L}{S}\right), \qquad N = 1, 2, 3 \text{ etc.} \qquad (4B\text{-}4)$$

This predicts that the pattern's size is larger than the wavelength of light by the factor (L/S). This enhancement factor is determined by the geometry of the experiment. For example, suppose $\lambda = 500$ nm, $L = 1$ m, and $S = 10^{-4}$ m. Then the first bright spot is found above the center spot by a distance $a = 5 \times 10^{-7} \times (1/10^{-4}) = 5 \times 10^{-7} \times 10^4 = 5 \times 10^{-3}$ m. Thus it is 5 mm above the center. The second bright spot is twice as far, or 10 mm; the third spot is three times further, or 15 mm; and so on. It is therefore quite easy to see this pattern, because it is so nicely spread out.

This derivation illustrates that the *angle* α determines where bright regions (and the intervening dark regions) fall. Therefore the further away we put the viewing screen, the more spread out is the pattern. This is predicted by the distance L appearing in the numerator of Eq. 4B-4. On the other hand the larger we make the slit separation S, the smaller the value for α to have q equal one wavelength, because S appears in the denominator. The distance between bright spots is thus proportional to the distance to the screen and inversely proportional to the slit separation. From this we can deduce a general feature of interference phenomena: that the pattern will be appreciably spread out only if the slits are quite close together compared to the viewing distance.

Interference patterns are interesting both because they demonstrate the wave aspect of light and because they provide a means to measure the wavelength of light. By measuring a, L, and S in Eq. 4B-4, we can deduce λ. In this way the knowledge of three conveniently measured numbers permits us to deduce the value of a much smaller number that cannot be directly measured.

QUESTION

4B-1 Suppose that Young's two-slit interference experiment is performed to measure the wavelength of light coming from a laser. Measurements show that the separation between the slits is 0.6 mm, the distance to the viewing screen is 10 m, and the distance between adjacent bright fringes is 1 cm. What is the wavelength of the laser's light?

8A: ILLUMINANCE AND LUMINANCE OF AN IMAGE

PREREQUISITE: Chapter 6

In photography and display work it may be necessary to estimate how much light forms the image that is produced by an optical system. This is conveniently given by the illuminance E_v of the image. The illuminance will be proportional to the subject's luminance L_v, which specifies how much luminous flux per steradian heads in the direction of the camera or optical system from each square meter of perceived area of the subject. The luminous flux Φ_v intercepted by the lens from an area A_s of perceived surface is $\Phi_v = \Omega_s A_s L_v$, where Ω_s is the solid angle of rays from the surface that is intercepted by the lens (fig. 8A-1a). A lens area of $\pi d^2/4$, where d is the lens' diameter, represents a solid angle of $\Omega_s = \pi d^2/4s^2$ where s is the subject distance. The intercepted luminous flux illuminates an area of $M A_s$ square meters of the image, if M represents the linear magnification provided by the lens. The illuminance of the image is correspondingly the incident flux divided by the area it covers:

$$E_v = \frac{\Phi_v}{M^2 A_s} = \frac{\Omega_s A_s L_v}{M^2 A_s} = \frac{\pi d^2 L_v}{4M^2 s^2} = \frac{\pi}{4}\left(\frac{d}{i}\right)^2 L_v \tag{8A-1}$$

where i is the image distance.

This equation tells us that the illuminance of the image is governed by the ratio of the lens diameter to image distance. The illuminance is very sensitive to this ratio, because the square of this ratio appears in the formula. A greater illuminance is produced with a

(a)

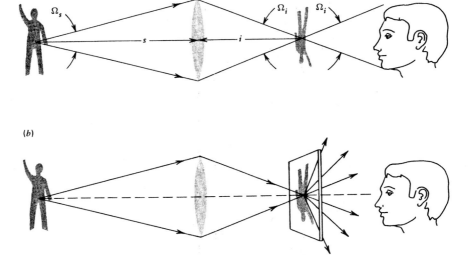

(b)

Fig. 8A-1. (*a*) Light emitted by a small area of the subject must be headed within the solid angle Ω_s to be intercepted by the lens, but on refraction is made to converge within the solid angle Ω_i. Continuing in straight lines through the image, the light emerges from the image position within the same solid angle Ω_i. (*b*) A diffusing screen at the position of the image causes the light leaving the image to be sent into a much larger solid angle.

large diameter and short image distance: the larger diameter gathers more light, and the smaller image distance produces a smaller image on which the light is concentrated. The ratio (i/d) is called the f-number of the optical system; so to achieve a bright image it is advantageous to have a small f-number.

It is interesting that an image can never have a higher luminance than the subject. To appreciate why, consider the situation when the subject of perceived area A_s emits a luminous flux Φ_v into the solid angle Ω_s intercepted by the lens as shown in Fig. 8A-1a. The luminance of the subject is $\Phi_v/\Omega_s A_s$ in the direction of the lens. This light is brought to a focus at the image position i and on continuing it spreads outward into the same solid angle Ω_i within which it arrived at the image: $\Omega_i = \pi d^2/4i^2$. If the luminous flux Φ_v comes from the image of area $A_i = M^2 A_s$, its luminance is therefore $\Phi_v/\Omega_i M^2 A_s$. Comparing this with the subject's luminance we have:

$$\frac{\text{Luminance of image}}{\text{Luminance of subject}} = \frac{\Phi_v/\Omega_i A_i}{\Phi_v/\Omega_s A_s} = \left(\frac{1}{M^2}\right)\left(\frac{\Omega_s}{\Omega_i}\right) = \left(\frac{1}{M^2}\right)\left(\frac{i^2}{s^2}\right)$$

$$= \frac{1}{M^2} M^2 = 1$$

(8A-2)

Thus this equation predicts that the luminance of the image is identical to the luminance of the subject. This equality comes about because the change in the solid angle (Ω_i/Ω_s)

is exactly compensated by the change in area (A_i / A_s). If instead the image is formed on a diffusing screen (Fig. 8A-1b), such as a ground glass, each portion of the image sends light into a greater solid angle; therefore the luminance is reduced. The image then appears dimmer than when viewed directly, and it is not as bright as the subject.

QUESTIONS

8A-1 Explain why the brightness of an extended subject is independent of its distance from us.

8A-2 When a diffusely reflecting flat surface is viewed from various directions, its brightness does not change. To explain this effect, the luminous intensity of each small portion of the surface must vary with the angle from the perpendicular to the surface. How does it vary with this angle?

9A:
CHARACTERISTICS
OF FILM EXPOSURE

PREREQUISITES: **Chapter 9 and Advanced Topic 2A**

To master the subtleties of photography it helps to understand how characteristics of the film affect the final photograph. All films misrepresent the actual scene. For example, the range of luminance that our eye can register in a typical scene greatly exceeds the range provided by a monochrome print. A film is incapable of simultaneously resolving subtle detail in brightly illuminated regions and in deep shadows, whereas the eye has this capability when scanning a scene. Some films have greater capability than others. This characteristic and other features of a film's response to light are taken up here.

A photographic negative is a diffuse filter, and the blackening of each portion can be described by its *optical density*. The notion of a filter's density was introduced in Advanced Topic 2A. Density is a measure of the degree of opacity, defined technically according to the transmittance of the particular region of the negative. The density is high when the transmittance is low. The formula that specifies the optical density D in terms of the transmittance T is:

$$D = \log \frac{1}{T} \tag{9A-1}$$

In this formula, T must be expressed in decimal form and not as a percentage. The common logarithm as used here increases by "1" for a decrease in transmittance by a factor of 10. This illustrates the rule that log 10^N, where N is any number, is simply equal to N.

For example, suppose that a region of the film is perfectly transparent: $T = 1$. Then the density is specified as $D = \log 1 = \log 10^0 = 0$. The density has the value zero. This is never observed in practice, because film is always slightly exposed all over, owing to imperfections in the film base and to light reflected within the camera. This overall exposure is called *fog*. It may typically produce a density of 0.02 or so.

To see what this means, suppose we inspect another region of the negative that had received much more light and after development transmits only 10% of the incident light. Then the density is specified as $D = \log(1/10^{-1}) = \log(10^{+1}) = 1$. Or in another region perhaps only 1% is transmitted, and $D = \log(1/10^{-2}) = \log(10^2) = 2$. A density of 3 corresponding to a transmittance of 0.1% is rarely encountered in photography. At the other extreme end of the density scale, a density of 0.02 commonly associated with fog corresponds to a transmittance of about 96%.

One reason why optical density is a useful measure of the transmission properties of a negative is simply that the eye judges differences in brightness nearly as the logarithm of the luminance of a surface, as explained in Chapter 6. The second reason is that the density of a negative is rather simply related to exposure, as we shall discuss in the next section.

It is remarkable that photographs with a much narrower range of luminance than the eye is capable of resolving in natural surroundings succeed in suggesting a full range of tones. The reflecting surface of a photographic print is even more limited than the range of transmittances that is produced by a negative. A reflecting surface on a print appears white if its reflectance exceeds about 70% and black if less than 5%. Consequently an adequate rendition is provided if the reflectance spans a range of only a factor of 20. The negative from which the print is made therefore need not have a much greater span in transmittance from one place to another. This is why a span of 10^2 in transmittance or a span of 2 in density is adequate for most purposes.

Characteristic Curve of a Film

How the density increases with exposure for a particular type of film is described by the *characteristic curve* of the film (Fig. 9A-1). First studied in the late 1800s by Ferdinand Hurter and Vero Charles Driffield, it is sometimes called the "H and D" curve. The horizontal scale across the bottom of the figure is arranged so that the distance from the left increases as the logarithm of the exposure H_0. The shape of the curve at its left end indicates that the exposure must exceed a certain threshold value (point A) before appreciable blackening occurs above the fog level. The "toe" of the curve (A to C) shows that the density increases progressively more rapidly for greater levels of exposure.

The central region (C to D) is particularly interesting, because the straight line indicates that here the density increases in proportion to the logarithm of the exposure. The straight line variation indicates that in this region a factor of 10 increase in H_0, causes a factor of 10 decrease in transmittance. When faithfully translated into the final positive print, this correspondence produces values of reflectance across the print that range upwards and downwards in proportion to the luminance of various subjects in the original scene. This is why photography gives us a realistic rendition of the relative brightness seen by the eye directly. The central region of the characteristic curve is the desired exposure range for a "normal" photograph.

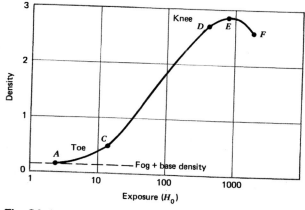

Fig. 9A-1. Example of a characteristic curve for monochrome film.

Still greater exposures (D to E) cause the density to increase less rapidly with exposure, and the curve displays a "knee." There all of the silver halide crystals produce development centers during exposure, and the density cannot be further increased. Surprisingly, additional exposure (E to F) triggers a number of complex reactions that cause the curve to turn downward! In this overexposed region where density decreases with increasing exposure a positive image is produced, known as a *solarized image,* which is occasionally used by photographers for special effects.

The linear portion of the characteristic curve (C to D) embraces a range of exposures called the *latitude* of the film. When a picture is correctly exposed, all regions of the negative will be represented by points on this portion. If some region has insufficient exposure so that it is represented on the toe of the curve, detail in the scene will be lost, because density varies only slightly for large differences in exposure. The same is true for overexposed regions in the knee. The latitude of a typical film can render all details when the brightest portion of the scene provides about 50 times more exposure than the darkest. In a natural setting the human eye can easily span a range four times greater. The more limited latitude of a film means that detail seen by the photographer at both ends of the range in brightness will generally be lost when recording on film.

Contrast

The slope of the linear portion of the characteristic curve indicates how rapidly the density varies with the logarithm of exposure. The number indicating the steepness of the slope is denoted by the Greek letter gamma (γ). Figure 9A-2 gives several examples of curves with different slopes. To have a photograph correspond with how we see the scene directly, γ should be close to 1. A film with $\gamma = 2$ has its density increase much more rapidly with exposure and is said to have a "high contrast." Dim regions in the scene appear much darker in the final print in comparison with light regions. Thus the film artificially enhances contrast between light and dark regions. On the other hand a film with $\gamma = 0.5$ is said to have low contrast. A film with high contrast requires a careful choice of exposure because its latitude is comparatively limited.

Development procedures can affect the shape of the characteristic curve. Extending

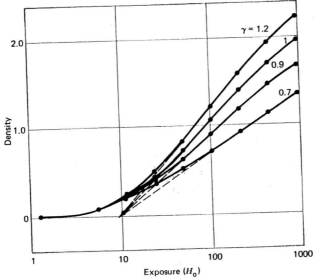

Fig. 9A-2. A family of characteristic curves for a film, with increasing values of γ produced by longer development times.

the development time beyond the recommended duration increases the steepness of the slope of the linear portion, enhancing γ and contrast. But γ reaches a maximum value for very long development times, and this value is denoted as γ_∞ ("gamma infinity"). Information about γ is provided by manufacturers of each film.

Film Speed

Correct exposure of a film must take into account its sensitivity to light, indicated by a number called the *film speed*. The DIN system popular in Europe is based on a reference exposure that gives a density of 0.1, The ASA speed recommended by the American Standards Institute uses a slightly different reference exposure labeled as H_m in Fig. 9A-3. This is the exposure that increases the density by 0.1 above what is contributed by the film base and fog. This definition of H_m is appropriate for a film that has been developed to the point where the density increases by 0.80 when the exposure is increased by 1.30 logarithmic units above H_m.

For an average scene outdoors, a light meter when set for the specified ASA number is calibrated to recommend the following exposure H_0:

$$\text{Monochrome negative film: } H_0 = 9.4\,H_m \qquad (9A\text{-}2)$$
$$\text{Color negative film: } \quad H_0 = 7.5\,H_m$$

These numbers giving the "standard exposure" were selected on the basis of experience with many tests under varying conditions. Color film and the special case of color reversal film will be discussed in the next section. Exactly how a numerical value is assigned to

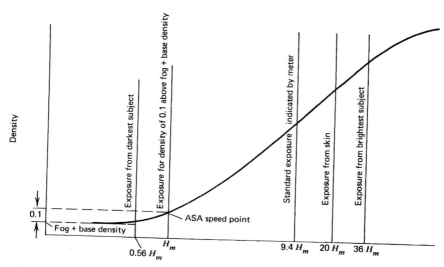

Fig. 9A-3. The characteristic curve (*D* versus log *H*) for a monochrome negative showing the exposure H_m on which the ASA speed of the film is based. Also shown is the standard exposure recommended by a light meter for an average outdoor scene and the resulting exposure from representative subjects. (Adopted from A. Stimson, *Photometry and Radiometry for Engineers*, Wiley, New York, 1974.)

film speed will also be explained there. Longer development time can "force" or "push" a film to give it a greater speed, but at the expense of having greater contrast. A relatively fast film with an ASA number of 400 can be pushed as high as ASA 3200 in this way.

9B: DEFINING FILM SPEED

PREREQUISITES: Chapter 6 and Advanced Topic 8A

The reading of an exposure meter of the reflected light variety tells us how to set the camera so that the average exposure falls in the central portion of the characteristic curve for the particular film. Specifically it indicates the proper exposure H_0, once its scale is adjusted for the ASA speed of the film. The position of the meter's needle indicates which combinations of f-number F of the lens and exposure time t for the shutter are correct. A reflected light meter responds to the average luminance L_v of the scene in its field of view. The discussion in Advanced Topic 8A shows how to predict the illuminance E_v falling on the film, and the result is given by Eq. 8A-1:

$$E_v = \left(\frac{\pi}{4}\right)\left(\frac{L_v}{F^2}\right) \tag{9B-1}$$

The luminance L_v of the scene is expressed in lumens per steradian per square meter. The average outdoor summer scene provides a luminance of about 3200 lm/sr·m². If a lens with low transmittance is used with the camera, the right-hand side of Eq. 9B-1 should be multiplied by its transmittance. With modern cameras a typical value for the transmittance is 90%, so the right-hand side would be multiplied by 0.9. It should also be noted that this expression is appropriate for the illuminance at the center of the scene. Somewhat off axis, the illuminance is reduced; thus a more appropriate value for the image illuminance would be decreased by an additional 8% from the prediction of Eq. 9B-1.

To predict the exposure of the film, the exposure time t must be chosen. If a reflected light meter is used to determine the exposure time, the calibration provided by the manufacturer must be relied on. The meter is calibrated by the manufacturer for optimal results when directed toward an average scene. This calibration takes into account the f-stop F of the lens and the speed S of the film. The meter is designed to indicate the exposure time according to the following formula:

$$t = \frac{10.76}{\pi} \cdot \frac{KF^2}{SL_v} \tag{9B-2}$$

This calibration incorporates a number, denoted by K, which takes into account the size of the film. For 35 mm films its value is $K = 3.3$. In the case of incident light meters the calibration formula has a similar form but K is replaced by the symbol C. This number takes into account the average reflectance of a typical scene. It has a value of about $C = 22$. Also for incident light meters the symbol L_v in Eq. 9B-2 is replaced by E_v, representing the illuminance that the meter detects.

With the image illuminance predicted by Eq. 9B-1 and the exposure time specified by Eq. 9B-2 we can calculate the exposure given by the expression $H_0 = E_v t$. Multiplying E_v and t, we find that both the f-stop F and the luminance L_v are eliminated from the final expression:

$$H_0 = \frac{10.76}{4} \cdot \frac{K}{S} \tag{9B-3}$$

We find what we had expected: the recommended exposure decreases for films with higher film speed S, and the exact value of the exposure is fixed by the meter's internal calibration number K. The manufacturer specifies film speed by where H_0 should fall on the characteristic curve. From Fig. 9A-3 it is clear that the lower the value of H_m, the more sensitive is the film. The ASA speed is therefore defined by introducing still another constant, denoted by k:

$$S = \frac{k}{H_m} \tag{9B-4}$$

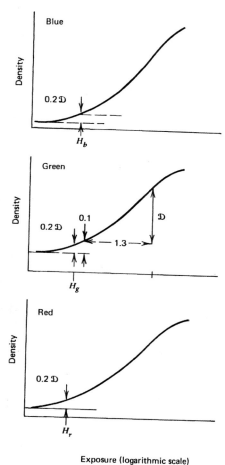

Exposure (logarithmic scale)

Fig. 9B-1. The three characteristic curves for the blue, green, and red sensitive layers of color negative film and their corresponding reference exposures. The references lie where the density is 0.2 $\mathcal{D}$ above the level contributed by fog and base, where the value of $\mathcal{D}$ is the measured increase in density of the green curve when the exposure is increased by 1.30 logarithmic units from the point where the density is 0.10 above fog and base.

The value of k is chosen by the manufacturer to take account of the shape of the curve. For instance, for monochrome negative film the value of k is chosen to be 0.8.

Color negative film must be assessed differently, because the sensitivity of each of the three layers responding to different portions of the spectrum should be considered. Figure 9B-1 illustrates three characteristic curves for a typical color negative film. First the individual reference exposures H_b, H_g, and H_r are determined when the film is exposed to white light. As for monochrome film these lie where the film density is increased by 0.1 above the level contributed by fog and the film base. Then the overall reference is defined by the geometrical mean of these individual levels: $H_m = (H_b\, H_g\, H_r)^{1/3}$. Best results for the finished photograph are obtained by using the value $k = 1.0$ when determining the speed in Eq. 9B-4. Consequently for color negative film, the speed S is very close to the inverse of the reference exposure H_m when H_m is expressed in the units of lumen-second per square meter of film area.

From Eqs. 9B-3 and 9B-4 we can deduce by how much the recommended film exposure

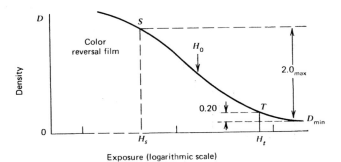

Fig. 9B-2. Color reversal film such as Kodachrome when developed has its density decreasing with increasing exposure, so that more light is transmitted where subjects in the scene were bright. The reference exposure is defined in terms of the exposures at the shoulder (S) and toe (T), and H_t lies where the density is 0.2 above the level provided by fog and the base of the film.

exceeds the reference exposure H_m. This is given by the factor in parentheses in the following expression:

$$H_0 = \left(\frac{10.76}{4} \cdot \frac{K}{k} \right) H_m = \left(\frac{8.9}{k} \right) H_m \qquad (9\text{B-}5)$$

The value in the numerator of the factor in parentheses should be reduced by about 10% to take account of the fact that the typical transmittance of a camera's lens is only 90%. Furthermore, there should be an additional 8% reduction to predict a more representative image exposure away from the optical axis where the image is slightly darker. With these two factors taken into account the final expression for the exposure becomes:

$$H_0 = \left(\frac{7.5}{k} \right) H_m \qquad (9\text{B-}6)$$

Thus for monochrome film the recommended exposure becomes $7.5/k = 7.5/0.8 = 9.4$ times greater than H_m. And for color negative film H_0 is $7.5/1.0 = 7.5$ times greater than H_m.

The case for color reversal films such as Kodachrome is quite different from that of negatives, because the characteristic curve slopes the other way, with density decreasing as exposure increases. By convention two reference exposures are defined (Fig. 9B-2) with H_t at the toe, where the density is 0.2 greater than that of the base plus fog, and H_s at the shoulder. Then by convention $H_m = \sqrt{H_t H_s}$. The ASA speed is chosen for optimal results by letting $k = 10$. Then $H_0 = 0.75 H_m$, which puts the recommended overall exposure close to the geometric mean of the reference points at shoulder and toe.

Index

SOME CONVERSION FACTORS

Physical Measures

Mass

1 kilogram (kg) = 2.21 pounds (lb)

Length

1 meter (m) = 39.4 inches (in) = 3.28 feet (ft)
1 kilometer (km) = 0.621 mile (mi)
1 nanometer (nm) = 10 Ångstroms (Å)

Angular measure

1 radian (rad) = 57.3 degrees (deg)

Energy

1 joule (J) = 0.239 calorie (cal) = 0.738 foot-pound (ft-lb)

Power

1 watt (W) = 0.00134 horsepower = 0.738 foot-pound per second (ft-lb/s)

Psychophysical Measures

Luminous flux

1 lumen (lm) = $\frac{1}{683}$ watt at 555 nm

Illuminance

1 lux (lx) = 1 lumen per square meter (lm/m^2) = 0.0929 footcandle (fcd)

Luminous intensity

1 candela (cd) = 1 lumen per steradian (lm/sr)
= 1 candlepower

Luminance

1 nit (nt) = 1 candela per square meter (cd/m^2) = 0.2919 footlambert (fL)